Comstock Women

Wilbur S. Shepperson Series in History and Humanities

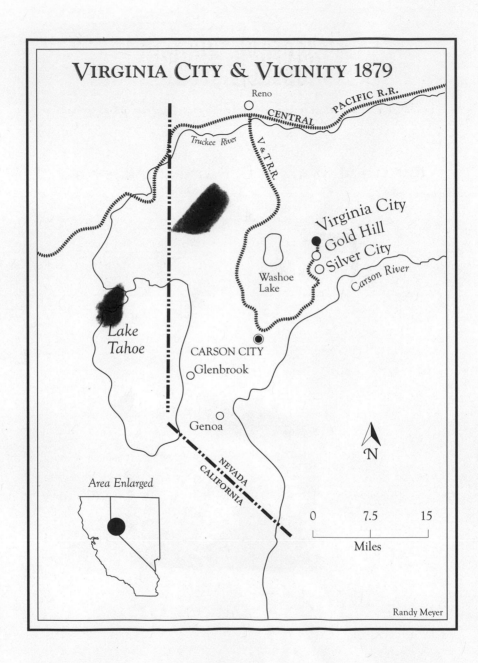

Virginia City & Vicinity 1879

Reno

CENTRAL PACIFIC R.R.

Truckee River

V & T.R.R.

Virginia City
Gold Hill
Silver City

Washoe Lake

Carson River

Lake Tahoe

CARSON CITY

Glenbrook

Genoa

NEVADA
CALIFORNIA

N

Area Enlarged

0   7.5   15

Miles

Randy Meyer

Comstock Women

The Making of a Mining Community

EDITED BY

RONALD M. JAMES & C. ELIZABETH RAYMOND

UNIVERSITY OF NEVADA PRESS

RENO / LAS VEGAS

Wilbur S. Shepperson Series in History and Humanities
Series editor: Jerome E. Edwards

This book was funded in part by a grant from the Nevada Humanities
Committee, an affiliate of the National Endowment for the Humanities.

University of Nevada Press, Reno, Nevada 89557 USA
Manufactured in the United States of America
Book design by Carrie Nelson House
Library of Congress Cataloging-in-Publication Data
Comstock women : the making of a mining community / edited
by Ronald M. James and C. Elizabeth Raymond.
p.   cm. — (Wilbur S. Shepperson series in history and humanities)
Includes bibliographical references and index.
ISBN 0-87417-297-7 (pbk. : alk. paper)
1. Virginia City (Nev.)—History.   2. Women—Nevada—Virginia City—
Social conditions.   3. Comstock Lode (Nev.)   4. Mines and mineral
resources—Nevada—Virginia City—History.   I. James, Ronald M.
(Ronald Michael), 1955– .
II. Raymond, C. Elizabeth.   III. Series: Wilbur S. Shepperson series in
history and humanities (Unnumbered)
F849.V8C65   1997   97-11958
979.3′56—dc21   CIP
The paper used in this book meets the requirement of American National
Standard for Information Sciences—Permanence of Paper for Printed
Library Materials, ANSI Z39.48-1984. Binding materials were selected for
strength and durability.

First Printing
06   05   04   03   02   01   00   99   98   5   4   3   2   1

In memory of the women of the Comstock

CONTENTS

Illustrations

Figures

Tables

Appendices

Maps

PART ONE

❧ Introduction ❧

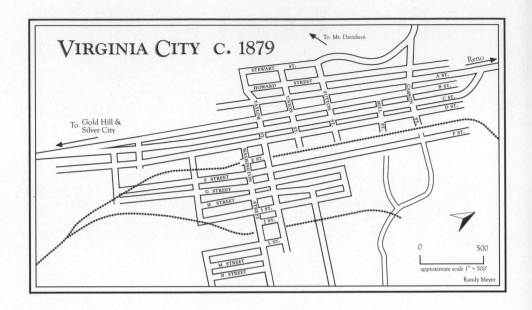

VIRGINIA CITY C. 1879

To Mt. Davidson

Reno

STEWART ST.

HOWARD STREET

A ST.

B ST.

C ST.

D ST.

To Gold Hill &
Silver City

TAYLOR ST.

UNION ST.

SUTTON ST.

MILL ST.

CARSON ST.

F ST.

WASHINGTON STREET

E ST.

F STREET

G STREET

H STREET

I ST.

J ST.

L ST.

M STREET

N STREET

0   500

approximate scale 1" = 500'

Randy Meyer

❧ 1 ❧

"I Am Afraid We Will Lose All We Have Made"

Women's Lives in a Nineteenth-Century Mining Town

C. Elizabeth Raymond

John is superintendent of the 'Great Bamboozle' now, and is besides a member of the Legislature, so of course we move in the best society. I spent a week with him in Carson [state capital] a little while ago . . . the 'best society' gave us some excellent dinners, and placed their fastest teams at our disposal.   LOUISE PALMER, 1869

We were half-starved coming over [from New Zealand]. The biscuits were full of worms, the potatoes rotten and the salt meat skinny, and when they killed a sheep we only got a taste of it.

We were here nearly three months without work.   GEORGINA JOSEPH, 1866, 1869

Both Louise Palmer and Georgina Joseph were residents of Virginia City, Nevada, during the period of its fabulous mid-nineteenth-century boom, from 1860 to 1880. They lived in an initially remote mining town in western Nevada that rapidly developed into one of the world's richest gold and silver mines. Unlike the gold mining camps of California, which were small settlements scattered north and south for 200 miles along the gold-bearing lode, the Nevada deposits were concentrated. They gave birth to an industrial marvel, a city of perhaps 20,000 people in what seemed like the middle of nowhere. Their fame was sufficient to obtain statehood for Nevada in 1864, long before it had the required population to qualify. For a time, the Virginia City mines—located literally beneath the city streets—were on virtually every western tourist's itinerary.[1]

Georgina Joseph wrote home from Virginia City to her in-laws in New Zealand, because her husband was illiterate and couldn't. Her letters chronicle a working-class family's daily struggle—husband Frank's frequent illnesses that kept him from work, the feeling of accomplishment when they finished paying their debts, her dismay when they lost the savings they had invested in mining stocks, her joy at his purchase of life insurance. Hers was a life, above all, of uncertainty. Louise Palmer wrote for publication in the prestigious *Overland Monthly*, turning her free time and her literary gift to fame and profit as she archly described for public consumption her life as a social lion of the world-renowned Comstock Lode. Hers was a world of luncheon parties and choral societies, a place where women competed with each other by means of fashion.

These two women led disparate lives. Louise Palmer, an upper-class woman of leisure, sought amusement. She was able to travel in pursuit of pleasure. Her article recounts a trip to the top of one of the Sierra Nevada peaks, with a stunning view of Lake Tahoe "amid the peaks upon peaks that rise to the clear blue sky." The sojourners celebrated the view and their climb with a bottle of champagne and a picnic lunch. Georgina Joseph, by contrast, stayed put in Virginia City. She had trouble making ends meet, pointing out for the benefit of relatives that, "in California everything is very cheap, but up here in the mountains they are dearer, for it is a long road to bring them." Her satisfactions were modest ones, as she reported in December 1869: "We have our winter provisions in and a fine pig in the corral so we are all right for the winter."

It is difficult to imagine the two women inhabiting the same social space. Yet both were present at the same time to witness and bear testimony to the spectacle of a wealthy western silver boomtown. Their divergent experiences can tell us much about the texture of daily life in Virginia City, the other side of a mining bonanza that is traditionally associated almost exclusively with the labors and recreations of men.

Fascinating though they are, Palmer and Joseph don't adequately represent the diversity and complexity of women's lives on the Comstock Lode. Just as men's situations varied in the mining town, so, too, did women's. Not only wives but also single women lived in Virginia City during the boom years. A few women of color and numerous European immigrants mingled with native Paiutes and with Chinese women. Women religious and prostitutes both found a calling in the overwhelmingly male mining town. This collection seeks to recover their experiences as well, to reveal the myriad ways that women's lives shaped and were shaped by residence on the Comstock Lode. We hope to recount more stories than just Georgina Joseph's and Louise Palmer's.

Recovering those stories, of course, is difficult. Along with Mary McNair Mathews, a feisty widow who published a book about her adventures on the Comstock Lode, Palmer and Joseph are the only resident women known to have left personal accounts of life in Virginia City. Other women come and go from public records such as newspapers and court decisions, from the published accounts of men like Samuel Clemens and J. Ross Browne, and from private diaries like journalist Alfred Doten's; but they seldom speak in their own voices. Scholars who seek to recover women's lives are thus forced to rely on other tools.[2]

In the case of Virginia City they are fortunate to have the resources of the Nevada Census Project, a computerization of all the individual U.S. census records for Storey County, to draw on.[3] This database allows for a level of precision not possible in places where demographic sampling techniques must be used. Because Virginia City was sufficiently productive to attain status as a genuine city and not simply as a transient mining camp, it lasted long enough for census data to be collected at crucial points in its development. By accident of circumstance, the decennial federal census was timed to capture Virginia City at its very inception, in 1860, when word of the initial silver discovery was just beginning to spread; again in 1870 as the town was growing toward its peak; and in 1880 when the richest ore was mined out and the population was beginning to decline.

Combined with a special 1875 state census, which enumerated the Comstock Lode close to its peak, these surveys provide scholars with invaluable data for reconstructing the entire population profile of a successful western mining town. As historian Elliott West points out, "Mining camps were among the most unstable and transient human gatherings in the history of the unsettled young republic." Virginia City was no different. Observers constantly remarked upon the bustle in its streets and the transience of the wealth-seeking population. Settlers came and went, but the settlement itself persisted. Several of the scholars whose essays are included here draw insights based on demographic data made available through the Nevada Census Project.[4]

Others bring to bear different techniques and tools that are necessary to understand the circumstances of women who left no written records and may have been misrepresented even in the census. Paiute women, for example, are not listed in some censuses, and the data is notoriously unreliable for Chinese residents. In cases such as these, anthropology and historical archaeology can provide insights unavailable in traditional documentary form. Census information about women's occupations is also erratic; so careful, contextual interpretation of the census data is imperative, as several of the essays point out.

The women's experiences catalogued here are literally part of the other

Women strolling below the Virginia City fire lookout. (Courtesy of Fourth Ward School, Virginia City)

side of bonanza. At the time of the Comstock boom, women were depicted as remarkable because of their relative rarity. Virginia City was a man's playground, a stage for the pursuit of wealth and power. To subsequent historians women became virtually invisible except in the guise of familiar stereotypes, as either prostitutes or millionaires' wives. Although their collective efforts were essential to establishing and maintaining the mountain mining community, women were and remain absent from popular depictions. Even the television show, *Bonanza,* ostensibly set just outside Virginia City in its heyday, depicts a ranch family entirely without women. In this volume we strive to reincorporate the full panoply of women into Comstock history.

Because of the limitations of the evidence, some of the contributions to this volume are necessarily suggestive rather than comprehensive. Some things that we might like to know—about the circumstances of the few African American women who lived on the Comstock, for example, or about the Chicanas who were among its first settlers—are virtually impossible to ascertain. Collectively, however, these essays greatly enlarge our knowledge of the lives of women in nineteenth-century mining settlements. We see beyond Louise Palmer and Georgina Joseph to glimpse many others.

This collection also mingles the voices of academic historians and anthropologists with those of museum professionals and demographers. The work benefits from the perspective of historic preservation and architectural history, as well as from the enthusiasm of interested nonprofessionals. For some essayists Virginia City is not simply a subject but also a working environment or even a place of residence. Their range of interests and methodology is reflected in the individual essays.

The volume is divided into sections by topic. An introduction by Ronald M. James and Kenneth H. Fliess provides general background on the history of Virginia City during the nineteenth century and on the structure of its female population during those decades. Their data pointedly remind us that nineteenth-century Virginia City was a predominantly male environment. Although women were more numerous than later popular perception would suggest, they were always a numerical minority.

Subsequent sections examine women "At Home in a Mining Camp," a kind of settlement simultaneously notorious for its upheaval and desirous of permanence. "Occupations and Pursuits" includes four essays on various ways that Virginia City women spent their time, while "Ethnicities" explores three distinctive subcommunities within the heterogeneous population of the Comstock Lode. The final section, "Image and Reality," includes two overviews. One explores the historical construction of popular imagery relating to Virginia City women, and the other suggests ways that historical archaeology could further illuminate the history of those women.

In many respects, Virginia City was anomalous. Its size, wealth, rapid growth, concentrated population, level of capital investment, and national prominence, as Ronald James and Kenneth Fliess point out, all distinguished it from other nineteenth-century mining towns. Unlike the California Gold Rush that preceded it by a decade, the "Rush to Washoe" was primarily a west-to-east movement, as disappointed California miners, Chinese laborers, and speculating merchants hastily relocated themselves to the dry basin on the eastern slope of the Sierra Nevada. Many Comstock residents, both men and women, were thus already westerners when they arrived in Nevada.

The barren slopes of Mount Davidson were a more hostile environment than California had been, however. Temperatures sometimes fluctuated by as much as 50 degrees in one day. Summer temperatures could exceed 100 degrees, while winter nights dipped below zero. Scouring afternoon winds known as "Washoe zephyrs" were a frequent and disagreeable feature of the mountain town, located 6200 feet above sea level. In the midst of such forbidding conditions, Julie Nicoletta observes, the new residents set about reproducing familiar physical spaces and social systems.

Women of means built middle-class homes with ample space for both family privacy and public display. Women who relied on their living quarters as a source of economic support built boardinghouses to accommodate the influx of bustling newcomers seeking opportunity amidst the mining excitement. In such dwellings the parlor, a special room set aside as a place to socialize within the home but outside the private chamber, was a significant feature. Nicoletta argues that the lodging-house keepers of the Comstock were, collectively, "Redefining Domesticity." Louise Palmer inadvertently confirmed that this might have been the case when she remarked that, on the Comstock, despite the fashionable display, "persons are not judged by the places they live in."[5]

Other women, of course, enjoyed no such luxuries as Nicoletta's subjects. Less well-to-do, they lived crowded with families in rented rooms or survived alone in small shacks, members of the more spatially constricted world of Georgina Joseph. Marion Goldman documents the varying circumstances and general poverty of Comstock prostitutes in *Gold Diggers and Silver Miners,* but Kathryn D. Totton and Sharon Lowe point to other groups of women who also failed to realize the nineteenth-century domestic ideal in Virginia City.

Totton surveys the divorce records of the Comstock Lode and concludes that conditions in Virginia City facilitated divorce. A fairly lenient law, combined with opportunities for divorced women to support themselves, a generally liberal interpretation of grounds by judges, and the relative absence of stigma attached to divorce, served to make divorce more common on the Comstock Lode than it was in eastern cities. Totton finds the high rate of transiency associated with a mining settlement to be confirmed by the frequency with which Comstock women's divorce suits cited desertion as a cause.[6]

Lowe explores the world of drug addiction, both in the lurid form of opium smoking and in the less visible form of women's self-medication with laudanum and other opiates. In "The 'Secret Friend,'" her exhaustive search of court, hospital, and institutional commitment records provides important correctives to popular stereotypes. First, she dispels the contemporary nineteenth-century myth of a Chinese population helplessly addicted to opium smoking. Second, she reveals the growing popularity of opium smoking among Euro-Americans and its predictable legal prohibition in 1876, once middle-class youths were publicly revealed to be participants in opium culture. Lowe's account of women whose experimentation with medicinal opiates proved disastrous further demonstrates that not only Virginia City's "criminal classes" were seeking solace in mind-altering drugs. Collectively, these three essays suggest the tremendous variety of domestic circumstances, some of them quite chaotic, in which Comstock women found themselves "at home."

Indeed, the circumstances of Comstock women were far more diverse than conventional images might suggest. In "Occupations and Pursuits," the authors discuss ways that various Comstock women spent their time. Working women labored for wages at essential and traditional female occupations. Women religious worked for the spiritual and social good of the community. Other women engaged in recreational or political activities. All of these pursuits helped to constitute and to maintain the community of Virginia City just as much as did mining, its principal industry.

In "Creating a Fashionable Society," Janet I. Loverin and Robert A. Nylen consider the different activities pursued by women within the sewing trades. Needleworkers ranged from elite milliners, through skilled dressmakers and ordinary seamstresses, to the sewing girls who performed rough bulk work. Loverin and Nylen offer a valuable comparative perspective, noting, as expected, a lower concentration of needleworkers in mining states than in East or West Coast centers such as California or Massachusetts. They also observe, however, that Virginia City had a higher proportion of personal clothiers than did the larger city of Denver, which serves to confirm anecdotal accounts, like Louise Palmer's, about the Comstock as a center of fashion and sophistication during the nineteenth century.

As Loverin and Nylen conclude, the first women to arrive in Virginia City were not all unskilled. Among the earliest residents were stylish and proficient practitioners of a demanding trade who made their living purveying fashion essentials to other women. Virtually from the beginning, Virginia City was understood to be a women's city, not simply a male mining camp.

The mining boom also attracted women whose motives were not eco-

Children enjoying an outdoor class in Virginia City. (Courtesy of California Historical Society)

nomic. As Anne M. Butler explains in "Mission in the Mountains," the Daughters of Charity were an important source of social and spiritual services in the burgeoning community. Arriving in 1864, the convent community established a boarding school and hospital and even functioned for a few years as a state orphanage. Largely Irish or Irish American, these women religious established important ties with the non-Catholic community, as well. They raised money to support their ventures through a variety of alliances with local residents who sponsored charity balls and fairs or donated needed supplies.

Secular denizens of Virginia City led a hectic life, a heady combination of great fortune and devastating loss as mining stocks rose and fell in value and terrible mining accidents alternated with stagnant periods of unemployment. In the midst of so much instability, the Daughters of Charity, as well as following their own spiritual vocation, offered solace to the families and individuals who were casualties. Butler's essay takes an important step to redress what she calls the "historical invisibility of women religious."

Other Comstock pursuits were less nobly disinterested, though surely no less related to the spectacular uncertainties of life in a mining boomtown.[7] Bernadette S. Francke's overview of other varieties of women's spiritualism in "Divination on Mount Davidson" points out the continuing popularity of these activities.

Women in Virginia City practiced spiritualism and fortunetelling both as a form of social recreation and as a profession. Paid lecturers extolled the virtues of spiritualism, or "spirit rapping," as it was sometimes known. Beginning in 1867, Virginia City even had a formally organized Spiritualist Society. Claims by a fourteen-year-old Catholic girl that her dead father had contacted her from the spirit world were hotly debated both within and beyond the church. Professional fortunetellers advertised their services in newspapers, while amateur practitioners preserved their disinterested status by refusing to charge for their readings. Billing herself as the "Washoe Seeress," former millionaire Eilley Orrum Bowers was the most famous, but clearly not the only Comstock woman willing to explore—and perhaps to exploit—connections to the supernatural world.

In "The Advantages of Ladies' Society," Anita Ernst Watson, Jean E. Ford, and Linda White explore the thoroughly respectable realm of middle- and upper-class Comstock women. Their daily activities represent a catalogue of women's traditional sphere varying little from their well-studied sisters in the urban northeast. These women, the wives of mine superintendents and attorneys, of doctors and merchants, seldom participated in paid employment. Instead they raised money for poor relief, participated in the organization and financial support of churches, campaigned for temperance and against the chaotic male world of the saloon, and arrayed themselves on both sides of Nevada's 1871 suffrage debate. Watson, Ford, and White tell a story of women's civic involvement that suggests that, at least for well-to-do women, continuity rather than disruption of social roles characterized the move to Virginia City.[8]

For Chinese women, by contrast, *Yin-shan*, or "Silver Mountain," represented disruption of virtually all expectations. Traditional family roles, along with familiar cultural practices like foot binding, eroded and disappeared in the United States. Yet the Chinese immigrants were perpetually marginalized within American culture, where they were reviled as barbaric and unclean by Euro-Americans who resented their economic competition. Opium smoking was identified as a Chinese vice, and Chinese women were almost invariably dismissed as prostitutes.

Sue Fawn Chung explains that the Chinese women who came to the Comstock actually filled a variety of roles. Some were prostitutes, serving the general sexual needs of a majority male population in a society where intermarriage between Euro-Americans and Chinese was defined as miscegenation. Others were "second wives," or concubines, women whose alliance with a single man was an accepted role in Cantonese society. Still others were wives and unmarried relatives, women who accompanied men and then ran households, raised families, or worked at businesses. Some remained in Virginia City for forty years. In all cases, because of the small size of the community to which they came, Chung argues that their adjustment to U.S. society was more rapid and more far-reaching in Nevada than it was in larger Chinatown districts such as San Francisco's. In "Between Two Worlds," Chung depicts women whose lives necessitated flexibility.

The Comstock's indigenous population was placed under similar pressures with the advent of Euro-Americans. Mining in Virginia City introduced permanent settlements and large-scale appropriation of natural resources into a delicately balanced system of seasonal hunting and foraging that had sustained the Northern Paiute for generations. Eugene Hattori argues, perhaps surprisingly, that Paiute culture survived this onslaught rather well.

Due in large part to the activities of Paiute women, who adapted their traditional activities of foraging and food preparation to the new Euro-American economy, families remained intact, birth rates continued high, and the traditional cultural identity of the Northern Paiutes remained unimpaired. The American Indian women of the Comstock followed patterns similar to those they had observed before the Comstock boom, neatly incorporating relevant elements from Euro-American material culture, including children's toys. Late in the century, when ore yields declined and the invasive Euro-Americans moved on to inhabit new towns, these women remained behind with their families in the place that had been their home all along.

"Erin's Daughters on the Comstock" were among those Euro-Americans who came, stayed for a while, and then moved on to new and more promising locales. Ronald James suggests, however, that this pattern of transience belies the underlying continuity created by the Irish community on the Comstock. The Irish were always a pronounced presence on the Lode, making up one-third of the population in 1880. James's careful analysis of census data and neighborhood reconstruction reveals that the Irish congregated in distinct neighborhoods, worked among and hired members of their own group, and tended to express their ethnic identity by means of the Catholic Church and ritual occasions such as St. Patrick's Day. Irish families had more children than did other groups, and even women who married non-Irish men tended to settle in Irish neighborhoods. When they moved on, these people moved within communities of Irish friends and neighbors.

This revelation of the persistence, the continuity, and the stability of the Irish community, as James points out, recontextualizes notions of the transience of the mining frontier. Serial communities, reconstituted at successive physical locations, may characterize the mining frontier more than does the notorious mobility attributed to single locales such as the California gold camps.

The three essays in the section on "Ethnicities" surely do not exhaust their important topic. In many cases, however, traditional historical sources concerning women of color or women from other ethnic subcommunities are meager or missing. Chung, Hattori, and James all demonstrate that census data may be combined inventively with other evidence to recover the history of women who did not produce written records of their own lives in Virginia City. Their essays provide more insight into these three groups of women than might otherwise have been thought possible.

In the book's final section, "Image and Reality," the authors employ different lenses to examine the position of women on the Comstock Lode. Andria Daley Taylor's historiography of Virginia City mythmakers probes the literary legacy of the 1940s to find the origins of contemporary Comstock myths, including the tremendously popular television series, *Bonanza,* and the twentieth-century veneration of murdered prostitute Julia Bulette. In "Girls of the Golden West," she points out that Virginia City is today a place of "congealed history," where those few women whose stories are told have been forced into "edited lives." Among the villains, she argues, are folklorist Duncan Emrich, newspaper editor Lucius Beebe, and novelist Vardis Fisher.

Donald L. Hardesty's conceptual exploration of ways that gender might be manifested in Comstock archaeology suggests that systems of meaning are located in artifacts as well as in texts. The middle-class world of taste and behavior described by Nicoletta and Watson, Ford, and White, should have physical counterparts on the ground, according to Hardesty. Gardens and floor plans, discarded tableware, and decorative objects are further manifestations of middle-class life and of the Comstock's connection to the world market, facilitated by completion of the transcontinental railroad in 1869.

As yet, this archaeological investigation has not been completed in Virginia City, although Hattori's essay on the Northern Paiutes relies in part on archaeological evidence. In "Gender and Archaeology on the Comstock," Hardesty demonstrates the rich contribution that this field can make to the exploration of gender in a nineteenth-century mining town. Occasionally, as he points out, the material record can be a valuable corrective to idealized written descriptions.

The literature of western women's history is a large one. In addition to numerous monographs, two landmark collections are *The Women's West*, edited by Susan Armitage and Elizabeth Jameson, and *Western Women: Their Land, Their Lives*, edited by Lillian Schlissel, Vicki L. Ruiz, and Janice Monk.[9] The subject by now has a well-defined historiography. Earlier studies tend to concentrate especially on the circumstances of nineteenth-century Euro-American women who traveled overland to some part of the West defined at the time as frontier.[10] More recent work has moved beyond this nucleus to include women who already resided in the region when the Euro-Americans arrived and women who came from other places and cultures to inhabit the West. These works speak of encounters rather than frontiers and suggest a more diverse population of women in the West.[11]

Partly as a result of the development in recent years of New Western History, scholars are reexamining both men's and women's history. Recent work displays greater attention to the intersections of race, region, ethnicity, and class with gender.[12] No longer does western history concentrate so exclusively on the Euro-American "pioneering" period of the nineteenth century.

In all of this explosion of work on the area, however, the historical worlds of interior mining communities are still relatively obscure. Historians have resurrected women's experiences in the placer mining gold camps of California and on the farms of the Great Plains, but studies of women's lives in hard-rock, deep mining communities such as Virginia City are more recent. The degree to which the contours of women's lives in these industrial communities repeat those of other regions is not yet clear.[13]

In *The Female Frontier*, Glenda Riley argues that location was essentially irrelevant for women in the American West. Although her book compares only women on the agricultural prairie and plains, she concludes that women's experiences on the American frontier were everywhere so similar, and so distinct from those of men, that life on a ranch would not have differed in its essential details from life in a mining town: "Despite locale or era, women's experiences exhibited a remarkable similarity, which was shaped largely by gender and its associated concept of 'women's work.'"[14] Her formulation is problematic for areas like Nevada's Comstock Lode. There, during Virginia City's boom years, the nature of the mining economy dramatically affected the texture of women's lives.

The heady decades of successive boom and bust created an urban outpost that was both bustling and transient. Rather than isolation and agricultural settlement, urban crowding and impermanence were the prevailing conditions. As Mark Twain described the place in 1863, "It claimed a population of fifteen thousand to eighteen thousand, and all day long half of this little army swarmed the streets like bees and the other half swarmed among the drifts and tunnels of the 'Comstock,' hundreds of feet down in the earth directly under those same streets." As residents, Virginia City women experienced the bustle in all its plenitude.[15]

Unlike the placer mining period in California, extensive capital was required in order to successfully exploit the gold and silver buried in Mount Davidson. Labor alone was not enough. Even potentially valuable claims became worthwhile only when combined with the capital and technological resources necessary to exploit them. An active San Francisco stock market grew up to finance the operations of the Comstock mines, linking the town early and tightly to a larger economic network. Its fortunes rose and fell with the price of shares. The women of Virginia City entered this exuberant economy as entrepreneurs of every sort, as both Mathews's memoir and the list of occupations in James and Fliess's appendix attest. Even Georgina Joseph invested unsuccessfully in mining stocks.[16]

Theirs was not a rural world of women confined to traditional domestic and agricultural endeavors. Virginia City's work force was an industrial one, organized into relatively powerful unions. The miners' expensive tastes in food, drink, and entertainment bespeak a generally high level of consumption during flush times.[17] An elaborate commercial infrastructure grew up to supply their desires. This was also a world of sudden and frequent reverses. Miners lost limbs and lives in accidents. Families as well as service providers suffered when the mines shut down operations from time to time. Women, perhaps seeking refuge from such uncertainties of life in a mining town, divorced or disappeared into the hazy worlds of opium and laudanum.

People came to the Comstock to make money through wages or through speculation. They stayed only so long as there was money to be made. In a mining economy, land was of little value in and of itself, and agriculture was a negligible activity. Mark Twain observed that the Nevada legislature spent $10,000 on an agricultural fair to display $40 worth of pumpkins. When mines or stock markets played out in Nevada, men and women who could do so moved on to more promising locations. Often these were other, newer mining camps.

Locations and circumstances thus took on a temporary quality. When business on the Comstock fell off, for example, Mary McNair Mathews went down to California to do sewing, and ultimately returned to New York. Those like Georgina Joseph, who could not afford to move on, took in boarders to make ends meet and hoped for change in the form of good health and better times. Others, like the Paiute women in Six-Mile Canyon, were more remote from both bonanza and borrasca. At home in a place that others encountered as newcomers, the Paiutes adjusted their daily patterns and waited out the boom. They continued to live among the gaudy remnants after most of the Comstock population had moved on to other mining rushes.

The circumstances of the various women whose lives intersected in Virginia City, Nevada, thus bore little resemblance to Riley's Euro-American agricultural frontierswomen. The fact that, as one California man put it, "we come and go and nobody wonders and no Mrs. Grundy talks about it," suggests the anonymity emerging in large eastern urban centers more than the tight-knit social worlds of small, interior, agricultural western towns.[18]

Those women who were followers of mining booms, like their male companions, lived lives of uncommon mobility. Attracted to Virginia City by the lure of its riches, they were perpetually vulnerable to rumors of bigger workings somewhere else. In each new community they encountered a suspension of traditional social circumstances and expectations that could be disquieting or exhilarating, depending on individual perspective. In Virginia City they resided amidst a polyglot, primarily male population of unsettled habits and uncertain future. The interlude offered them both the opportunity, and the sometimes stark necessity, for considerable autonomy.

By rendering these women's experiences on the side of a mountain in Nevada more visible, both in their commonality and their uniqueness, the editors of this volume hope to expand our general knowledge of women on the nineteenth-century western mining frontier. We hope that *Comstock Women* will serve to recover not only Louise Palmer and Georgina Joseph, but some of the thousands of other women who participated in the brief, flamboyant history of the nineteenth-century Comstock Lode.

❧ 2 ❧

Women of the Mining West

Virginia City Revisited

Ronald M. James & Kenneth H. Fliess

Augusta Ackhert, a thirty-nine-year-old German immigrant, lived on Virginia City's commercial C Street in the midst of the Comstock Mining District during the early 1880s. Widowed a few years earlier, she worked at a candy store and took in lodgers to support herself, her eleven-year-old Irish American adopted daughter, and two younger children. She had left New York in about 1875 and eventually arrived in Nevada, coming by way of California. Her renters included an African American family of seven. Concurrently, down the hill in Chinatown, Ty Gung of China cared for her infant daughter and kept house for her sixty-year-old husband, a cook who had been unemployed for nine of the previous twelve months. Emma Earl, a fifty-one-year-old miner's wife born in Pennsylvania, lived in the residential neighborhood on the north end of town. She augmented her household's finances by working as a dressmaker.

These profiles, based on the tenth U.S. manuscript census of 1880, incorporate one of the best sources of information available on a remarkable corner of nineteenth-century America. Although thousands of women lived there and helped build it into a place of importance, the history of the mining district typically focuses on men.

In 1859 prospectors in the western Great Basin found the Comstock

Lode, an incredibly rich ore body of gold and silver. In the ensuing rush, hopeful would-be millionaires streamed into the mining district, establishing the boomtowns of Silver City, Gold Hill, and Virginia City, frequently referred to as the "Queen of the Comstock." Roughshod camps quickly transformed into cosmopolitan centers of industry and commerce with a diverse international array of inhabitants. At its peak in 1875 the combined communities of Gold Hill and Virginia City had about 25,000 people, making this an urban center rivaling all but the largest cities on the continent. From 1860 to 1882 the Comstock produced $292,726,310 in precious metals, and this during a time when its miners regarded themselves as well paid at four dollars a day.[1] The Comstock quickly won renown because of its productivity but also because its miners, engineers, mill men, and banking executives defined hard-rock mining at the time. The district gave the American West some of its finest examples of corporate mining, and it also created an industrial prototype that others used throughout the region.[2]

Historians have frequently celebrated the technological and industrial aspects of the Comstock, paying less attention to the families that built the place into a community. Ironically, one of the most discussed episodes of Virginia City history affected everyone, women as well as men: on October 26, 1875, the core of Virginia City burned to the ground. This "Great Fire," as it is called, is a pivotal event in Comstock history, and yet of far more significance was the rebuilding. Within months the community rose from the ashes, a phoenix grander than before. The speed of the resurrection provides graphic evidence of Virginia City's vitality at the time.

In spite of the astonishing amount of wealth that the Comstock produced, its days of bonanza were destined to end. By 1880 the mining district had begun a long downhill slide into depression that ultimately reduced its population to less than 1,000. Despite desultory revivals of mining in the intervening years, the local economy has depended for decades on tourism, promoted especially by the popular television show, *Bonanza*, which ran from 1959 to 1972.

Given this history, it should come as no surprise that treatments of the Comstock's past have usually focused on the district's men. Engineers and miners designed and built the mines that brought fame, and historians have written about them for over 100 years. Famous residents like Samuel Clemens, John Mackay, William Sharon, James Fair, and Adolph Sutro have been the object of repeated study. Still, Augusta Ackhert, Ty Gung, and Emma Earl, together with the thousands of other women from all over the world, also made the Comstock home and helped form its character. This volume attempts to tell some of their stories. An analysis of the women of the Comstock reveals that they, as much as the men, created a community and established the infrastructure needed to support the local industry. The women complemented the diversity and complexity of the mining district. Any history that predominantly focuses on the men of the Comstock ignores a vital force that helped shape the place.[3]

Some of the most powerful tools to study women on the Comstock are the federal manuscript censuses. Isolated census entries, while providing glimpses into the lives of people, offer too little information to allow much progress to be made. Together, however, they can suggest general trends, laying a foundation that is much more solid than that which can be constructed through the use of other sources. A comparative overview based on census records provides, therefore, not only a means to organize the existing data, but also a basis for further research. In this way, it is possible to place the Augusta Ackherts, the Ty Gungs, and the Emma Earls in context and to understand them and their counterparts better.

Statistical summaries of data on Comstock women are easily obtained but are not without their own problems. Enumerators invariably missed some people. This may have been more common in Chinatown and the red-light district, for example. In these places more people may have wished to remain anonymous, or they may have discouraged enumerators from lingering. In addition, the U.S. census mandated that Native Americans be recorded separately with little more than a simple head count. Detailed information is consequently not available for most of these people. A few Northern Paiutes appear in the 1880 manuscript census against the enumerators' instructions, but such records are anomalies. In spite of these limitations, the federal censuses of Storey County provide information on trends within the female population. To this end, the Nevada Census Project undertook the comprehensive encoding of all federal manuscript census data for Storey County from 1860 to 1910.[4]

This and many of the following chapters employ data from a first phase of the Project. Although the results are preliminary, it is clear that the Project affords researchers an unprecedented opportunity. A comprehensive computerization of all census records for a sizable community does not exist elsewhere. The implications of such an undertaking are far-reaching. With these records, it is possible to pursue calculations related to demography without concern for sampling error made due to the fact that only a

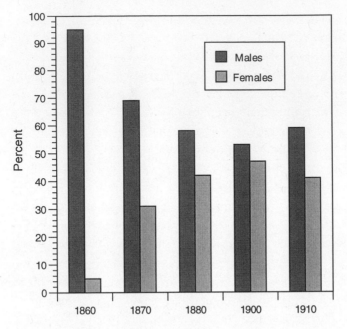

*Fig. 1. Percentage of population in Storey County, by sex, 1860–
1880 and 1900–1910.* (Source: U.S. Manuscript Census)

part of the population was examined. The value of computerized census
material to historical researchers, as is indicated by many of the chapters in
this volume, suggests that at this time we can only begin to understand the
ramifications of this resource on subsequent studies.[5]

Figure 1 graphically illustrates the percentage of males and females for
the first five federal census years (see also appendix 1, table 1.1 for the statistical summary of the same data). Clearly, the female portion of the population, at approximately 5 percent, was quite small in 1860, while the sexes
were most closely balanced in 1900, when females comprised more than 47
percent of the population.

When using census data for more than raw numbers, it is important to
approach them with healthy skepticism, because situations were rarely as
simple as the enumerator made them out to be. For example, women could
pursue a number of occupations simultaneously, a juggling act perhaps less
frequently required of men. Mary McNair Mathews, who lived in Virginia
City in the 1870s, worked as a teacher, a nurse, a seamstress, a laundress,
and a lodging-house operator, and yet she would have appeared in the

manuscript census as having just one pursuit. The census provides only the occupation that a given woman, her neighbors, or the enumerator chose to place at the top of the list. Presumably, that occupation was usually the type of employment that dominated the woman's time, but this may not always have been the case. A woman might have declared the occupation she regarded as most prestigious even if she pursued it infrequently. Neighbors or the enumerator may have assessed a woman's principal occupation as the one that was most obvious. In addition, one should always be on guard against the simple stereotypes that have dominated much of the way in which Western writers have portrayed women. As Elizabeth Jameson points out, "The interplay of gender, family roles, and economics which shaped Western women's options is more complicated than the images we have imposed on them."[6] In spite of these problems, however, census records are a valuable tool in the effort to understand the women of Virginia City.

The eighth U.S. manuscript census occurred in 1860, only thirteen months after the discovery of silver on the Comstock. It provides a snapshot of the community during its first boom. It was the first census to record the area now known as Nevada, which at the time was still the western part of the Utah Territory.[7] The census allows for a demographic profile of the few women present and challenges commonly held views of what boomtown life was like in the mining West.[8]

The myth of the Wild West incorporates the idea that the earliest period of a boomtown's social development was dominated by dance-hall girls and prostitutes who led the community in an unrestrained stint of abandon. People imagine this era to have ended when respectable women arrived and held the licentious proclivities of humanity in check. This has resulted in a twist on the Frederick Jackson Turner story: substituting frontierswoman for frontiersman, some see westering white women (the bad and reckless followed by the upstanding) taming the frontier. The manuscript census records no prostitutes in Virginia City and Gold Hill in 1860 and careful examination of the information concerning the documented 111 adult women suggests that prostitutes were either extremely rare or nonexistent at the time.[9] Sally Zanjani alludes to a similar phenomenon in her study of the mining frontier as it arrived in the turn-of-the-century boomtown of Goldfield, Nevada. She points out that one of the first women there "was not a stereotypical frontier harlot, with or without a heart of gold, but rather an independent entrepreneur as tough, unscrupulous, and aggressively ambitious as any of the men surrounding her."[10]

Virginia City's Hurdy-Gurdy Girls of the early 1860s. (Illustration by J. Ross Browne; courtesy of Nevada Historical Society)

Few women declared occupations on the Comstock during the 1860 census, and, given the already low numbers, developing a statistically valid demographic profile for any profession is impossible. The schoolteacher, the three seamstresses, the laundress, the milliner, and the two saloon keepers seem unremarkable and merely serve as precursors of those who followed. Emma Rigg, a seventeen-year-old native of Nova Scotia, claimed "theatrical" as an occupation; she was recorded along with three men, including her husband, all with the same profession. The diversions of 1860 were certainly growing in complexity, but the census is more tantalizing than informative in this as in so many other cases. Nonetheless, women who may have been less than respectable by the standards of the day were in the minority. Of the 111 women, 83 were living with husbands in the two communities, and 43 of these were looking after more than 100 children.

Thus, the 1860 census clearly indicates that the roots of a family-based community on the Comstock were established early. Certainly, some women were living on the wild side, but they were relatively few. This observation concurs with that of J. S. Holliday, who, in his monumental study of the California Gold Rush, points out that single women tended to come West by ship, while women who traveled overland usually had husbands. In addition, he maintains that "the women who landed in San Francisco stayed in that metropolis or settled in Sacramento City, for there was little reason to go to the primitive mining towns and camps."[11] That tendency appears also to hold true for the early Comstock.

The earliest period of the Comstock differs in one important way from that of the California Gold Rush. The latter featured a growing population scattered over a large territory with little infrastructure to support itself. Historian JoAnn Levy, for example, documents that women at the time frequently became successful entrepreneurs. They cooked, cleaned, and managed boardinghouses, often charging exorbitant rates in response to the scarcity of the commodities offered. They were able to do this for months or even years in some places, because it took considerable time for services to catch up with demand.[12]

The Comstock, on the other hand, relied on a concentrated ore body, discouraging far-flung settlements. In addition, the strike occurred ten years after the discovery of gold at Sutter's Mill, and the region was ready to respond more quickly and effectively with an established infrastructure. The Comstock consequently did not have a long period during which people, and women in particular, could charge exorbitant rates because their services were so rare. That is not to say that inflationary costs did not remain a problem throughout the mining West, but extreme price gouging was possible for a much shorter period on the Comstock than was the case in the early days of California's boom. The women who arrived first on the scene were able, therefore, to enjoy the halcyon days of capitalism and profits only briefly.[13]

An important element of life in a mining community was its fluidity. Change occurred constantly and sometimes suddenly. Census data after 1860 provide an opportunity to assess how some of this change occurred. Figures 2 through 6 graphically report the percentage in each age interval of females for each census year (see appendix 1, table 1.2 for a statistical presentation of the female age structure from 1860 through 1910). Figures 2 and 3 indicate a high percentage of females between the ages of twenty and thirty-five for 1860 and 1870, while Figure 4 indicates a high percentage of females between the ages of twenty-five and forty in 1880. These three figures also show a high percentage of females age five and below, especially in 1870. This may indicate that many of the twenty- to forty-year-old women gave birth soon after migrating to the Comstock, while the lower percentages of females between five and fifteen years old represent girls born before migration. There are also few women in the older age categories in 1860 and 1870, and it is not until 1880 that there is a sizable population older than age fifty.

Figures 4 and 5, which represent 1900 and 1910, respectively, show another pattern. Both are more pyramidal than are the figures for 1860 through 1880. There are no age groups that stand out for their large or small membership, though for 1900 the groups of children ten years old and younger are somewhat reduced. This is, most likely, the result of sampling error in a small population. The figure for 1910 indicates one anomaly: the relatively large percentage of females between the ages of sixty and seventy. This too is most likely the result of sampling error in a small population, although it may correspond to the exaggerated numbers of young women found thirty and forty years before.

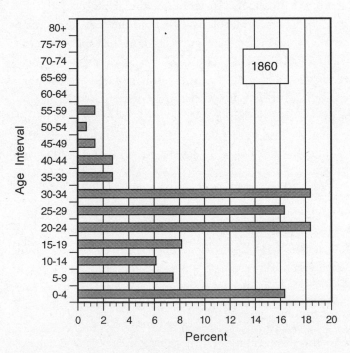

Figs. 2–6. Percentage of females by age, Storey County, 1860–1880 and 1900–1910. (Source: U.S. Manuscript Census)

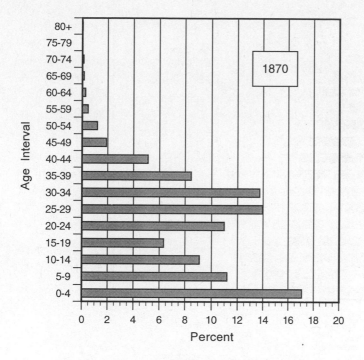

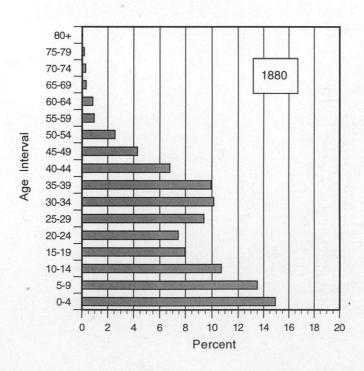

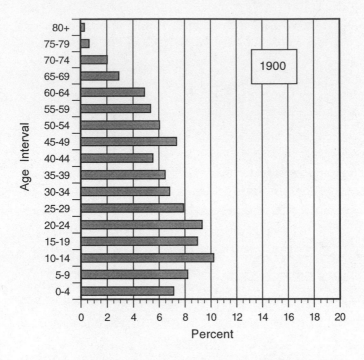

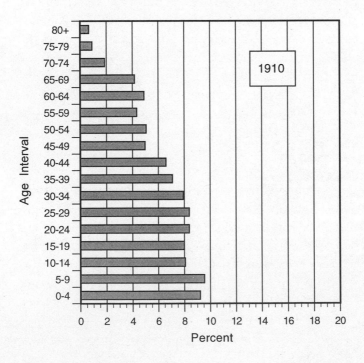

This type of graphic representation yields a clear picture of the age structure for each census year. While it is important to keep fluctuations in population in mind when reviewing these figures, comparing percentages makes it possible to determine and understand differences and similarities across time.

The Comstock was noted for its international population, but its complexion also changed over time. Table 2.1 reports the ethnic structure of females between 1860 and 1880. It is evident that the preponderance of females in all census years are categorized as "white," a category that made up anywhere from somewhat over 96 percent to 100 percent of this subpopulation, depending on the census year. The diversity among Comstock women was chiefly that of nativities from throughout Europe as well as North America. Appendix II reports the place of birth for females from 1860 through 1910. It also lists the places of birth for fathers and mothers for 1880 through 1910. Together, this information dramatically illustrates the range of possibilities. While it is desirable and appropriate to define the stories of groups such as African Americans, Asians, and American Indians, their representatives are so few in number that it is hard to arrive at generalizations and to maintain a clear focus.[14] This also makes it extremely difficult to draw conclusions concerning these groups based on census data. Similarly, appendix III reports the various occupations that women pursued during each census year. Again, the nature of the community changed dramatically over time.

Several trends are evident in the Storey County population during the fifty years in question. First, the female component of the Comstock increased substantially, especially when it is viewed as a percentage of the total population. It grew from about 5 percent in 1860 to between 30 and 47 percent for all subsequent census years. Second, the age structure of females on the Comstock transformed over time from a population dominated by women in early adulthood to a more regular pyramidal pattern. This is probably an indication that the population on the whole was beginning to develop in place without relying heavily on migration to build and sustain itself. Third, it is clear that the population was dominated by females of Euro-American ancestry. Those of other ethnic identities are relatively rare. In addition to this, the proportion of those born in the United States is quite high, ranging from just under 60 percent to over 75 percent. Still, a large number of native-born Euro-American women and girls' parents were born elsewhere. Fourth, women's employment changed over time.

TABLE 2.1   Percentage of Females by Ethnicity, 1860–1910

	African American		Chinese		American Indian		White	
Year	N	(%)	N	(%)	N	(%)	N	(%)
1860	0	0.00	0	0.00	0	0.00	147	100.00
1870	25	0.71	103	2.94	0	0.00	3,370	96.34
1880*	35*	0.52	41	0.60	65	0.96	6,641	97.92
1900	1	0.06	3	0.18	1	0.06	1,680	99.70
1910	4	0.33	1	0.08	3	0.16	1,220	99.43

*Includes fourteen females listed as "m" or mulatto.
U.S. manuscript census for the years 1860–1910.

Even if occupational categories are collapsed, more and more types of occupations are listed as the district matured. Given these conclusions, it is apparent that women became important as they spread throughout the community and helped diversify the economy of the Comstock.

While not disregarding other census records, it is particularly useful to look at the tenth U.S. manuscript census of 1880. Although mining was in decline at that time, there were still about 10,000 people in Virginia City, and much of the society's structure remained intact. The 1880 document is the first to include street addresses, and it thus yields precise identification of locations. Information on parental nativity, also first recorded in 1880, allows for a better definition of ethnicity.[15] In general, the 1880 census reveals a diverse community, with women figuring into many aspects of the economy and society. Tables in appendix 1 report data from the 1880 document that complete a statistical profile of the women of Virginia City. They show that marital status, age, and ethnicity influenced the choices women made when deciding how to earn money.[16] A discussion of each occupational group and of these factors underscores the complexity of women's participation in a nineteenth-century mining town.

For a variety of reasons, such an investigation must begin with prostitution. This was not the most important occupation for women on the Comstock. Indeed, relatively few of them pursued it. The profession must be addressed first, however, because it requires a clear definition before one can begin sorting out other groups. Sexual commerce raises several issues regarding the demography of Virginia City. Prostitutes were the group most likely to deceive a census enumerator regarding their occupation. Sociologist Marion Goldman discusses the resulting inaccuracies in the census in her *Gold Diggers and Silver Miners,* a crucial, pioneering study of the Comstock.[17] The author's consideration of a varied body of primary data makes her contribution a useful reference for this line of research. Unfortunately, it appears that Goldman includes more women as prostitutes than the data indicate as having existed.[18]

In evaluating the occupations of Virginia City's women, Goldman considers age, marital status, household, and place of residence. Such an assessment is necessary not only to understand prostitutes, but also to discuss women truly pursuing other employment, and so it is crucial that this task be undertaken as carefully as possible. Evaluating women for possible association with prostitution is difficult. Goldman reasonably assumes that a young single or divorced woman living with or near prostitutes but claiming another occupation was probably also involved in sexual commerce.[19] This seems likely, but there are several problems with the approach. First of all, women fitting this profile may have been nothing other than what they professed to be. In addition, a prostitute may also have pursued other occupations, and a seamstress, for example, may have worked occasionally as a prostitute.[20] And finally, prostitutes who lived away from the traditional red-light district and claimed other professions will remain unnoticed given this standard for reclassification. Still, a conservative reassessment of the 1880 census using these criteria increases the number of Virginia City women involved in prostitution from the forty-seven who reported it as their occupation to seventy-seven. Employing this approach before 1880 is not possible, because earlier U.S. censuses do not record addresses. For this reason, Goldman focuses on the 1880 document, an emphasis repeated here.

The 77 identifiable prostitutes in 1880 fall far short of the 134 women Goldman identifies as pursuing this profession in the Virginia City census. A careful examination of Goldman's methodology reveals several difficulties that misdirected her analysis.[21] Ultimately, it is possible to concur with fewer than half of her identifications of alleged prostitutes. Because of this, Goldman's work with the census must be regarded as seriously flawed.[22]

In contrast to Goldman's methodology, strict use of the same criteria with census data can isolate a group of women more reliably associated with the profession. This method also reduces the likelihood that prostitution will affect the evaluation of other occupations. With this in mind, it is possible to use the 1880 manuscript census to arrive at conclusions regarding prostitution as well as regarding other occupations involving women on the Comstock.

In 1880 prostitution was apparently the fourth most common occupation for Comstock women, following keeping house, working as a servant, and needlework. Of course, it is not possible to identify all prostitutes from the census, but this ranking is defensible.[23] In 1870, however, prostitution appears to have been the most commonly pursued wage-earning occupation listed in the manuscript census. Using these two documents, which are separated by ten years, it is possible to arrive at an understanding of prostitution in the mining district. The average prostitute was young, was often divorced, and was rarely widowed or married. She usually had no children and lived with fewer people than other women did.[24]

Some ethnic groups dominated Comstock prostitution. Although the census lists many Asians and Hispanics as prostitutes,[25] even more Asian and Hispanic women declared keeping house as an occupation. In most cases there is no reason to believe, as Goldman apparently does, that some of these women were also prostitutes.[26] In 1870 exactly the opposite may have been the case. It is more likely that some Asians and Hispanics categorized as prostitutes were not engaged in sexual commerce. Because of prejudice, the enumerator may have assumed that some of these women were prostitutes when they were not.

Much has been written on the relative status of ethnic groups in red-light districts. This is particularly true for Asians but is also the case for Hispanics. Society generally restricted both groups to the lower rungs of the socioeconomic ladder within the community of prostitutes.[27] It is clear that racial prejudice and circumstances of immigration limited the choices of these women, many of whom played important roles in brothels and cribs. Nonetheless, sweeping conclusions about Asian and Hispanic women and prostitution can be inaccurate. In spite of the limited options and opportunities available to them, many apparently found work outside brothels.

Although relatively few of the Comstock women born in the United States became prostitutes, the ones who did were sufficiently numerous that they filled the rank and file of the red-light district. Unencumbered by prejudice, these Euro-American prostitutes occupied the full range of available statuses. They appeared as everything from madams to the most destitute of prostitutes. Euro-Americans were often the preferred prostitutes on the Comstock, but those who professed French nativity could most easily strive for higher status.[28] Several prostitutes claimed to be French, but women of that nativity were so few in general in both 1870 and 1880 that generalizations would be unfair. Still, their ethnically based prestige afforded them more attention and status than their number might indicate.[29]

Four African Americans of Virginia City can be identified as prostitutes in the 1870 census, but given that there were only twenty-three African American women in the city at the time, generalizations are problematic. Similarly, three of twenty-six African American women were prostitutes in Virginia City ten years later. Nonetheless, it is possible to make a few observations about these women. In 1880 two prostitutes claimed to be mulattos, an ethnicity that had the potential to increase the price commanded by an African American prostitute. In 1870 the four African American women identified as prostitutes were born in Virginia, Massachusetts, Nova Scotia, and California. Ten years later, yet another Californian worked the line, joined by women from Mexico and Panama. African American women coming directly from the southern states to the Comstock almost always pursued other careers.[30]

With the ranks of prostitutes carefully identified, it is possible to provide an overview of other occupations undertaken by Comstock women. Most women on the Comstock were involved in domestic work in one form or another. Three quarters of the women on the Comstock in 1880 claimed to be keeping house. This may be misleading, because some may have earned money at a variety of tasks such as sewing, teaching, or washing clothes. In addition, the 1880 census clearly indicates that many of the women who claimed housekeeping as an occupation managed households that included some lodgers. Women listed with no occupation were usually young, were living with parents, and apparently were not required to contribute to the household finances. It is reasonable to assume that most families expected them to help their mothers until the young women married and began keeping house themselves. They were, therefore, part of the cycle of domestic work and should also be considered here. For the most part, only Euro-American women born in the United States could afford the luxury of not pursuing an income.

A few women reported that they maintained lodging houses. As JoAnn Levy points out in her study of women of the California Gold Rush, those who rented rooms were involved with a wide variety of domestic tasks including cleaning, cooking, and otherwise caring for their families and tenants.[31] Besides lodging-house operators, housekeepers, and those listing no occupation, there were also many servants in 1870. Ten years later, there were more women in domestic service than in any other income-producing profession.

Not surprisingly, the majority of women keeping house were also married. Some widows maintained the title of housekeeper, but more of these were certainly engaged in moneymaking activities than were their married counterparts. Few of the women who kept house were divorced. Housekeepers also had, on average, the most people in their families and were the least likely to cohabit with large numbers of lodgers. This occupation is related to the enormous topic of child rearing, the history of which is also not often studied and could potentially offer valuable insights into Comstock society.[32]

A series of letters to her parents from Georgina Joseph, an immigrant from Auckland, reveals a great deal about women involved in domestic activity on the Comstock. Joseph arrived with her husband in 1867 and described her efforts to care for her household, including the two babies she delivered in the following three years. Writing home, she pointed out that domestic work could be profitable and that "wages here for girls are very good, from 30 to 50 dollars." She went on to write that she could secure a position for a friend of hers "in the same house with me to cook and clean for three single men at 35 dollars a month."[33] It is likely that many women who listed "keeping house" as an occupation had an income based on caring in this way for one or more roomers.

Women who claimed to keep a lodging house tended to be older, and they were more likely to be widowed than married. The census lists as married many who did not have husbands living on the premises. A larger percentage of divorced women than of women as a whole ran lodging houses. No doubt this was because many had established households, which represented one of their few assets, while at the same time they felt a keen need to raise money for day-to-day existence.

Affluent housekeepers on the Comstock frequently sought to hire servants to deal with the more tedious aspects of caring for a house and family. The luxury of such assistance was also an expression of wealth. Lodging-house operators, particularly those running larger establishments, employed servants as a necessity. The earliest Comstock sources point to the difficulty of hiring servants. In 1860 only one woman from Virginia City or Gold Hill—twenty-two-year-old Bridget Deobin—appears to have been a servant. Although the enumerator listed her with no occupation, she was living in the same household with a machinist and his wife, to whom she appears otherwise unrelated.[34] J. Ross Browne, in his *A Peep at Washoe,* comments on how difficult it was to hire a servant during his 1860 excursion to the Comstock.[35] A booming economy and diverse markets offering well-paid employment in many fields proved to be obstacles when it came to attempting to lure women into domestic service. Nonetheless, many households eventually employed servants, and, as is mentioned above, they gradually became numerous on the Comstock.

Mary McNair Mathews discusses the topic at length in her observations from the 1870s. She was concerned about competition between women and Chinese men among the ranks of servants, and although her racism kept her guarded, the topic generated what may have been her only positive comment about Asians.[36] Quoting a woman who employed a Chinese servant, Mathews pointed out: "They will do things for us I would not like to ask a white person to do; besides, they never tell any family affairs like white girls do."[37] The cheap source of labor offered by the Chinese community was a constant threat to non-Asian women who wished to work as servants. To balance the economic efficiency and hard work of Asians, Euro-American women could count on the preference that their ethnicity afforded them. Although more costly and reputed to work less, a Euro-American servant provided an employer with more prestige than an Asian one did.[38] As a result, many of the wealthier households employed one of each: a Chinese cook, who was also responsible for the heavier and dirtier household tasks, and a woman (usually Irish) who cared for children and guests and who was involved in the more social activities.

The 1880 manuscript census for Virginia City reveals a community with ample opportunities for the employment of servants. There were 156 women and 43 men among their ranks that year. A few generalizations concerning the women so occupied are appropriate here. Almost all were single, and most were young.[39] Half of the female servants on the Comstock at the time were Irish, supporting the Victorian-era stereotype of Bridget the parlormaid. In general, the more affluent homes up the hill from the commercial corridor were the ones most likely to employ servants. Businesses along C Street also provided a market for these workers. Residences and businesses outside this central core employed few servants, and those they did hire tended to be Irish. This was particularly true of the Irish neighborhoods themselves (see page 250).

By contrast, there are no Chinese, Hispanic, French, or Jewish women appearing in the 1880 census as servants.[40] Chinese men, on the other hand,

"Visiting French actresses" on a mine tour. (Courtesy of Nevada Historical Society)

were commonly employed in that capacity, forming the chief source of competition to Irish servants. In fact, servants, restaurant workers, and laundresses, who together represent the vast majority of wage-earning women on the Comstock, all competed with Chinese men. Nevada historians cut their teeth on stories of railroad workers and miners who excluded the Chinese from their industries for fear of competition, yet most wage-earning Comstock women appear to have been in direct economic rivalry with the Chinese. In order to survive in a marketplace with Chinese workers willing to live on lower wages, Euro-American women established themselves in specialized niches, but they were not able to exercise exclusive control of their industries in the way their male counterparts did.

The absence of mass-produced clothing, the difficulty of home laundering, and a Victorian preference for elaborate attire created many jobs for women making clothing and for those serving as milliners and laundresses.

Euro-American men who worked as tailors or as wash-house owners and laborers provided some competition, but the vast majority of those involved in these trades were women. The many Chinese men who operated laundries were again a remarkable exception.

Nearly a hundred seamstresses, dressmakers, and sewing women worked in Virginia City in 1880. Some of the more affluent households kept one on staff as they would a domestic servant.[41] Other women making clothes operated independently, depending on piecework. Most were young and single, although a few older widows returned to the occupation when faced with financial need. The thirteen who worked as milliners in both 1870 and 1880 fall into much the same pattern, except that the millinery business attracted more married women than it did other garment workers. Few occupations were as heavily dominated by women born in North America as was the clothing industry.

Most women involved in laundry work were Irish. As in the case of the female servants, the prime competitors of the Irish women in laundry work were the Chinese. This subject has been treated comprehensively

The interior of a Virginia City house, depicting the Goodmans—a Jewish family in 1869. (Courtesy of Nevada Historical Society)

elsewhere, but a few general observations bear repeating here.[42] Washing clothes was an arduous, time-consuming process in the nineteenth century. One of the first luxuries purchased with disposable income was laundry service. Not surprisingly, most considered the profession to be among the least desirable. Large, mechanically run laundries owned by Euro-American males handled most of the commercial washing on the Comstock. An assortment of facilities owned and operated by women and by Euro-American or Chinese men satisfied the private market.

Although it was backbreaking work, washing provided women a means of support when none other was possible. Competing with Asians, however, appears to have been difficult. Mary McNair Mathews, who worked on the Comstock as a laundress for a brief time, bitterly railed against Chinese labor.[43] Nonetheless, fifteen women in Virginia City found a niche in the industry during the time of the 1880 census.

Many other women, including those listed as keeping house and operating lodging houses, certainly did washing for money on occasion. It is impossible to identify such cases, but the few women listed as running laundries usually fit a well-defined profile. They were typically older than those involved in other occupations, and they were all widowed or living without a spouse because of divorce or separation. Most had children. The Irish dominated the industry, a fact made all the more striking when the location of Irish neighborhoods is compared with that of Irish laundries. Irish laundresses appear to have taken advantage of their neighborhoods to protect themselves from Chinese competition.

Besides the women involved in the more common occupations, there were a few who pursued alternatives. Daughters of Charity, teachers, and nurses were relatively rare on the Comstock. In 1880 only sixty-six women of Virginia City claimed these three occupations. Because of the custom of employing single women as teachers, members of that group were predictably young. They also tended to be Euro-Americans of North American nativity. The Comstock did not strictly follow nineteenth-century prohibition against married teachers, so several married women taught in Virginia City. This perhaps reveals a necessity born out of the scarcity of women in the mining West. In addition, there was considerable opportunity for private education in Virginia City. Mary McNair Mathews, a widow with a school-age child, also briefly ran a private school.[44] It appears that any literate woman could pursue such an option.

In contrast to the youth of the teachers, the nurses of the Comstock were usually older, and they were often widowed.[45] It appears, therefore, that women undertook nursing, like laundry work and lodging-house operation, because of financial need later in life. Presenting a similarly older demographic profile, the Daughters of Charity had a thriving order in Virginia City. The sisters worked as teachers at their orphanage and convent between H and I streets and as nurses at St. Mary's Hospital down Six-Mile Canyon.

About 1 percent of the women pursued options rarely available on the Comstock, and by their exception they help define the whole. They tended to be older and more likely to be widowed than other women were. Census manuscripts record such women as working as merchants, waitresses, and cooks, and as operators of saloons and restaurants; there was also a florist, an opium-den operator, a bookkeeper, a book folder, a fortuneteller, a peddler, an upholsterer, a hairdresser, a house painter, a dairy operator, and a laborer. These unusual careers date to the earliest period of Comstock development, as is evidenced by Emma Rigg, the actress of 1860 mentioned above. In 1863 Madam Schroder conducted business as a diviner, an astrologer, and a folk healer.[46] Three years later, Susan Carroll and Mary Conway, two needleworkers, recognized an opportunity and opened a "Female Employment Office." According to the *Territorial Enterprise*, they could assist women "in search of respectable employment . . . and accommodations to families in the way of obtaining female 'help.'"[47]

In spite of the diversity of occupations available to women on the Comstock, they were almost entirely excluded from mining and milling, the district's largest source of employment. Mine managers frequently set aside the traditional prohibition against women entering a mine for numerous visitors to the progressive industrial community, but convention was sufficiently strong to deter their employment there.[48] In 1871, however, a group of women opened a mine, digging an adit on B Street by themselves. Small-time operators on the Comstock numbered in the thousands, usually representing the attempts of the common man to stumble into a bonanza and instant wealth. For women to found and excavate a mine was unusual, if not unprecedented. Like most of these limited endeavors, the women's mine probably remained active for only a short time, but the *Territorial Enterprise* felt it remarkable enough to justify at least two articles on the subject. One included a supportive editorial comment: "We do not see any reason why women should not engage in mining as well as men. If they can rock a cradle, they can run a car; if they can wash and scrub, they can pick and shovel. Although some gentlemen friends of the ladies are attempting to persuade them from continuing work, they are determined, and we are pleased to see it."[49]

Unfortunately for women, this enlightened sentiment appears to have been as rare as the mine that inspired it. On the other hand, additional research may reveal that women were more frequently involved in mining, particularly with smaller, family-run operations. The Nevada State Historic Preservation Office gained insight into this possibility in 1990, while recording a small rathole mine on the Comstock. A collapse of the mouth of the mine had sealed it, together with a large assortment of nineteenth-century artifacts, for at least seven decades. One of the most tantalizing of the artifacts found was a woman's basque, or close-fitting undergarment, found beneath what was probably a man's vest. Perhaps it was only a rag used by men, but there is the possibility that it belonged to a woman who was helping out in a family enterprise.[50] Conclusive answers are unattainable in this case, but clearly the question begs for more research on the involvement of women in the mining industry, and it may be archaeologists who ultimately shed light on the topic.

Eilley Bowers and Mary McNair Mathews serve as good examples of women trading in mining stocks and claims, but the extent of this activity, as with many others that women engaged in, awaits further research.[51] In addition, much of this activity will require a sorting out of interests and efforts by spouses. Georgina Joseph wrote of stock investments she made with her husband. It is clear that she took an active interest in the market. From her detailed account, it appears that she participated in decisions, but again the nature of men's and women's roles in family undertakings requires more study.[52]

This brief overview of women as they appear in the census manuscripts clearly suggests that age, ethnicity, and marital status each played a role in the choices women made, or were allowed to make, with regard to occupation. It is equally clear that the Comstock was a place of diversity, where women of widely varied backgrounds lived and worked. In spite of their value, summaries of groups lack the life breathed into the past by individuals. The definitive treatment of women on the Comstock will include both.

In his memoirs of Virginia City, the impresario David Belasco recalls a poignant incident involving a woman and her dying daughter: "The mother was holding [her sick child], and I knew when I glanced at the flushed face that there was little hope. The faro men were calling outside the window, music came in from a dance hall near by, and rowdies were swearing in a saloon across the way. But here was a mother with the mother love in her eyes and all the tragedy of it too, watching the life ebb from the small body held close to her heart."[53]

Too often, the image of women of the mining West evokes only prostitutes and bonanza queens. It is important to remember that the experiences of women in the region were diverse and complex. While earlier histories of the Comstock have given attention to Eilley Bowers, Julia Bulette, Marie Louise Mackay, and a few others, the broad cross section of women must be understood in the context of their communities, their occupations, and their ethnicities. The history of the Comstock's women should be given in the context of full lives, experiencing happiness, achievement, and mother love, as well as frustration, sadness, and tragedy.

In the letters from Georgina Joseph, for example, one sees the gamut of loneliness, sorrow, joy, and hope. She took delight in her newborn children as well as in the daughter she brought with her. She worried about her husband's incessant illnesses that kept him from work. She expressed pride in her family's ability to do well economically, but she fretted over its losses in the stock market. Sadly, she could not write the final chapter of her Comstock story. The Gold Hill Cemetery Records show that in 1870 she died at age twenty-six while delivering the third of her Comstock babies, which in turn succumbed shortly afterward.[54] For all that her life was and was not, her story survives in the letters she penned and the tale she told. As historian Susan Armitage points out, "ordinary lives are the true story of the West, for men as well as women."[55] We may never know much about individuals such as Augusta Ackhert, Ty Gung, and Emma Earl, described at the beginning of this essay, but we can attempt to uncover their circumstances, piecing them together with the stories of Mary McNair Mathews, Georgina Joseph, and others, to understand something of their opportunities and motivations. It is with this ambition in mind that the authors of this volume pursued their work, beginning, at least, to paint the portraits that Comstock history has too often neglected.

PART TWO

❧ At Home in a Mining Town ❧

As a western mining town, Virginia City conjures up an image of prostitutes and bachelor miners living in a rough-and-tumble society. This stereotype is easily challenged by statistics showing that the community grew quickly into a stable one in which there were many families. Nonetheless, a thorough understanding of what it was to be "at home" in a mining town requires further explanation. The following essays portray three facets of a complex jewel.

The sudden development of the Comstock caused it to grow explosively. Available housing often lagged behind need; in such a young, opulent economy, many women saw that providing lodging could be profitable. Similarly, chaotic conditions and relaxed enforcement of traditional mores made divorce a more common option than it was elsewhere in the United States. Not all women took in lodgers, and not all of them pursued divorce as an alternative to a bad marriage, but both practices affected home life throughout the community. The same is true of the use of opiates. Drug use occurred at home in the form of the consumption of accepted medications and in opium dens. Wherever women used opiates, the drug affected the domestic realm. It was occasionally even implicated in divorce cases.

The Comstock invented itself, combining eastern middle-class standards of behavior with a range of international influences in an environment far removed from the rest of the nation. The result was a new community reminiscent of other industrial towns but with its own, unique western stamp. Much remains to be done to understand the nature of home life in this mining district. The prevalence of widowhood, the scarcity of servants, and the effects of poverty and wealth are yet to be researched thoroughly. Nonetheless, the following chapters add to our understanding of important facets of domestic life in Virginia City.

❦ 3 ❦

Redefining Domesticity
Women and Lodging Houses on the Comstock
JULIE NICOLETTA

D omestic architecture and labor on the Comstock is a topic that has received little attention from historians. While engineering feats such as
the Sutro Tunnel and the mines themselves have captured the imagination of contemporary writers and twentieth-century historians, virtually no research has been devoted to housing in Virginia City. The man-
sions of Bonanza kings James Fair and John Mackay and of the town's
other wealthy inhabitants were heavily publicized in the local newspapers
at the time they were built and have been documented to some extent since
then.[1] Yet the buildings housing less-known people—the bulk of the men
and women who came to the Comstock to take part in the gold and silver
boom—remain largely unstudied.

By focusing on women who worked as keepers of lodging houses, a type
of building that accommodated hundreds of Virginia City inhabitants, we
can see how women used a specific occupation to make a living. This chapter considers these women and the lodging houses they ran from 1860 to
1910 as a means to understand why women operated these businesses and
to what extent they controlled decision-making concerning lodging-house
construction, furnishing, and use. In addition, it analyzes the change over
time of lodging-house forms, the makeup of their households, and the

TABLE 3.1   Population Numbers for Lodging-House Keepers by Sex in Storey County, 1860–1910

Total Population		Male	Females	Male Lodging-House Keepers	Female Lodging-House Keepers
1860*	6,841	6,104	737	18*	6*
1870	11,359	7,864	3,495	34	32
1880	16,115	9,294	6,821	21	80
1890	8,806	5,144	3,662	—	—
1900	3,673	1,933	1,740	6	28
1910	3,045	1,781	1,264	1	7

*Nevada portion of Utah Territory. Numbers for lodging-house keepers include Virginia City, Silver City, and Gold Hill only.
— means no data available.
Sources: Eighth U.S. manuscript census for the Nevada portion of the Utah Territory, 1860; ninth U.S. manuscript census for Storey County, 1870; tenth U.S. manuscript census for Storey County, 1880; eleventh U.S. census records for Storey County, 1890; twelfth U.S. manuscript census for Storey County, 1900; thirteenth U.S. manuscript census for Storey County, 1910.

relationship these buildings had to the economy of the Comstock. It concludes by putting female lodging-house keepers in the context of nineteenth-century ideals of domesticity, arguing that these women extended the boundaries of traditional work roles in order to expand their economic opportunities in the West. Despite some historians' claims that women had broader opportunities in the West than they did elsewhere in the country, women lodging-house keepers still worked within domestic spaces. To better understand their roles, we need to understand these spaces.

Studying women lodging-house keepers is somewhat difficult, because they and their boarders left few records. In addition, many of the nineteenth-century lodging houses noted in written documents no longer stand. Nevertheless, a variety of sources contain information about lodging houses on the Comstock. City directories and newspaper advertisements and articles provide names and addresses of some lodging houses and their proprietors. In some cases, articles describe more prominent houses, offering a sense of the exterior and interior appearances of these buildings. The manuscript records of the U.S. census from 1860 to 1910 contain useful information, such as age and marital status, about individuals listed as "lodging-house keeper," "boardinghouse keeper," or "hotel keeper"[2] (see tables 3.1 and 3.2). These records also provide valuable lists of household members of lodging houses. Contemporary accounts of life in lodging houses offer a rare view into a world that no longer exists. Finally, studies of women in other western towns contribute a comparative context for lodging-house keepers in Virginia City and the surrounding area.

This chapter deals with five distinct economic periods between 1859 and 1910. The events of each period had an effect on women on the Comstock and on their lodging houses. The first period, from 1859 to 1860, covers the time of initial settlement, immediately after the discovery of gold and silver. The second, from 1860 to approximately 1863, covers the first boom period. During the third period, from 1863 to 1872, the Comstock experienced several cycles of depression and prosperity. The fourth period, from 1873 to 1878, was the time of the Comstock's second great boom. During the fifth period, from 1878 to well into the twentieth century, the Comstock fell into a profound depression. As the region saw economic decline, the lodging-house profession became increasingly feminized as men searched for more profitable occupations.

Based on the census records and written accounts of lodging-house keepers on the Comstock, it appears that the vast majority of women in this profession were white and middle class. Relatively few people of color lived on the Comstock. For example, the 1860 census recorded seven African American men and no African American women. The 1870 and 1880 censuses include fewer than 100 African American men and women in total populations of 11,359 (1870) and 16,115 (1880). In general, the number of African American men was double the number of African American

TABLE 3.2   Female Lodging-House Keepers, 1860–1910

Year	Total	Average Age	Average Household	Married	Widowed	Divorced	Single
1860*	6	38.8	4.3	1	—	—	—
1870	32	37.3	5.6	13	—	—	—
1880	80	37.0	8.3	30	32	6	12
1890	—	—	—	—	—	—	—
1900	28	57.6	4.4	6	18	0	4
1910	7	52.9	5.3	3	4	0	0

*Includes Virginia City, Silver City, and Gold Hill only.
— means no data available.
Sources: Eighth U.S. manuscript census for the Nevada portion of the Utah Territory, 1860; ninth U.S. manuscript census for Storey County, 1870; tenth U.S. manuscript census for Storey County, 1880; eleventh U.S. census records for Storey County, 1890; twelfth U.S. manuscript census for Storey County, 1900; thirteenth U.S. manuscript census for Storey County, 1910.

women during this era. These proportions reflect population figures in the West, where, as historian Lawrence B. De Graaf shows, "in relation to whites, black women, like men, made up a minuscule portion."[3] The proportion of Chinese women to Chinese men living on the Comstock was similar to that of African American women to African American men, and, again, these proportions are similar to those in the West as a whole. The census does not include Native Americans until 1870 and even then counts only Native Americans on reservations and in towns. Nevertheless, women of color who ran lodging houses do appear in the census records, though in much smaller numbers than do white women.

A word must be said about the differences between the terms *lodging house* and *boardinghouse*. Until 1900 or so, the word *boardinghouse* applied to a place where both lodging and meals could be obtained, whereas a *lodging house* was only a place to stay. The term *hotel* was used interchangeably with *lodging house*, though the former generally applied to larger, more expensive establishments. The 1878–79 Virginia City directory cross-lists boardinghouses, lodging houses, and hotels. By 1900, however, the term *lodging house* seems to have replaced the term *boardinghouse* in the census; this change may reflect a shift in usage of the terms in society at large. For the purposes of this study, boardinghouses, lodging houses, and hotels are all examined, since they were often run by women, and since all provided temporary housing for a variety of people working on the Comstock.

Canvas tents, which offered temporary shelter for those who arrived in Gold Hill shortly after the discovery of gold in 1859, functioned as the earliest lodging houses. Nick Ambrosia, called "Dutch Nick," set up the first such house in Six-Mile Canyon. He simply erected a large tent and ran it as a saloon and boardinghouse, charging customers fourteen dollars per week for board.[4] Although men far outnumbered women on the Comstock in the winter of 1859–60, Eilley Cowan, who later made and lost her fortune as the wife of Sandy Bowers, ran one of the first boardinghouses and restaurants in Gold Hill; her establishment may even have been the very first of its kind in the area.[5] According to writer Eliot Lord, the first hotel in Virginia City was a canvas tent measuring fifteen by fifty-two feet. The communal lodging room, at fifteen by thirty feet, accommodated thirty-six guests who paid one dollar each per night.[6] An early visitor to the Comstock in 1860, J. Ross Browne, perhaps exaggerated for comic effect when he noted that Gold Hill's "Hotel de Haystack or Indication," had 300 guests crammed in together.[7] Both men and women operated their businesses out of tents in these months of initial settlement. Lodging houses, like other early buildings on the Comstock, were crude structures built for a mining camp rather than for an established town.

At this time, few women ran lodging houses in the Nevada portion of what was then the Utah Territory; at least women do not appear in large numbers in the 1860 census as lodging-house, boardinghouse, or hotel keepers. Out of a total of 111 adult women in Storey County, only 6 (5 in Virginia City and 1 in Silver City), all of whom were white, appear in those categories. Their average age was thirty-nine years. In contrast, 18 men appear as keepers of boardinghouses, lodging houses, or hotels located in Gold Hill, Virginia City, and Silver City. The total population of what would become Nevada in 1860 equaled 6,841—6,104 men and 737 women. The 6 women lodging-house keepers represented approximately 5 percent of the adult females living in Storey County. Other female professions listed in the census include domestic service and sewing, both traditional occupations that women could perform within the home for payment.

The 1860 census does not list street addresses, so it is difficult to pinpoint the location of houses in the early days of the Comstock boom. Most of these "houses," however, were probably tents or temporary, movable structures that had no permanent foundations. Thousands of people, mostly men, came to the Comstock right after the initial discovery of gold and silver in 1859. Building construction lagged behind the demand created by this rapid influx. In addition to tents, other lodgings consisted of small family houses with rooms that owners rented to miners or laborers. This small-scale practice continued throughout the late nineteenth and early twentieth centuries, even though large lodging houses were also built during this period.

According to the census, households of lodging houses kept by women did not vary widely in 1860. These households ranged in size from two to nine individuals; the average size consisted of approximately four people. In general, households encompassed the keeper, the family members, the lodgers, and the servants. The majority of lodgers consisted of single men—miners or workers related to the mining industry. Other lodgers, however, included families—either married couples and children or single women with children. A few lodgers were single women. As more people settled the Comstock, more women and children appeared in such households. The average size household for men running hotels, lodging houses, or boardinghouses was approximately eight people. The reason for the relatively smaller size of women's households may have been economic. Perhaps in 1860 these women, newly arrived, had just entered the business of lodging-house keeping, or perhaps wealthier male hotel keepers were able to attract more clients with larger accommodations. No evidence exists, however, to indicate that lodgers preferred male keepers over female keepers or vice versa.

During the first boom period the streets of Virginia City changed considerably. Upon his return to the Comstock in 1863, J. Ross Browne noted a great difference in accommodations compared to what he had observed three years earlier. He secured rooms in a private lodging house kept by a widow and considered them good rooms with clean beds.[8] The streets were no longer dominated by canvas tents. As Browne commented, "large and substantial brick houses, three or four stories high, with ornamental fronts, have filled up most of the gaps."[9] He also observed that Virginia City builders were striving to give the burgeoning community a sense of style, however dubious: "The oddity of the plan, and variety of its architecture—combining most of the styles known to the ancients, and some but little known to the moderns—give this famous city a grotesque, if not picturesque, appearance, which is rather increased upon a close inspection."[10] The city had grown so quickly that buildings took on various shapes and forms, without following a comprehensive plan. Similar rapid development occurred in many other western cities after the initial settlement period.

Both before and after the Great Fire of 1875, which destroyed much of the center of the city, many structures adopted aspects of currently popular forms such as the Greek Revival, the Italianate, and the Queen Anne styles. The major difference between these two periods manifested itself in building materials. Until the Great Fire a large number of buildings were constructed of wood. After 1875 many property owners built with more fire-resistant materials such as brick, stone, and cast iron. Numerous commercial buildings lining C Street—the main road through Virginia City—as well as larger dwellings, reflected the Italianate style, with large bracketed cornices and bay windows. Residential structures often exhibited a mixture of styles. The C. J. Prescott House (1864), a single-family house at 12 Hickey Street, had a front-facing gable roof and plain window surrounds characteristic of the Greek Revival, which was then at the end of its period of popularity. Other parts of the house, such as the front porch, with its simple ornamentation, reflected elements of the Italianate style, which was near its peak of popularity in the United States. After the Great Fire, other

The C. J. Prescott house in Virginia City dates to 1864 with additions dating to the 1860s and 1870s. (Photo by Bernadette S. Francke; courtesy of Nevada State Historic Preservation Office)

styles became popular as the city rebuilt many of its structures. These styles included ornate ones such as Second Empire, with its towers and mansard roofs, and Queen Anne, with its decorative finials, brackets, and spindles.[11] Regardless of the style or styles used in a structure, the end result was often a simplified version of the high-style manifestations that occurred in large cities like New York and San Francisco. Nevertheless, the variety of forms used on the Comstock represented a wide range of knowledge of contemporary architectural styles among builders.[12]

In the early years, erecting any building on the Comstock was a difficult and expensive task. The lack of lumber made it necessary to haul building materials from places as far away as the eastern Sierra and even San Francisco. Until the Virginia and Truckee Railroad, running from Carson City to Virginia City, was completed in 1869, teamsters and packers provided the only means of freight transportation to the Comstock. Although some shops produced finished carpentry work, many lumber yards provided builders with ready-made, mass-produced architectural elements from larger urban centers. Browne noted that lime, brick, ironwork, sashes, and doors had to be brought in at great expense and cost three to four times what similar articles cost in San Francisco.[13] Yet there were local quarries, kilns, and foundries in the region early in the settlement period that provided at least some locally produced building materials.

The 1870 census shows that, after ten years of Euro-American settlement, an almost equal number of women and men, thirty-two and thirty-four respectively, ran lodging houses on the Comstock, representing a large increase in the number of women in the profession. As was the case with most, if not all, western mining towns, men initially settled the area in large numbers; women followed. The male-dominated work structure of mining towns provided great opportunities for women to earn money by offering domestic service including lodging.[14] Furthermore, although some mining companies ran their own lodging houses, most accommodations were small operations, allowing women to easily enter the market with little capital. The census recorded 11,359 people, 7,864 men and 3,495 women, living in Storey County in 1870. Despite its first depression beginning in about 1863, the Comstock economy flourished again in the early 1870s as new ore discoveries opened more mines and brought more workers into the area. Buildings of this "Big Bonanza" period were more substantial and were constructed of wood or sometimes of brick. Of the thirty-two women identified as lodging-house keepers in 1870, thirteen were married, while the status of the other nineteen was unknown or not listed. In addition, the average age of this group, at thirty-seven years, was slightly younger than was that of the 1860 group. By 1870 economic growth over the past decade had attracted more women who sought their fortunes in the region. Only four other occupations outnumbered the keeping of lodging houses for women; these were keeping house, domestic service, seamstress work, and prostitution.

Between 1860 and 1870 the average household size for women lodging-house keepers increased from approximately four to six individuals per household. With this increased size came more diverse occupants and more women and children. In 1870 these households ranged from one woman with no lodgers to a woman whose lodging house accommodated her husband and child and thirteen lodgers.

It is certain that more women ran lodging houses during the 1860s than the census indicates, because the first economic peak occurred in the early 1860s, after the census had been conducted. Mrs. Wylie, for instance, was a prominent lodging-house keeper in this decade. She ran the Oro Hotel, a three-story, wood-frame building on D Street, and she advertised frequently in the local newspapers. She received a great deal of attention when fire heavily damaged the Oro on January 1, 1867. Despite this disaster, Wylie recovered quickly. Two weeks after the fire, the *Daily Union* noted that the Oro "has been rebuilt, and in three or four days will be finished and refurnished throughout, better than of yore. Such enterprise as [Wylie] has displayed in so speedily rebuilding and refurnishing her hotel should be encouraged by liberal patronage."[15] On the same page of the *Daily Union,* Mrs. Dirks advertised her house, the Occidental Hotel, a wood-frame building on South C Street, stating that it "has been thoroughly overhauled, repaired, refurnished, papered, painted, etc., making it one of the most pleasant houses in the City of Virginia."[16] A late-1870s photograph of Virginia City shows the Occidental, a white, one-and-a-half-story, wood-frame building with a front-facing gable roof. Though smaller than the Oro, the Occidental provided both double and single rooms, suitable accommodations for most lodgers. By the late 1860s women had taken an active role not only in running but also in building and furnishing lodging houses.

The records identify no women of color specifically as lodging-house keepers; however, a close reading of the census for 1870 reveals that one African American woman, Eliza Lawson, age thirty-six, kept house with three white men. Lawson was economically well off, with $6,000 in real estate and $5,000 in personal property. The men who lived with Lawson are not listed as lodgers, though it is likely that they paid her rent. In any case, Lawson's wealth was unusual for a woman taking in a few lodgers. In another instance, Elizabeth Vincent ran a millinery store and lodged two white women who may also have worked as shop assistants. In addition, the census shows a cluster of African Americans living in separate but adjacent dwellings. Most of the individuals are single men, but families also lived in the area. Unfortunately, the 1870 census does not list street addresses, so it is difficult to pinpoint the location of these dwellings or even to ascertain whether these separate dwellings may actually have been a large lodging house.

Although Virginia City was immortalized by eastern writers such as Mark Twain and others as a rough, western mining town, it was actually quite cosmopolitan.[17] Its architecture reveals that its residents had a knowledge of current building styles and fashions in furnishings. Both Wylie and Dirks made sure that their lodging houses were carpeted and wallpapered. Owners who could afford the expense purchased furniture from producers

Virginia City in the late 1870s with the location of the Occidental so indicated.
(Courtesy of Nevada Historical Society)

A note on the back of this nineteenth-century photograph reads: "Parlor and dining rooms of Our First Home—St. Paul's Rectory, Virginia City, Nevada." (Courtesy of Comstock Historic District Commission)

in San Francisco and other more established urban centers so their buildings would have the appearance of civility. Many lodging houses also included parlors, which were becoming fashionable in the mid-nineteenth century. The parlor, in contrast to the sitting room, provided a more public meeting space for inhabitants and guests, as opposed to the private family areas in a house. Parlors also became common in single-family houses throughout the United States in both rural and urban areas, as builders and homeowners sought to delineate more clearly domestic space from the outside world. For lodging houses, the parlor served as a communal place for inhabitants to meet outside their own lodging rooms but within their "homes" instead of in saloons or restaurants. Parlors may also have functioned as rooms in which the owner and his or her family could meet with lodgers rather than using the family's private quarters for this purpose.[18]

The second great economic boom in Virginia City encouraged the establishment of more lodging houses in the 1870s, even though the majority were small, with fewer than four or five lodgers. Most of these buildings were probably single-family houses adapted to accommodate lodgers. Despite the overall prosperity of the area, many women did not have the means to provide larger quarters, a situation that was typical of urban mining centers throughout the West.[19] Some women, however, did run larger houses and played a prominent role in their construction.

The 1870s saw a relatively large explosion of bigger and more elaborate lodging houses than had previously existed. In addition to the mining boom, the Great Fire of 1875, though terribly destructive, forced the immediate reconstruction of the city, and it was rebuilt on a grander scale. In November 1875 the *Sutro Independent* announced that Mrs. Cooper "will immediately begin the re-erection of her block of palatial residences on Taylor street. She has all the lumber engaged, and the contract will probably be let before this reaches the public."[20] Less than a year afterward, Cooper placed an advertisement in the *Virginia Evening Chronicle* to sell her new sixty-seven-room building, known as the Loryea House, because she was moving to Europe.

Other lodging houses built after the fire demonstrate the lengths to which some owners went to create elegant, first-class buildings. In April 1876 the *Mining Reporter* stated that Mrs. Dalles was completing the construction of a "very prominent building" on B Street and Sutton Avenue. The architect and contractor, John Thexton, had designed mill buildings on the Comstock and was also currently at work on a large commercial and residential structure on C Street. Dalles, who had kept the boardinghouse at the Eureka Mill, had branched off on her own to build an opulent house on B Street with a thirty-four-foot front and a depth of seventy-eight feet. The building stood three stories high and had a total of twenty-nine rooms. The interior contained many conveniences, including six double parlors, two on each floor; well-appointed bathrooms; and closets. The building was constructed of redwood, and the floors were made of Oregon pine. Rooms were to be *hard finished,* a term indicating the type of plastering on the walls, and ornamented in the "latest architectural fashion."[21] Though it was unusually elaborate, a building of this kind seems to have represented all that was considered stylish in housing at the time in Virginia City.

Eliot Lord's description presents a view of the more typical lodging house, but it confirms the comfortable accommodations that the buildings generally provided: "They are plain buildings, two and three stories in height, with an average floor-surface of 45 by 75 feet. Most of the rooms are simple dormitories, averaging 10 by 12 feet in size, but the connected rooms or suites are considerably larger. . . . These apartments are all carpeted and comfortably furnished according to the requirements of the lodgers.

Water-closets near all the houses are well constructed and cared for."[22] For many lodgers, these houses offered the most comfortable and economical accommodations to be found.

Despite the availability of descriptions for some lodging houses, it is difficult to define a lodging-house type. Many buildings accommodating lodgers were simply single-family houses with rented rooms. Larger buildings had many rooms off central hallways but probably did not greatly differ, if they differed at all, from large single-family houses. If women did modify their domestic spaces to better adapt them to the function of a lodging house, only more research will provide evidence to support this.

Mary McNair Mathews, who wrote an account of her life in Virginia City during the 1870s, provided detailed information on many lodging-house types ranging from small cottages to large houses. Mathews, widowed in the East, arrived in Virginia City in 1869 to recover her dead brother's property. She ended up staying until 1878, finding that she could make money in the boom economy. To support herself and her young son, she engaged in a variety of professions, primarily sewing, teaching, and keeping lodgers, none of which took her outside the home. Mathews recounted her struggles to make a living and in the process described the various houses in which she lived. She started out in a rented room and eventually constructed her own three-story lodging house on C Street.

Mathews's first day in the city may be representative of that of other women in the thriving mining town of the late 1860s. Upon her arrival she immediately began searching for lodgings. She first stopped and secured a room on the third floor of Mrs. McKinney's house on B Street for fifty cents a night.[23] She then began looking for work as a seamstress, stopping on her way at the lodging house of Mrs. Beck, a woman who eventually became a close friend. Married to a man who had a prosperous furniture dealership, Beck may have run her house to earn additional income for her family. Her house, on 21 North A Street, was a neat, white, two-story building with French windows and green blinds. Large, white pillars supported the porch with a veranda above. Two bird cages hung from the veranda ceiling. A white fence surrounded the front yard, which contained fruit trees, rose bushes, and young wheat.[24] In short, the exterior of the building and its yard reflected a neat, orderly appearance—not one stereotypically associated with mining town lodging houses. The house's main hallway led to a richly furnished sitting room, with a piano, table, whatnots, brackets with books, shells, vases, and ornaments. Mathews described

A two-story wood house in Virginia City with ample room for lodgers. (Courtesy of Nevada Historical Society)

the room as having "every luxury and comfort."[25] Indeed, Beck's house had the accoutrements that every fashionable middle-class home had at the time.[26] Although no photographs exist of Beck's house, images showing contemporary domestic interiors provide an idea of the appearance of these spaces.

Mathews's early experiences in Virginia City may have paralleled those of other recent arrivals. She and her son moved frequently from various lodging houses. When she made trips around the Comstock and as far as San Francisco to investigate her deceased brother's assets, she vacated her rooms, finding new accommodations when she returned. Her mobility reflected the transience of much of the Comstock's population.

When Mathews rented her own furnished house on G Street, she managed to obtain some of the objects required to make it a home. The building, which she rented for fifteen dollars a month, included a kitchen with a table, chairs, a stove, a cupboard full of dishes, two washboards, three tubs, and flatirons. She even had a sitting room with a three-ply carpet, a horse-hair sofa, six cane-bottomed chairs, a large rocking chair, a table, a bureau, a spring bed, and bedclothes.[27] Like many working-class American homes in the late nineteenth century, Mathews's house did not have numerous specialized rooms.[28] Instead her two rooms had to serve multiple functions; her kitchen acted as a dining room and work area, while her sitting room served as a parlor and bedroom. Although it was humble, at least she had her own house rather than rented rooms.

Mathews first became a landlord when she rented a four-room house on A Street for twelve dollars a month and then sublet two of the rooms for twelve dollars a month each.[29] Many women entered the business by taking in lodgers and boarders, because they did not have the capital to buy real estate. Instead, they could simply rent a number of rooms and then sublet them at a profit. After making some money speculating on mining stocks, Mathews decided to buy her own house. She purchased a small, three-room structure on C Street and soon decided to build a large lodging house with a store below on the ground floor.[30]

From descriptions of other large lodging houses erected at the same time, Mathews's house seems fairly typical. The wood-frame structure stood three stories high. Instead of being hard finished, or plastered, on the interior, Mathews finished the walls herself by stretching canvas over the rough boards and then applying wallpaper onto the canvas. Many families decorated their homes in western mining towns using this method of covering the walls, because it was quick and relatively inexpensive, yet it gave the interior a finished appearance. When the building was finally completed it had carpets, running water, and brick chimneys. It could accommodate twenty-six lodgers.[31]

As a widow running a lodging house, Mathews supported herself the way many single, white, middle-class women, mostly divorced or widowed, did on the Comstock. She took on various domestic jobs that could be performed in her home or in the home of another family. Running a lodging house was considered a respectable occupation, among the relatively few a woman like Mathews could pursue, because it kept her in the home.

Women in other parts of the West, particularly in mining towns where the population density was high, found that lodging-house keeping allowed them to ascend the economic ladder. This was significant for women looking for work, since female inhabitants of mining towns had few employment opportunities. These towns typically had no industries, such as textiles, that traditionally employed women, so women had to find other types of work.[32] Because in the early years of settlement there was often a large proportion of men to women, women found they could provide domestic services such as lodging, cooking, sewing, and laundry in their own homes and could receive good wages for them.[33]

On the Comstock, as in other western mining towns, the lodging business rose and fell with the boom and bust cycle of the mining economy. Mary McNair Mathews's business started out successfully; her lodging rooms were always occupied. Her house survived the Great Fire of 1875, but gradually the bonanza period on the Comstock waned. After 1877 the region began to slide into a deep economic depression, putting many miners out of work and forcing some families to beg for food. The downturn lasted well into the twentieth century. In January 1878 Mathews and her son moved back to her parents' house in New York. She left her lodging house in the care of a female friend until her agent could rent it. She also rented out the store on the first floor and the small house at the back of the lot.[34] After making a second trip to Virginia City, Mathews was still trying either to rent or sell her house.

Mary McNair Mathews left Virginia City too soon to be included in the 1880 U.S. census. Despite the declining economy, Storey County had a sizable population in that year; there were 16,115 people in the county, 9,294 of whom were men and 6,821 of whom were women. A search of manuscript census records for people listed as boardinghouse, lodging-house, or hotel keepers reveals that eighty women and twenty-one men ran such houses in 1880. The average age of these women, thirty-seven years, was the same as it had been in the previous census. For the men, the average age of forty-eight years was higher than it had been in 1870. That women in the profession outnumbered men four to one is not surprising. By 1880 the ratio of women to men had increased dramatically on the Comstock, and running lodging houses remained a viable profession for women. On the other hand, working as a hotel keeper seems to have been a predominantly male profession. The census for Storey County shows that men outnumbered women in that field. The rapid rise in the number of women running lodging houses suggests that the occupation grew feminized as opportunities for profits declined. This is similar to the situation in other western mining towns. For example, the number of women running boardinghouses in Lewis and Clark County, Montana, rose from four in 1870 to seven in 1880 to forty-seven in 1900.[35]

Although lodging-house keeper ranked fourth after keeping house, servant, and seamstress for women's professions on the Comstock in 1880, many widows turned to lodging-house keeping as a means of supporting themselves and their families. Of the eighty women running houses, thirty-two were widows. While some of them were like Mary NcNair Mathews, who arrived on the Comstock already a widow, others lost their husbands after their arrival. A miner's job was a dangerous one; many writers of the period, including Mathews and Eliot Lord, discuss the high mortality rate among men working in mines.[36] Running a lodging house was a relatively easy and logical way for a once-married woman who stayed at home to use the housekeeping skills she already had in order to earn a living for herself and her children. Of the remaining female lodging-house keepers, six were divorced, twelve were single, and thirty were married. In a few cases, women teamed up to run a house together. In most of these instances, at least one woman was single or widowed but had entered into business with a married or single female partner.

Households of female lodging-house keepers continued to grow over the years. By 1880 their average size was approximately eight people, and they ranged in size from one to thirty-eight people. While the makeup of most households consisted of the keeper and lodgers, who were usually single men, the makeup of others was different. Bridget Whiston, a widowed Irishwoman with three children, ran a lodging house on South G Street in Virginia City. Although she is listed as a lodging-house keeper, Whiston had no lodgers at the time. Another woman, Sarah Pottle, appears in the census as a lodging-house keeper, even though her household, made up of three prostitutes, eight gamblers, and four lodgers, resembled a brothel.

Many lodging houses had live-in servants. Seventeen households, nearly half the total number of boardinghouses, had at least one servant. Usually these servants were men from China who worked as cooks. Sometimes lodging-house keepers could afford to hire a female Irish or American servant, which increased the keeper's status. The number of servants in a lodging house did not necessarily correspond to the number of lodgers. It often depended on the income of the keepers. For example, the house of Maggie Bonner and Mary A. Lewis had thirty lodgers and four family members and only two Irish servants. Rosa Conners, a widow with no children, had fifteen lodgers and one African American servant.

For the first time in the census records, Chinese women appear as lodging-house keepers in 1880. Two single women, Chu Suh and Gee Pan, ran a house together with one Chinese cook and two Chinese lodgers at 98 South G Street. Farther north along the street, at 7 South G Street in

Chinatown, Ho Sing, also single, ran a house with three Chinese lodgers and a Chinese cook. Chinese women running lodging houses on the Comstock, or in the West for that matter, were rare, primarily because such a small number of Chinese women lived there. Furthermore, Chinese men tended to run lodging houses for male Chinese lodgers.

Other women claiming to be lodging-house keepers may have been hoping for better times. In 1880 approximately one-fourth of the seventy-seven lodging houses on the Comstock had no lodgers or only one lodger. Family members outnumbered paying tenants in these households. These situations, however, were most likely temporary, as lodgers came and went with great frequency depending on the business cycle of the local economy.

The 1880 census is useful, because it lists street addresses for each person. With this information it is possible to locate houses on the 1890 Sanborn Perris Fire Insurance Map for Virginia City as well as to connect women with advertisements in newspapers and with other citations. For example, the 1880 census lists M[argaret]. J. Fraser, a divorced woman of forty-four years, as running a boardinghouse at 199 Main Street, Gold Hill. This building was known as the Comstock House. Fraser figures prominently in Alfred Doten's journal during the 1880s, first for her nasty divorce from Owen Fraser in February 1880 and then as the Dotens' neighbor. Fraser initially ran the Comstock House with her husband[37] but apparently won ownership of the building as part of her divorce settlement, since she continued to run it by herself throughout the decade. Doten's wife's account of spending the night at the Comstock House is not a flattering one. Before moving to Austin, Nevada, she and her children slept at Fraser's house one night. "The noises, stinks etc, of the Comstock-Fraser House," however, drove them back to their own house for their last night in Gold Hill.[38] Despite Doten's unpleasant description of Fraser's house, Fraser continued in the business for the next two decades. She appears in the 1900 census running a lodging house on West A Street in Virginia City.

Many women had one or more jobs, but the census enumerator only noted one profession per person for men and women alike. A random sample encompassing 10 percent of the 1880 manuscript census found that twenty-six married women were identified as "keeping house," even when their households included lodgers.[39] All these women lived with husbands who had a profession outside the home. These women are in addition to the eighty women already listed as lodging-house, boardinghouse, or hotel keepers. For example, Charles Steel, an engineer who lived on South D

Street, had a wife, Amelia, who is listed as "keeping house." Their household, however, included two female servants, two Chinese cooks, and fifteen boarders. Unmarried men who headed households with lodgers are never described as lodging-house or boardinghouse keepers if they were miners or had another occupation. This distinction suggests different perceptions of men and women in the lodging-house profession.

Married women living with their husbands who provided part of the family's income may not have been regarded as lodging-house keepers even when they took in lodgers. The majority of women who appear in the census as lodging-house keepers are single—either widowed, divorced, or never married. The census enumerators apparently listed married women as keeping house, whereas women not living with men were assigned another occupation. We know that census enumerators had to use specific categories to identify individuals. It appears, however, that certain terms, such as *keeping house*, identified some women, namely those who were married and lived with their husbands. As in the 1870 census, no African American women are listed as lodging-house keepers, yet three married, African American women (out of a total of twenty-six African American women) took in lodgers, as did their white counterparts. Unfortunately, the catchall term *keeping house* masked a number of economic activities performed by married women at home.

One of the most interesting revelations of the 1880 census is that many parts of Virginia City were somewhat integrated. A boardinghouse on North B Street, run by a white family, included three African American families and one single African American man among its lodgers. At 45 North C Street an African American family took in two male lodgers, one white and one African American. Next door at 47 North C Street stood a large lodging house run by Alfred Turner with only white lodgers and servants. Although many Chinese lived in Chinatown, located roughly along Union Street in the vicinity of G and H Streets, Chinese laundries stood throughout the city and housed Chinese men.[40] These buildings were never called boardinghouses or lodging houses, but they functioned in the same way, providing dwelling space to single men.

Few women appear regularly from one decade to the next in the census records for Storey County. The economies of western mining towns forced most settlers to be transient, following booms from place to place. Nevertheless some women, like Margaret Fraser, remained on the Comstock, working as lodging-house keepers for many years. Rose Sissa also ran a lodging house, first with her husband and then on her own after his death. In 1871 a small article in Virginia City's *Territorial Enterprise* noted that the Capitol Hotel, at 25–27 North B Street, run by Eugene Sissa, had one of the finest conservatories in the city.[41] By the late 1870s, however, Sissa's wife, Rose, was listed as the proprietor of the house, by then known as the Capitol Lodging House. According to census records, Sissa continued to run this lodging house by herself at least until 1900. Although this building no longer stands, the 1890 Sanborn Perris Fire Insurance Map of Virginia City provides some information about the structure. Two adjacent two-story buildings, each with a bay front, made up Sissa's lodging house. Small one-story structures, probably privies, stood behind the buildings. The 1910 census lists Sissa at another address, 33 North C Street, where she ran a house with two lodgers. The 1890 Sanborn map shows a three-story lodging house, which could have accommodated Sissa and her lodgers, at this address.

Although Rose Sissa was unusual in that she remained in Virginia City for nearly fifty years, her longevity in running a lodging house shows that women could make a career of this occupation. In addition, Sissa represents the changes facing Virginia City lodging-house keepers overall. As the years passed and the mining boom on the Comstock faded, the population dropped. Consequently, the community needed fewer lodging houses to accommodate lodgers. In 1880 eighty women appear in the census as boardinghouse, lodging-house, or hotel keepers. After this peak, the number of women running such houses declined precipitously, though their percentage as part of the total female population in Storey County remained relatively stable. The 1900 census records twenty-eight women running lodging houses. The total Storey County population in 1900 equaled 3,673—1,933 men and 1,740 women. The average age of women lodging-house keepers was fifty-eight years, and the average household size was approximately four persons. Only six men, with an average age of forty-nine, appear as lodging-house keepers in this census.

By 1900 the makeup of households had changed as well. Many women listed as lodging-house keepers had only family members—usually children—living with them, and no lodgers. Other households had only one or two lodgers. While miners or men with other mining-related professions still accounted for most lodgers, a larger number of families lived in lodging houses than had previously done so. The proportion of individuals who worked in non-mining jobs also increased. The most striking change is that none of these households had servants. Perhaps these lodging-house keepers could not afford hired help to assist with the maintenance of their houses. By this time women were running their lodging houses on a much smaller scale than they did in the 1870s and 1880s, when the Comstock economy had been stronger.

By 1910 only seven women were listed as lodging-house keepers for the census. All but one of these women were over fifty years of age. The total population of Storey County was 3,045—1,781 men and 1,264 women. The average size household for lodging houses, at approximately five persons each, was slightly larger than it had been in 1900. One woman lived with her husband and daughter and had one lodger. Another women had five children but no lodgers at all.

Without the rapid growth of a thriving mining economy, the demand for lodging dropped. Many buildings on the Comstock stood empty, abandoned by owners who had moved on to more prosperous towns. Though a few lodging houses remained, this type of housing became less viable as a business. Lodging houses increased in number out of an urgent need for relatively low-cost housing and housekeeping services in a densely inhabited area. With the population dropping every year, fewer lodging houses were necessary to accommodate even the most transient of residents.

The rise in the average age of female lodging-house keepers over the years reflects economic changes as well. Younger women who had the opportunity to move to a more vital area could do so much more easily than could older women. Owning property also made a difference. Women who could afford to walk away from their property, as Mary McNair Mathews did, fared much better than did women who could not afford to lose their homes. Married women were more likely to move away than to remain when their husbands decided to look for work elsewhere. That four of the seven lodging-house keepers listed in the 1910 census were widows reflects the difficulty older, single women had in moving away from depressed areas.

With limited professional opportunities in western mining towns, many women supported themselves, sometimes quite successfully, by running lodging houses. In a region driven by an economy based on male labor working in mines, female lodging-house keepers were able to grasp a small portion of the wealth created by the Comstock boom. Yet the number of women who took in lodgers, including women identified in the census as "keeping house," never amounted to more than approximately 5 percent of the total female population of the county. Other mining towns had higher percentages. For example, in 1880 27 percent of all females in Helena, Montana, kept lodgers. This percentage includes women who took in lodgers informally but not women listed as lodging-house keepers.[42] Paula Petrik, the author of this study on Helena, notes that this percentage is similar to those of other mining communities but is greater than those of cities outside the frontier.[43] Perhaps the lower percentages on the Comstock indicate that by 1880, after twenty years of development, Virginia City and its neighboring towns could no longer be considered the frontier. Indeed, the percentages of women identified as lodging-house keepers in the total female population vary only slightly from decade to decade in the area.

Scholarship on women in the West has sought to assess women's opportunities on the frontier and has posed the question of whether or not western women had more freedom than their eastern counterparts did. Historians have provided arguments that support both views. From this examination of lodging houses on the Comstock, however, it appears that although women did have more opportunities to make money in the urbanized area of the West than they did in similar areas in the East, they were still governed by ideas of women's place in society. Historian Sandra L. Myres argues that women broke free of their domestic sphere by attempting "to civilize raw Western communities" through the establishment of churches, schools, and social organizations.[44] The connection between women and lodging houses, a type of domestic architecture, fits in with the nineteenth-century belief that women should civilize the West. The editors of *So Much to Be Done: Women Settlers on the Mining and Ranching Frontier* argue that the "urban character of California and Nevada gold rush territory allowed women to act positively within traditional female roles, both as economic partners and as 'civilizers.'"[45]

The urban nature of mining towns like Virginia City paralleled the increasing industrialization of formerly rural eastern communities. This change helped spawn the Cult of True Womanhood—a white, middle-class ideology that decreed gender-separate spheres—one at home where women raised children and the other in the outside world where men entered into commerce.[46] Though these separate spheres often overlapped in the West, distinctions still existed. Census enumerators, using occupation categories determined by the federal government in Washington, D.C., identified women in certain ways. Thus, many married women who took in lodgers were classified as "keeping house" rather than as lodging-house keepers. On the other hand, married women who ran lodging houses with their husbands were sometimes identified as lodging-house keepers, as were their spouses. Sometimes they were not. No pattern for this variation in identification is apparent. Nevertheless, female lodging-house keepers seem to have been accepted because they filled a crucial role, providing the closest thing to a domestic environment that many lodgers could obtain in the West.

Yet these women were more than just symbols of domesticity. As other authors writing on women in the West have shown, the frontier provided women with the opportunity to distinguish themselves as capable individuals. In her history of women and prostitution on the Comstock, Marion Goldman sees lodging houses as a logical extension of women's traditional domestic roles but notes that they provided new opportunities for women as well. Because of their strong presence in the lodging industry, the Protective Organization of Hotel, Boarding House, and Restaurant Keepers of Virginia City and Gold Hill allowed women to join their group at a time when women "were not permitted to join any other professional association or union."[47]

Other factors determined who ran lodging houses on the Comstock and in the West. The population on the Comstock was quite homogeneous. For example, in 1880, of a total of 2,232 adult women, ninety-four were women of color. Since most women were white, it follows that most female lodging-house keepers were white. As we have seen, women of color had opportunities on the Comstock to take in lodgers and run their own businesses. African American women in other parts of the West found success in these kinds of careers as well. However, the dominance of whites and persistent racist attitudes may have increased the difficulty women of color had in entering relatively high-status fields such as the lodging-house profession.[48] Most women running lodging houses were middle class, because entering the profession required some capital necessary to open and operate a lodging house. In addition, middle-class women could draw from their knowledge of running domestic households and apply it to the business of lodging-house keeping. Finally, like teaching school or taking in sewing, running a lodging house was a respectable profession for middle-class women. Women could work for themselves as entrepreneurs while remaining in a domestic environment.

More research is necessary to determine whether women of color ever accepted and tried to conform to ideas of domesticity, or if this ideology was simply imposed upon them by white, middle-class reformers.[49] In addition, further study of domestic architecture in the West, including single-family dwellings and lodging houses, may provide additional insight by defining the nature of the family in mining towns during the late nineteenth and early twentieth centuries. Many households on the Comstock, like households throughout the United States, did not comprise nuclear families.[50] For primarily economic reasons, lodging houses as well as private domiciles housed many unrelated individuals. As more people moved from rural to urban areas, lodging houses and rented rooms accommodated the influx.[51]

Lodging houses provided a type of domestic space that was common at the time, and the women who ran them simply broadened their traditional roles of mother and nurturer. As lodging-house keepers, they were active players in the economic life of the community. They had buildings erected and then purchased many of the goods brought from the outside to the Comstock—items such as furniture, other household objects, and clothing. Even though men initially made up the majority of lodging-house keepers in the early settlement and boom years of the Comstock, women gradually overtook men in this profession. During the peak of the Comstock boom, women of color as well as white women entered the profession in larger numbers than had previously been the case. As the work of running lodging houses became less lucrative, men abandoned the occupation, leaving it almost entirely to women. Similar shifts occurred in other traditionally male occupations such as teaching school.

Much more work remains to be done on domestic labor and architecture on the Comstock. This study presents only a part of the story by focusing on lodging houses. More in-depth studies of the manuscript census records will undoubtedly reveal additional information. Also, more research is needed on the documentation of the buildings themselves. While many photos of the Comstock in the late nineteenth and early twentieth centuries exist, few specifically record houses and lodging houses of the middle and lower classes. Further study of this topic will require an examination of additional written records, such as newspapers, diaries, and account books; archaeological evidence; and a comparison of Comstock lodging houses to similar facilities in other western mining towns. Studying domestic buildings will provide more insight into women's various roles in the West. Further analysis of lodging houses—domestic spaces where men's and women's spheres overlapped—should offer particularly intriguing perspectives.

❧ 4 ❧

"They Are Doing So to a Liberal Extent Here Now"

Women and Divorce on the Comstock, 1859–1880

KATHRYN DUNN TOTTON

According to Myron Angel's *History of Nevada*, the first divorce in the Comstock region occurred in the summer of 1853, when an irate father refused to allow the consummation of a marriage contracted in his absence between his fourteen-year-old daughter and a miner. The community chose sides, and the would-be husband was prevented from abducting his bride as she was whisked away to California by her father. "Thus was accomplished the first ceremony of marriage in Nevada, followed by a swift-winged and effectual divorce," without the benefit of courtroom, counsel, judge, or jury.[1]

Within a few years judges and the legal system replaced vigilantes and irate fathers as the means of obtaining divorce on the Comstock, and the number of divorces in the area increased steadily. There was "hardly a week but there is an application for a divorce. Virginia City is truly the city of divorces," wrote Mary McNair Mathews in her account of Nevada life during the 1870s. Although Mathews tended toward hyperbole, by the end of 1880 Nevada judges had granted divorces to more than 800 petitioners, over 350 of them in Storey County.[2]

Neither Nevada nor Storey County divorce rates were exceptionally high. By the 1880s legal action terminated approximately one in every sixteen marriages in the United States, and the West had the highest per capita divorce rate in the nation in the last half of the nineteenth century. Nationally, women obtained approximately two-thirds of all divorces, and like their contemporaries, Nevada women utilized the legal system to end marriages far more frequently than men did.[3]

Wives petitioned for divorce nearly three times as often as did husbands in nineteenth-century Nevada, and on the Comstock the ratio was even higher. Of the 185 Carson County (Utah Territory) and Storey County divorces granted between 1859 and 1882 that are included in this study, women were the plaintiffs in 160. Two were divorced twice during these years, and another prevailed only on her second attempt. A favorable legal climate, economic opportunities, and a tolerant community made divorce an option for the Comstock's women. Depending on their individual circumstances, divorce offered women an alternative to remaining in an intolerable marriage or an opportunity to achieve independence.[4]

Records of divorce suits permit examination of the lives of these women. Court documents reveal life circumstances and marital behavior that led women to the courtroom to divorce or to be divorced. The content of the files varies, but each file generally includes basic information about a couple's life: the date and place of marriage; some justification for the alleged cause of divorce; a reference to children, if any, for custody purposes; and a listing of any community property.

Neither Utah Territory nor Nevada courts demanded extensive proof of charges, neither required a jury to hear the case, and the laws of both permitted decrees by default if a defendant failed to appear in court or to respond to the suit. As a result divorce documents for the Comstock, unfortunately, are rarely as rich in specific detail as are those for other western states. Nevertheless, the petitions demonstrate how the middle-class Victorian values of male legislators resulted in divorce and property laws that allowed women to end unsatisfactory marriages without undue difficulty. At the same time, they reveal the dichotomy between those middle-class ideals and the reality of married life for many women on the mining frontier. When used in conjunction with federal census records, the Nevada state census of 1875, and other legal documents, they can also be used to suggest women's situations following divorce.[5]

Although the court records provide minimal information about economic status and even less regarding race and ethnicity, the available evidence supplemented with data from other sources, such as census returns, indicates that women who experienced divorce on the Comstock were a diverse group. Husbands' occupations, as far as is known, ranged from physician, attorney, and banker through handiron merchant, cook, and miner, suggesting that economically the couples represented a cross section of the area's upper, middle, and working classes. References in the records to community property also support this conclusion. In the majority of the cases there was no community property to divide, but in almost 20 percent of the sample the pair had accumulated at least some property, from a few shares of mining stock or a small house to extensive holdings.[6]

The ethnic diversity of the Comstock, where in 1870, for example, 44.2 percent of the population was foreign born, is also reflected in these women. Birthplaces, for those women who could be located in a census return, included England, France, Germany, Ireland, Mexico, Peru, and Sweden. Whether any members of the Comstock's small African American community were included in this sample, however, cannot be discerned from the available sources. None of the petitions mentions race, and no African Americans appear among the divorced in census returns.[7]

Divorce crossed lines of class and ethnicity on the Comstock, and the community appeared to accept the divorced, even if it did not condone the practice. In Virginia City, Mary McNair Mathews observed, "in the best circles" marriages often ended in divorce, "and perhaps before three months have passed, each one has secured another partner, and they will both meet at the same ball . . . and you would never imagine they had been husband and wife." Louise Palmer, in 1869, remarked on a similar attitude in the town of Dayton, noting that "it would appear as if there were some hidden law compelling ladies there to obtain divorces from their first husbands and to choose others." Divorced spouses amiably socialized after remarriage "as if no unpleasantness had ever occurred between them."[8]

The experiences of Theresa Dirks and her daughter Leonora seem to bear out at least some of these assertions. In 1856 Theresa and her husband Leonard were residing in San Francisco with their three children, twelve-year-old Leonora, Camilla, age four, and George, three. That year Theresa and Leonard agreed to allow William Allers to marry Leonora with the understanding that the couple would not cohabit for two years. When the two years had passed, however, Leonora defiantly refused to live with Allers. In 1860 she joined her mother, who had moved to Virginia City, and on Christmas Eve, "Mrs. Dirks . . . [and] Miss Leonora Dirks" were among the guests at the "first ball" in Storey County.[9]

William followed, attempting to persuade Leonora to change her mind, but to no avail. The following year he filed for divorce. Theresa testified in court to the hopelessness of her daughter's marriage:

> From what I know of the parties and the facts and circumstances since their marriage, I am confident of the opinion that plaintiff will never succeed in prevailing upon the defendant to live with him as his wife, and I now believe that the difference of age and disparity of disposition if defendant should consent to the marriage would render their living together as husband and wife in peace and union extremely improbable, and that the welfare of both plaintiff & defendant requires a separation.

The judge concurred and issued the decree.[10]

By 1869 Theresa had relocated to Hamilton in White Pine County, where, as she had in Virginia City, she operated a lodging house, and where in April she wed James Dean. Shortly after, not far away in Treasure Hill, Robert and Johanna Charles were married. Three years later, in February 1872, Judge Richard Rising in Storey County awarded a divorce decree to Theresa Dean and on May 15 granted a divorce to Robert Charles. On May 20 Robert and Theresa recorded their marriage contract in the Storey County courthouse. Extant sources do not reveal, however, whether the parties involved continued any social relationships.[11]

Leonora, whose marriage to William Allers had officially ended in 1860, remarried in 1865 and filed for divorce after five years, charging her husband, William Walsh, with desertion and neglect. William had never supported her, she alleged, forcing her "ever since her marriage . . . [to be] supported entirely by her own industry and the generosity of her mother." In March 1866 William moved to California, abandoning Leonora and their infant daughter. After notification of the suit was published in the *Territorial Enterprise* for one month, Judge Rising granted the divorce, restoring Leonora's maiden name and awarding her custody of her daughter and restoring Leonora's maiden name. In September 1870 a census taker found Leonora, twenty-five years old, with a $5000 personal estate, keeping house for her two siblings in Virginia City.[12]

By 1880 Robert and Theresa had moved to Virginia City. The census for that year recorded them at 91 South C Street. Robert, forty-six, gave his occupation as banker, and Theresa, fifty-two, listed hers as keeping house. The household included Theresa's son George, a twenty-five-year-old clerk;

Josephine Welsh, a fifteen-year-old niece; eight male lodgers; and Johanna O'Neil, a servant. Leonora Dirks does not appear in Storey County during the 1880 census.[13]

Beginning in 1852, when the Utah Territorial Legislature passed its first divorce statute, Nevada—then the far western portion of Utah Territory—provided a favorable legal climate for divorce, particularly for women. The Utah statute, one of the most lenient in the nation at the time, provided that: "If the court is satisfied that the person so applying is a resident of the Territory, or wishes to become one; and that the application is made in sincerity and of her own free will and choice and for the purposes set forth in the petition; then the court may decree a divorce from the bonds of matrimony against the husband." Six specific causes for divorce were subsequently delineated: impotency; adultery; "willful desertion of his wife by the defendant, or absenting himself without a reasonable cause for more than one year;" habitual drunkenness; felony conviction; or "inhuman treatment so as to endanger the life of the defendant's wife."

The statute also permitted a probate court judge to grant a divorce not only on the basis of proof of one of the specified grounds but also in the event that "it shall be made to appear to the satisfaction and conviction of the court, that the parties cannot live in peace and union together, and that their welfare requires separation." If the defendant failed to appear in court after receiving merely a "proper and timely warning," a decree could also be awarded for default. That the lawmakers intended to make the law accessible to women is evident in the unusual use of the feminine gender and in the incorporation of a section permitting a husband to obtain a divorce "for the like causes, and in the same manner as the wife obtains a divorce from her husband."[14]

Promulgated by a Mormon legislature to fit the unique circumstances in Utah Territory, the statute suited the needs of residents—both male and female—of the Comstock region as well. Husbands made frequent use of the "no-fault" peace and harmony clause in the statute. This provision had allowed the judge to terminate the marriage of William and Leonora Allers, for example. And on June 29, 1861, Henry Zottman secured a divorce in Probate Court in Carson County from his wife, Margaret, by convincing the judge that the "serious disputes, disagreements and animosities [that] arose between them" almost immediately after their marriage, and "which at last settled into violent mutual aversion," precluded their ever living in peace and harmony together.[15]

Eighty percent of the divorces granted in Carson County, Utah Territory, between 1859 and 1861, however, were awarded to women, who cited other, usually complex, grounds. Women's petitions incorporated accounts of alcohol abuse, cruelty, neglect, and abandonment. Desertion, often coupled with factors such as alcohol abuse and neglect, was the most common complaint. Elizabeth Wilkins and her six-month-old son, for example, were abandoned by Henry, a heavy drinker and habitual gambler. Josiah Hayse left Olive in July 1859, "on account of difference of temper and disposition." Almost immediately after their 1859 arrival in western Utah Territory from Michigan, Esie Bristol vanished, leaving Rebecca to fend for herself and their fifteen-month-old daughter.[16]

Allison "Eilley" Bowers, according to her biographers, sought independence by refusing to return to Salt Lake City in 1857 with her spouse, but her 1860 divorce petition presented an altogether different case. Eilley married Alexander Cowan in Salt Lake City in 1853, and the couple emigrated to Carson County soon after. In September 1857, Eilley charged, Alexander deserted her "without cause or provocation" on her part, leaving her "in a helpless and destitute condition." Although he returned in January, he decamped again in August, without making "any provision whatever for your complainant's maintenance or support." As far as the court was concerned, Eilley was an abandoned wife.[17]

After desertion, the second most common complaint was cruelty, or inhuman treatment, although only one plaintiff, Kate Hilsaa, utilized it as the sole ground. Kate had emigrated with her husband Joseph from Pennsylvania to San Francisco, where, according to her testimony in Utah Territory probate court, she supported the family while Joseph, "a man of violent passions" and "ungovernable temper" became a habitual drunk and "on many occasions . . . addressed to her the most opprobrious epithets and threats of personal violence."

In March 1860 Kate fled to Virginia City. There she provided for herself and her children by taking in sewing and laundry and managed to save enough money to purchase part interest in a town lot and a building, which she turned into a boardinghouse. According to witnesses' testimony, Kate was a successful businesswoman, operating a reputable establishment and paying her bills promptly. Then, in November, Joseph appeared in Virginia City and soon took up his old ways. After he attempted to burn the boardinghouse one night and threatened her on another with a rifle, forcing her to take refuge in a hotel, Kate finally sought and won a divorce.[18]

Although the political status of the region officially changed in March 1861, when Congress passed the Nevada Organic Act providing for the creation of Nevada Territory, divorces continued to be granted under the Utah law until October, when the new territory's first legislature created a legal code, including a statute for marriage and divorce. The new law provided that absolute divorce could be granted to a petitioner who had resided in the territory for at least six months on any one of seven grounds:

1. impotency at the time of the marriage, continuing to the time of divorce
2. adultery since the marriage, remaining unforgiven
3. willful desertion of either party by the other, for the space of two years
4. conviction of felony or infamous crime
5. habitual, gross drunkenness, contracted since marriage, of either party, which shall incapacitate such party from contributing his, or her share, to the support of the family
6. extreme cruelty in either party
7. neglect of the husband for the period of two years, to provide the common necessaries of life when such neglect is not the result of poverty on the part of the husband, which he could avoid by ordinary industry

A judge could also award a decree by default if the defendant failed to respond to the suit, as long as the plaintiff had taken the requisite steps to notify the accused.[19]

Nevada legislators thus raised no major obstacles to divorce. Elimination of the peace and harmony clause, addition of a residence requirement, lengthening of the desertion period to two years, and the qualifiers such as inability to contribute in the case of drunkenness, made the process marginally more difficult, but an additional ground—neglect—was added.

The advent of statehood three years later occasioned little change in the statute except to make the legal climate slightly more hospitable to women seeking divorce. Lawmakers in 1864 added provisions allowing divorce proceedings to be held in closed court. And to enable a wife to carry on or defend a suit, judges were empowered, at their discretion, to require husbands to pay support during the pendency of the case and to provide funds for legal expenses. If necessary, the court could attach the husband's property or income to that end.[20]

The provisions of the statute indicate that Nevada's mostly middle-class, male lawmakers subscribed to the Victorian ideals of domesticity and companionate marriage. Mutual affection and respect precluded adultery, desertion, and cruelty on the part of either spouse. Each partner was expected to contribute his or her fair share toward the support of the family, and inability to do so due to drunkenness was grounds for divorce. However, only the husband could be charged with neglecting to provide for the family, and only the husband could be compelled to provide support and legal expenses. Clearly, in the eyes of lawmakers, the wife's role did not include wage earning or other remunerative activities.

The divorce records, however, reveal that the language of the statute and the wording of the legal petitions represent the values of legislators and attorneys rather than the reality of Comstock women's lives. Regardless of the grounds, as Susan Gonda also observed in her study of San Diego women and divorce, attorneys were careful to couch complaints in the language of the law. The documents have a similar sound and explain less perhaps about women's experiences than about how their attorneys believed cases could be won.[21]

Most begin by establishing that the plaintiff had been a resident of Nevada for more than six months immediately preceding the commencement of the suit and then state the date and place of marriage. Enumeration of the cause or causes for divorce follows, carefully couched to conform to the language of the statute.

Carrie Wood's petition, for example, alleged that "for more than one year immediately preceding the commencement of this action said defendant has wholly failed and neglected to provide plaintiff and his family with the common necessaries of life, and that such neglect was and is not the result of poverty which ordinary industry could not have overcome." Charlotte Robinson's suit charged William with "habitual gross drunkenness, contracted since marriage, which incapacitates him from contributing his share to the support of his family."[22]

When language in court documents deviated from the formula, however, it may be considered more accurately representative of women's experiences. In many cases the women provided supplementary details of incidents of abuse, drunkenness, adultery, or neglect. These details offer insight into the reality of their lives that is lacking in the formal language of the petitions.

Catherine Hodges's attorney stipulated that Henry "at divers times committed an assault upon her," the particulars of which included throwing a stick of wood and a chair at her, threatening to "cut [her] liver out," and calling her "a whore." Adolph Goffin tore an earring from Mary Anne's ear on one occasion, and "in his violence and rages . . . burned and destroyed almost her entire wardrobe." Before deserting her, Mary Ann Hall's husband publicly called her a "*whore* and a *prostitute*,'" and placed a notice in the *Virginia Daily Union* falsely accusing her of infidelity with Samuel Robinson.[23]

Patrick O'Byrne beat, kicked, and bruised Mary, and "drove her out in the street, tearing her clothes off her person, and calling her a common whore—compelling her to seek shelter and protection with her neighbors." In an addendum to her original complaint, Mary alleged that after forcing his way into their house, Patrick had taken two of their children and hidden them from her. Mary was able to ascertain that they were staying with Mrs. Rickert and that one child "was very sick." When she tried to retrieve them, Rickert informed her that, "The defendant came to her house at midnight . . . and woke the children up and took them from their bed out in the street and that she had not heard from them since. That at that time one of the children was all broken out with a rash, and he the defendant would not allow her to place a shawl about either of them before taking them away, and when she remonstrated with him he threatened to kill her."[24] The sophisticated vocabulary in some of the accounts clearly reflects the work of an attorney, but the incidents themselves are not products of the legal imagination. They are the details of marriages that did not conform to Victorian ideals.

The perspective of Samuel Chapin, Storey County delegate to the state constitutional convention of 1864, may also more closely represent the situation of many Comstock women than did most court documents. Chapin argued forcefully and successfully in the convention to eliminate a provision prohibiting sole traderships for married women. When a colleague replied that women who were saddled with "miserable intemperate husbands who are incapable of transacting business," should get divorces, Chapin responded, "They are doing so to a liberal extent here now." A sole trader provision, he explained, would:

> prevent a great number of divorces and attempts to obtain divorces
> from miserable husbands, who are only a nuisance in the family rela-

tion. There are noble, struggling wives, who are willing to maintain such worthless husbands, if they can only themselves have the control of affairs, in such a manner that their husbands cannot get hold of and spend all their earnings. . . . They are willing to bear the burden of worthless husbands, and struggle on in that way, if they can only have the privilege which should justly belong to them, of protecting their own property, in their own right.[25]

In making this argument, Chapin acknowledged that for many Nevada women the ideals of true womanhood and the companionate marriage were far from reality. Married women—especially working-class women—living on the urban frontier, whether in California, Wyoming, or Montana, worked to help support their families. On the Comstock, as Bonnie Ford found in early Sacramento, "A husband's financial responsibility for a wife—a key tenet of the middle-class ideology—was honored far more in the breach than in the observance."[26]

In 1873 lawmakers took another step, although perhaps not intentionally, to facilitate divorce for women. *An act defining the rights of husband and wife* permitted married women to own, control, and register separate property. It also required community property to be divided equally in the event of divorce (unless the defendant was guilty of adultery or extreme cruelty, in which case the judge could determine a just division). Provision for an ante-nuptial contract allowed women to prevent their property from becoming community property and therefore subject to the rules of coverture. Intended to protect a married woman's property from profligate or spendthrift husbands, the separate property and community property provisions also operated in some instances to provide women with the peace of mind to enter marriages and the economic security to end them if necessary, as the cases of Theresa Dirks and Mercedes Nevarra demonstrate.[27]

In 1869 Theresa Dirks and James Dean filed a marriage contract in Storey County District Court. Although Theresa reposed "full confidence in the integrity, ability and business capacity and habits of said James C. Dean," her property holdings in San Francisco, Virginia City, and Hamilton City had "been acquired by her own unaided industry and exertions," and she therefore deemed "it prudent to assume to herself the exclusive control of said Estate." James agreed that Theresa should have full control of the property she owned when they married and any she should acquire afterward. Theresa subsequently filed an inventory of her separate property, and after divorcing Dean three years later, convinced Robert Charles to sign a nearly identical contract before their marriage.[28]

After her spouse of ten years, Charles C. Ogden, "left the State of Nevada [and] never returned," Mercedes filed for divorce in May 1871 on the grounds of desertion. Judge Richard Rising issued the decree, restored her maiden name, and awarded her complete control of the two Virginia City lots that she had acquired after Charles departed, "by and through money and other property possessed by her before her marriage and by her own industry."[29]

In December of the same year she married Edward Reyes. Less than four months later, Mercedes fled their home to escape Edward's physical and mental abuse after he struck her, "threatened to mash her head to a jelly" with the butt end of a pistol, hit her in the back with an iron frying pan, and chased her through the house calling her "a damned bitch and whore." In her suit she charged Edward with extreme cruelty and further alleged that he had committed adultery and contracted a "loathsome disease" in a Carson City "house of ill fame." Once again, Judge Rising obliged with a decree granting the divorce, restoring her maiden name, and reaffirming her control of her property.[30]

Mercedes Nevarra's suits utilized the two most frequently cited grounds for divorce in Nevada during this period: desertion and cruelty. Between 1867 and 1886 cruelty was the leading cause for divorce in Nevada, accounting for 254 of the 1,128 cases recorded. When cases alleging cruelty in conjunction with other causes are added, the number rises to 436, with women as the plaintiffs in 398 of the suits. For the female petitioners in the Comstock cases included in this study, cruelty, often coupled with desertion and/or neglect, accounted for the largest number of petitions.[31]

Descriptions of physical abuse abound in the records, and in Storey County the court effectively continued the practice established under Utah Territory of expanding the definition of cruelty to include mental abuse, as well. This interpretation, which was gradually winning acceptance in courts in the United States, was another result of the transformation of the marriage relationship by the mid-nineteenth century. Words or actions that caused psychological suffering were sufficient grounds for divorce even in the absence of physical evidence of abuse. False accusations of infidelity, particularly by a husband against a wife, figured prominently in cases alleging mental cruelty.[32]

In February 1864 Wilhelmina Zimmerman filed a complaint in the court of Judge Leonard Ferris against her husband Simon. Not only had he abandoned her for over two years and neglected to support her and their sixteen-month-old son, whom she had supported "by her own labor and industry" since his birth, but he was also guilty of extreme cruelty. He had "openly and publicly" charged her with infidelity, called her a "dirty slut, a whore, . . . and other abusive insulting and obscene epithets too coarse and vulgar for repetition." Simon filed an answer to the suit concurring that he "may have used improper and perhaps unbecoming language to said plaintiff." Judge Ferris agreed and granted the divorce on the grounds of desertion, neglect, and extreme cruelty.[33]

In 1866 the Nevada Supreme Court, hearing a case appealed from Washoe County, clarified the grounds of cruelty: extreme cruelty must be judged on the character of the parties involved and the circumstances of the case. It need not involve violence "if it appear probable that the life of one of the parties will be rendered miserable by any character of misconduct on the part of the other." One act of violence, which was not likely to be repeated, and which was the "result of rashness rather than malignity," did not, however, constitute extreme cruelty, and a divorce could not be granted on the ground of extreme cruelty if the complainant "has wilfully provoked the violence or misconduct complained of, unless such violence greatly exceeds the provocation."[34]

Either spouse could seek a divorce on the ground of extreme cruelty, but for women the practical result of this ruling was double-edged. It permitted wives to secure divorces on the ground of mental cruelty, such as abusive language or threats of physical violence, while at the same time forcing them to endure repeated acts of mental or actual physical violence before allowing them to file for divorce.

A case in point is that of Mercedes LeGuen, who married Thomas LeGuen in Lima, Peru, in 1867. Three years later the census taker found the family—Thomas, a thirty-three-year-old, Louisiana-born physician; Mercedes, a twenty-year-old homemaker born in Peru; and their daughter, Emeline, a one-year-old California native—living in Virginia City. A graduate of the Medical and Surgical College of New York, Thomas maintained an office on South C, over the Young America Engine House. By the time Mercedes filed for divorce in early March 1871, an infant, then four months old, had been added to the family.

In court, Mercedes testified, through an interpreter, that Thomas had, at various times over the previous two years, threatened her with a pistol, struck her in the face, obliged her several times "to leave her home through fright," and accused her "of prostituting her body to other men, and called her by vulgar and obscene names." Additionally, she alleged, he had committed adultery numerous times, including in July 1870, when he spent five or six days with a woman at the hotel "at the Warm Springs near Genoa." Witnesses supported her allegations, and Judge Rising issued the decree and awarded her custody of the children.[35]

Despite the courts' liberal definition of cruelty, women rarely attempted to win a divorce solely on the grounds of mental cruelty. The one woman in this sample, Louisa Schenadeke, who was awarded a decree for the single cause of mental cruelty, did not seek divorce. The suit was filed by her husband, Frederick. Frederick and Louisa were married in Reno in September 1869 and moved to Virginia City shortly afterwards. There they rented a house that they shared with four male boarders. According to Frederick's complaint alleging adultery and extreme cruelty, Louisa had an affair with one of the boarders and contracted gonorrhea, which she knowingly transmitted to her husband.

Louisa responded to the suit by denying all of Frederick's allegations and charging him with impotency and extreme cruelty for "libelling and slandering her character." Her attorneys introduced as evidence two letters written by Frederick—one to Louisa's mother and one to her sister and brother-in-law—in both of which he accused her of infidelity with John Truscott, the boarder, and alleged that she had contracted gonorrhea. Three doctors—Thomas LeGuen, C. C. Green, and Dr. Heath—testified for Louisa, and although the record of their testimony has not survived, the court concluded on the basis of evidence from medical examinations that Louisa was a virgin. Judge Rising ruled: "That the letters of plaintiff referred to . . . were calculated, if true, to render defendant infamous, and that having been published apparently without any cause or foundation the charges made in said letters and declarations, constitute such extreme cruelty as will entitle the defendant . . . to a decree."[36]

Women's charges of mental cruelty were almost inevitably compounded with allegations of desertion, physical abuse, or drunkenness. The ideals of Victorian womanhood may have been firmly fixed in the minds of judges, but the court documents reveal that this was not the reality for many women. The circumstances of women's lives in this urban mining community included violence, desertion, and alcohol abuse.

The experience of Margaret Booth demonstrates the difficulty women faced in convincing the court to accept charges of cruelty and the importance of presenting sufficient detail. For Booth, winning a decree required two years and two attempts. The federal census taker enumerating residents of the Comstock in July 1870 found the Booth family living in Gold Hill. Thirty-six-year-old, Irish-born Margaret was keeping house for thirty-four-year-old Joseph—a miner born in Kentucky—and their four children. Two of the children, Ellen, age ten, and Thomas, age eight, were born in California. The younger two, Susan, age three, and Walter, age one, were born in Nevada.

In 1872 Margaret filed suit against Joseph after fifteen years of marriage, charging him with extreme cruelty and neglect. Since the time of their marriage, her petition averred, Joseph had treated her in a "cruel and inhuman manner." As evidence, her attorney cited one incident when Joseph "did beat and kick this Plaintiff in anger and with great violence . . . and did threaten to kill" her. He had also allegedly abused his children and neglected "to support his family though abundantly able to do so" as a miner earning a four-dollar-per-day wage. Joseph appeared in court with his attorney to deny all charges, and the judge dismissed the case.[37]

Fifteen months later Margaret retained another attorney and again filed suit on the grounds of extreme cruelty, alleging that Joseph "violently struck plaintiff under the left eye, inflicting a severe blow, and lifted plaintiff up by the hair of her head," that he "made threats of personal injury" against her, frequently "struck and assaulted plaintiff in a violent manner, and has called plaintiff opprobrious names." And, she asserted, "plaintiff really believes she cannot longer cohabit with said defendant without incurring great personal danger."

The children were living with and supported by her and should remain in her custody, she argued, because with his "dissolute habits and violent temper" Joseph was not a fit custodial parent. The family's home in Gold Hill, where she and the children were living, had cost $225, all but $25 of which had been "furnished by plaintiff from her own earnings." Judge Rising found the defendant guilty of extreme cruelty and granted the decree, awarding the property and child custody to Margaret.[38]

Christina Kruttschmitt was the only wife in these Storey County cases to be charged with extreme cruelty. Noting that he had always "conducted himself as a good and faithful husband and amply provided for the wants of his wife," her spouse, A. M. Kruttschmitt, successfully sued for divorce

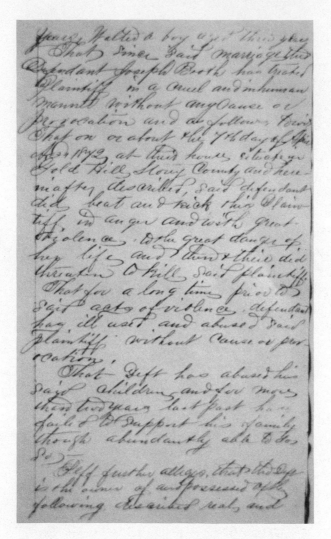

Court case featuring divorce from the Storey County records.
(Courtesy of Dangberg Microfilm Collection, Special Collections, Getchell
Library, University of Nevada, Reno)

in 1873, accusing Christina of physical and mental cruelty, cruelty to his
children by a previous marriage, and desertion for two months. For more
than three years, he alleged, Christina had "by violence and threats . . .
made it unsafe to live and cohabit with her." On one occasion she struck
him "with her fists and scratched and disfigured his face." On another she
attacked him "with a hammer, and then and there . . . threatened to mark

him with said hammer, 'so that he would remember it all his life,' called him all manner of opprobrious epithets, to wit a son of a bitch and a 'whoremaster.' That at said time he was compelled to call the assistance of lodgers in the house to restrain her violent and outrageous conduct."[39]

Christina, like other female defendants in this sample, clearly violated the standards of appropriate conduct for a proper Victorian woman. The case is unusual, though, for the charge of extreme cruelty based on physical violence. Men more commonly relied on charges of desertion, adultery, or drunkenness. Nevada husbands, for example, charged adultery twice as often as wives, although when adultery with other causes is considered, the ratio declines. In the Comstock cases examined for this study, twelve men and eleven women cited adultery as at least one cause for their dissatisfaction.[40]

Eight male plaintiffs succeeded in obtaining divorces with charges of adultery alone. One was Henry Jewett, a forty-six-year-old painter. According to Henry, Emily, his wife of six years, had committed adultery with E. M. Long in San Francisco for several months and subsequently on a steamer to Oregon, after which they had lived together in The Dalles and in Boise before returning to San Francisco.[41]

Two of the four who utilized additional grounds charged drunkenness. Robert Charles accused his wife Johanna of adultery and public cohabitation with Michael Lamb in "a wooden building situated on Main Street" in Treasure Hill, beginning ten days after their marriage; of adultery and public cohabitation with Thomas Daley in Pioche since August 1870; and with habitual drunkenness. Anna Porteous had been committing adultery with William Persy since 1869 and was living with him on South B Street when Samuel charged her with adultery and drunkenness in 1871.[42]

William Dunn cited desertion and adultery. Jane, he claimed, deserted him, committed adultery with one Mr. Russell in Port Wine, California, and bore a child as a result. George Rogers initially alleged only adultery, but in an amended complaint added charges of extreme cruelty, accusing Ida of slandering him and his family and of consorting with "persons who are beyond the pale of respectable Society and unworthy of the consideration of any good woman." As a result "his life has been rendered miserable."[43]

Middle-class strictures against extramarital sex apparently had little meaning for these women. Unfortunately, since none of them filed an answer to the suit, it is impossible to determine or even speculate on circumstances that might have led, or driven, them to form these liaisons.

Of the eleven female plaintiffs, six cited adultery as the sole cause of their dissatisfaction with their spouses, and five used a complex ground of adultery and cruelty. Because Victorian society was more tolerant of male promiscuity, plaintiffs and their attorneys may have believed that a complex suit stood a greater chance of success. The details of the LeGuen, the Megger, and the Legg suits also reveal, though, that adultery was often part of a male behavior pattern that included verbal and physical abuse. Ensebia Megger won her decree on the grounds of adultery and extreme cruelty after her adulterous husband Charles infected her with gonorrhea. Mahala Legg successfully charged that, besides committing adultery, James choked and kicked her and turned her out of the house in the middle of the night.[44]

One of those who succeeded in obtaining a divorce with the sole charge of adultery was a nineteen-year-old California native, Carmen Gomez. Judge Rising found her twenty-eight-year-old husband, Trihana, a miner, born in Mexico, guilty of adultery with Margareta Leon. Carmen received the divorce, obtained custody of their eight-month-old daughter, Aurelia, and reclaimed her maiden name.[45]

Alcohol abuse was, likewise, a contributing factor for both husbands and wives in the decision to divorce, but husbands again were far more likely to utilize it as a sole ground. Since excessive alcohol consumption, especially if it led to public drunkenness, fell into the realm of improper behavior for Victorian women by the standards of the court, husbands could rely on the charge of habitual drunkenness alone to win a divorce suit. Virginia City physician C. C. Green accused his wife Ellen of "habitual gross drunkenness contracted since marriage which compasitates [sic] her from contributing her share to the support of the family, and from attending to her household duties." Ellen declined to respond to the suit, and the decree was entered.[46]

Far more complex was the case of John and Sue Rothenbucher. The couple was married in Virginia City in 1867, and census returns show them living together in Virginia City in 1870. John, forty years old, was a jeweler with $2,500 worth of real estate and a $5,000 personal estate, and Sue, twenty-four, was keeping house. The Nevada state census conducted in 1875 recorded only John, however, because according to her testimony in this case, Sue at that time was living and working in San Francisco.

This convoluted case is one of the few in this study in which the parties requested a jury trial. When John sued for divorce, he charged Sue with habitual drunkenness. She countered by accusing him of adultery and physical and verbal abuse and argued that her use of alcohol was the result of her affliction with a "female disease." To support herself while seeking treatment for her disease, she was forced, she stated, to take a housekeeping job in San Francisco in 1875, which explains her absence from the 1875 Storey County census. Allegations of female drunkenness were apparently less tolerated and more convincing than were unsubstantiated charges of adultery and abuse, because in the end the jury ruled for the plaintiff, and John received his divorce. In 1879 Sue was committed to the Nevada State Hospital.[47]

A woman's excessive consumption of alcohol, especially among members of the upper and middle classes such as the Greens and the Rothenbuchers, occasioned social disapproval that often resulted in divorce. Husbands' drinking, on the other hand, frequently led to mental and physical abuse of their wives and to desertion and neglect. As a result, Comstock wives rarely cited drunkenness alone as grounds for divorce. Harriet Glander's seven-year marriage to John, a carpenter, ended in 1863, when she charged him with gross drunkenness, harsh epithets, extreme cruelty, and turning "her out of her house to give room for Chinamen." Anna Daley accused Henry of habitual drunkenness and "choking and striking her with his fist." James Dill, his wife Sarah alleged, was extremely cruel and a "habitual drunkard."[48]

Neglect was the second most common reason for which Comstock women sought divorce, although plaintiffs generally incorporated additional charges of desertion, drunkenness, or abuse. In the case of Ellen and Frank Burns, the additional charge was extreme cruelty. According to the complaint Ellen filed in February 1873, she and Frank had married in Virginia City in October 1871. If this date is accurate and is not an error on the part of the court clerk, the couple married after living together five years or more and having two children, because in 1870 a census taker recorded Frank, a butcher, and Ellen, a homemaker, both thirty-four, living in Virginia City with five children ranging in age from ten to two.

By the time Ellen filed for divorce a sixth child had been born, so that Ellen had, as she stated, "six children depending on her for support," because "defendant has not supported, nor assisted in supporting any of said children or said plaintiff." Three of these children had been born to her and

Frank during their marriage. Furthermore, Frank had on numerous occasions, she averred, beaten and kicked her and left her in fear of her life. The decree was granted, and Ellen was given custody of the children and sole possession of the family home on Silver Street, which she had purchased before their marriage.[49]

Desertion, the third most common complaint, seems also, at first glance, to be a cause for seeking divorce cited more often by husbands. Indeed, statistics show that when desertion alone was charged, husbands were the plaintiffs more than twice as often as wives were. Some women, like Leonora Allers and Margaret Garhart, simply refused to remain with their spouses. Allers was rebelling against an arranged marriage. As a witness in the case testified, "she wanted to make her own choice of a husband." According to William Garhart, Margaret deserted him in 1865, "without any cause or provocation on his part."[50]

If every case in which desertion was a factor is considered, however, women in Nevada utilized desertion nearly twice as often as men did. Among the cases considered in this study, the ratio is almost three to one, with desertion rarely constituting the sole ground for divorce. Abandonment was usually preceded by some combination of abuse, neglect, and drunkenness. Such complex suits may have provided, as Paula Petrik contended, a means of circumventing the prescribed waiting period for desertion in some instances. Most reflect, though, the transiency of the mining frontier and the tensions and strains in these marriages.[51]

Before Margaret Snyder's husband abandoned their four-year-old daughter for the mines of Diamond City, Montana Territory, "he was often very much intoxicated, and for some months prior to their separation . . . was so constantly intoxicated as to almost incapacitate him from attending to ordinary business." During that time he became "exceedingly abusive." Twenty-seven-year-old, Prussian-born Margaret was forced to "rely on her own personal exertions and the assistance of relatives."[52]

For John Adams, the Beaver Head Mines of Washington Territory proved irresistible. During the five-month course of their marriage, however, he was frequently physically and mentally abusive to his wife, Mary. He kicked her, struck her in the face with his fist with such force that she required a doctor's care for two weeks, threatened her with a "bowie knife," and called her "a whore and made use of other opprobrious language towards her."[53]

In 1876 Mary Gardner charged her husband Thomas with neglect and desertion. Despite his ability to provide for his family "either by the practice of his profession as an Attorney and Counsellor at law, or in the pursuit of his avocation as an editor, or by manual labor as a working man," he had, "for a greater portion of the time during the eight years last past," failed to provide for his family, forcing her to support herself and their son. Three years earlier he had abandoned them completely and was living in San Francisco.[54]

Divorce, regardless of the cause, frequently left Comstock women like Margaret Snyder and Mary Gardner solely responsible for the support of their children. Women in the West received custody of their children in a divorce at a rate more than three times that of the national average. Of course, this figure must be attributed in part to the high rate of desertion in western divorce cases, which left the court with little or no alternative to awarding custody to the mother.[55]

The Comstock women plaintiffs considered in this study also had a high success rate in the eighty-four cases—45 percent of the total—in which custody was an issue. By the mid-nineteenth century, courts in the United States had developed the tender years principle, and this theory that the mother was best suited to care for children under the age of seven increased a woman's chance of obtaining custody of her young children in a divorce. Despite this precedent, and although mothers failed to win custody in only ten of the suits, the records indicate that wives were not sanguine about their chances of receiving custody. Women petitioners frequently made reference to the defendant's parental unsuitability. Stephen Randolph, his wife Susan averred, was an unfit parent because of his "dissolute and drunken habits," while William Walsh's "improvident, immoral, and degraded" habits made him unfit to have child custody.[56]

Nor did a favorable court ruling necessarily ensure success. In the decree ending the marriage of Mercedes and Thomas LeGuen, Judge Rising awarded custody to Mercedes and ordered Thomas to "surrender said children [ages twenty-two months and four months] to said plaintiff." Presumably, Thomas did so, since there is no court record of further action.[57]

Mary Derby, however, was not so fortunate. In February 1876 she received her divorce from Solon, her husband of twenty-one years, on the grounds of extreme cruelty but could not gain custody of her children because they were not in the court's jurisdiction. Nearly a year previously

Solon "against the wishes and consent of plaintiff and clandestinely took the two youngest children to California and has continued to keep them out of this State (although he resides here) and refuses to inform plaintiff as to their whereabouts or to let her see them and she has not seen them since they were carried away."[58]

Women defendants, on the other hand, were rarely successful in obtaining child custody. A woman whose conduct caused her spouse to seek divorce could scarcely be considered an appropriate custodial parent. In the case of Samuel and Anna Porteous, for example, when Samuel sued for divorce, Anna did not appear in court to contest the suit but did file a petition for custody of their year-old son. After hearing testimony of witnesses for both sides, Judge Rising awarded custody to Samuel with the provision that Anna "have the privilege of visiting it at least once each week, at any reasonable time."[59]

When Johanna O'Neil sued for divorce, she charged her husband Dennis with neglect, gross and habitual drunkenness, and extreme cruelty, and requested custody of their three-year-old son. Dennis responded by denying her allegations and asserting that she had committed adultery for several years with John McHugh at the house of Mrs. Farry, was habitually drunk, and had beaten and neglected their child while drunk. Because of her "habitual & gross drunkenness, her moral delinquencies and bad associations" she was "unfit to have the care and custody of the infant son." Johanna did not respond and failed to appear in court. The judge concurred with the defendant, ruling that Johanna had left Dennis "voluntarily and of her own accord" and was guilty of gross and habitual drunkenness. Custody of the child was awarded to Dennis.[60]

Johanna O'Neil lost her suit and her son; Susan Gelzter won her case but lost a son. Gelzter sued her husband Charles for divorce on the grounds of extreme cruelty. He had on one occasion accused her of being unchaste, "called her a bitch and a strumpet and said she was an actress and all actresses were whores." On another occasion he accused her of being "a common prostitute," of having "had illicit commerce with strangers and that her youngest child was a bastard." On yet another occasion, "without any cause or provocation [he] committed a violent assault and battery upon [her] by beating and bruising her severely, telling her at the same time that she was only a thing to use for his own convenience." Charles' attorneys filed an answer denying all charges. Judge Rising issued the decree to Susan and ruled that she should have custody of three-year-old Frederick, since Charles was "unfit to have custody," but without explanation awarded custody of seven-year-old Edward to Charles.[61]

When child custody was contested on the Comstock the doctrine of tender years was generally ignored, as the Porteous, O'Neil, Gelzter, and Legg cases demonstrate. Of course, Anna Porteous and Johanna O'Neil had been deemed morally unfit. Susan Gelzter and Mahala Legg, however, convinced the court of the veracity of their complaints and won their suits, yet they lost custody of their young children to their former spouses.

In the case of Mahala and James Legg, Mahala sued, charging extreme cruelty—both mental and physical, neglect, and adultery—and requested custody of their daughter, Ella Alice. James denied the charges, alleging that she had left him and that, further, she was not the proper person to have custody of Ella Alice. The child, he claimed, was not receiving proper care, and neither the sex nor the age of the child was a reason to give custody to the mother: "She has not any trade profession or occupation whereby she can properly provide for said child or properly educate her: that Plaintiff has not any means of her own but is dependent upon her relatives for support and maintenance." After four months of legal wrangling, Judge Rising issued his decision: Mahala received the decree, but James was to have "care and custody of the child . . . until further order of this court." Mahala was to be "permitted, without hindrance, to visit said child as often as she may choose so to do."[62]

The Legg case raises another issue that is important to understanding the roles of divorced women on the Comstock. Whether a woman sought divorce to achieve independence or because desertion or an abusive spouse left her no alternative, the result was the same: She had to rely on her own resources or on family and community support for survival. Alimony and child support were seldom requested and rarely awarded in these Comstock cases.

Of course, as their petitions indicate, many of these women had been self-supporting, or they had at least been contributing to the family income, long before filing suit. Since the high cost of living and the unpredictability of mining employment often created circumstances that necessitated a two-income family, particularly among the working class, a large number of married women extended their household work to increase the family income. Like their contemporaries in other western states and territories, some took in washing and sewing or housed and fed boarders. Others worked outside the home in some capacity. Thus, if desertion, neglect, or divorce occurred, many women had a means of self-support in place. Attempting to determine the specific means by which these women and their families survived, however, is frequently a frustrating task.[63]

Several factors contribute to the difficulty in tracing divorced women on the Comstock. During the 1870s 299 couples were divorced in Storey County, but only 46 divorced women appear in the 1880 census of the county. In part, this may be attributable to inaccuracy in the returns due to human error or deception. The unmarried servant, Johanna O'Neil, living in the Dean household in 1880, for example, may be the same Johanna O'Neil who lost her divorce suit and custody of her son in 1872. In just one segment of the community, which was made up of people who were Irish born or of Irish descent, 81 women listed in the 1880 census as married had no husband present in the household, which also suggests the possibility of at least some under-reporting of divorced status through design or error.[64]

Mining frontier mobility presents another problem as well as a partial explanation for the census statistics. Since most married women in the Comstock community had moved at least once, and perhaps several times, it is not unrealistic to assume that a divorced woman would do so once again when seeking employment, family support, or the opportunity for re-marriage. Working-class women with no property stake in the area, in particular, could be expected to relocate, but Theresa Dirks, who owned property in San Francisco and Virginia City, moved to Hamilton and then back to Virginia City, keeping her property in Hamilton as well. A move of just a few miles to Silver City or Dayton, even temporarily, would effectively exclude an individual from a Storey County census enumeration.

The sources do provide sufficient evidence to confirm certain points about women's experiences after divorce. Occasionally, as in the case of Mary and Francis Lacy, the court record provides clues. Married in Boston in 1853, the couple emigrated ten years later to Gold Hill, where Mary "supported the family consisting of Defendant[,] herself & one child by washing." In September 1863 Francis was legally declared to be "of unsound mind . . . and sent to the Insane asylum at Stockton," and in December, seven months after arriving in Nevada, Mary sought and won a divorce on the grounds of extreme cruelty. More often, however, court documents include a reference to a petitioner's supporting herself or her family through her own industry and labor but do not specify the means.[65]

Evidence in the records of the effects of the separate property and community property provisions of the law also provides hints. When Mary

Anne Goffin divorced Adolph in October 1863 on the grounds of extreme cruelty, Judge Leonard Ferris awarded her the boardinghouse and saloon at 85 B Street, where she had "labored for him constantly ever since their marriage" in July of that year. When Ellen Green's physician husband divorced her on the grounds of drunkenness, she received $1,000 in lieu of their house and property between C and D streets in Virginia City. Hotel keeper Christopher Miller retained his real property when he divorced Maria, but she received ten shares of Belcher Silver Mining Company stock and $600 in gold coin as her share of the community property.[66]

Census returns can also be of some assistance, although in terms of occupations, especially of women, they can be misleading. A woman who housed boarders, did laundry, and took in sewing, for example, could list only one of these occupations for the enumerator. Nevertheless, useful patterns can be determined.[67]

Of the forty-six divorced women who appear in the 1880 federal census for Storey County, the majority were performing "women's work." The nature of the Comstock society, in which men outnumbered women by approximately two to one, created a demand for housing and domestic services. The most common occupation, professed by twelve women, was that of seamstress or dressmaker. Nine were "keeping house," an amorphous term that sometimes meant boardinghouse operator, sometimes meant homemaker, and occasionally served as a euphemism for prostitute. Seven women were proprietors of lodging houses. Other occupations included milliner, music teacher, laundress (two), fortuneteller, authoress, waitress, servant (two), and prostitute (two). Two of the four women without a stated occupation were living with family. One was a patient in the county hospital, suffering from "debility," and officially listed as a pauper. The other was Sue Rothenbucher, twenty-seven years old, living alone on Mount Davidson, and listed as head of house, occupation "none."[68]

Marital status, unfortunately, was not included in the 1870 census returns. When the data from the census corresponds to information from the court documents, though, it is possible to gain additional information. Eliza Cleveland, for example, who divorced Valentine on the grounds of neglect and conviction of the felony of forgery, supported herself and their five- and three-year-old sons as a schoolteacher, according to the 1870 census. After her drunken and abusive husband deserted her for the mines of Diamond City, Montana in 1866, Margaret Snyder supported herself and daughter Barbara by working as a dressmaker, which we learn from the census, and enlisting the aid of relatives, as she informed the court. Mary O'Byrne exercised the option of returning to her maiden name of McCabe when she divorced Patrick in 1876, but the 1880 census lists Mary O'Byrne, divorced, a seamstress living on G Street in Virginia City with three children whose names and ages correspond to those provided in the court documents.[69]

But in other instances the evidence, while suggestive, is less conclusive. In her complaint, Catherine Hodges noted that she "has been and still is supporting" the couple's four children, aged two through eight, although she did not add neglect to her charge of extreme cruelty against Henry. The judge granted the divorce and awarded the saloon business on the west side of South C Street to Catherine. The Virginia City census return for 1880 includes Catherine Hodges, a thirty-two-year-old divorced woman with a seven-month-old son keeping house at 133 So. D Street. Had a fourth child been born after the divorce? Were the other three children away at school or living with relatives when the census enumerator visited? Did the income she received from the saloon permit her to stay at home with her infant son? Or were there coincidentally two divorced women named Catherine Hodges in Virginia City? The Nevada state census of 1875 includes a forty-six-year-old female, V. Mendosa, born in Mexico, a housekeeper—apparently unmarried—with $800 in real estate. Could this be Vicenta Meza, who divorced Nicholas in 1869 after a year of marriage, and resumed her maiden name of Mendoza? Unfortunately, the sources are too vague, and definitive answers remain elusive.[70]

Although the extended family was not a common feature of the extremely mobile mining frontier, some women, like Leonora Dirks, Mahala Legg, Margaret Snyder, and Kate Bonafous, were able to turn to relatives for assistance. Kate and Eugene Bonafous lived on the Comstock for the duration of their thirteen-year marriage. The census enumerator in 1870 found them keeping a boardinghouse in Virginia City. At that time they had two children, Pauline, age three, and one-month-old George.

In 1877 Kate sued for divorce, charging that she had endured more than five years of cruelty and neglect. The precipitating incident occurred in January 1875, when Eugene struck her with enough force to knock her out of a chair, ordered her to leave, and locked her out of the house on a cold winter night, forcing her to take refuge with her sister. Since that time she had not lived with him, and, she alleged, he had contributed nothing to the children's (by now numbering three) support for seven years. She "was

compelled to support herself and her children by the charity of her relatives." And in fact he "frequently . . . threatened to abandon Plff. [sic] and
leave her to depend solely upon her relatives for support." The majority of
women, however, lacking the support of family, were forced to rely on their
own resources.[71]

Yet another option was remarriage. The demographics of the Comstock
made it a strong possibility, which no doubt accounts for some of the divorced women who do not appear in the 1880 census, that a divorced
woman would marry again. Leonora Dirks and her mother both exercised
this option, as did Mercedes Nevarra and Kate Bonafous: In June 1878 a
marriage license was issued to W. H. Sheridan for his marriage to Mrs.
Kate Bonafous.[72] Bessie Gallagher also chose to remarry shortly after her
divorce. Bessie married Thomas Muckle in Virginia City in 1871, when
both were in their teens. By 1875 Thomas was working as a mason, and
Bessie was keeping house and caring for their two young daughters.

The following year, in May, Bessie sought a divorce. Thomas, she
averred, was addicted to drink and had treated her cruelly. One day in February that year, as she was recovering from a serious illness, he broke the
door and the window in their house, smashed dishes, and when she resisted his efforts to force her out of the house, removed the stovepipe and
attempted to smoke her out. In April he "threatened to blow plff [sic]
brains out if she attended the Catholic Church," and later that month he
slapped her face and "otherwise abused her." The court found Thomas
guilty of extreme cruelty and awarded Bessie the divorce, the family home,
and custody of the children, although Thomas was permitted to visit them
"once every two weeks during the day time." Bessie resumed her maiden
name of Gallagher and six months later the county court recorded a marriage license for Thomas Muckle and Bessie Gallagher.[73]

Evidence also suggests that efforts to legally terminate a marriage, especially in the case of desertion, may often have begun only when remarriage
became a possibility. For a woman dependent on her own labor for survival, there was otherwise no clear reason to incur the expense required to
obtain a divorce. Alexander Cowan purportedly deserted his wife in 1858,
yet Eilley waited two years before suing for divorce and marrying Sandy
Bowers. Annie Lowden wed E. H. Smith in Gold Hill on May 11, 1876,
eight days after receiving her decree and resuming her maiden name. According to her suit, her husband, John Thompson, had abandoned her in
1873. Although Nevada law permitted divorce two years after desertion,

and lawmakers had reduced the waiting period to one year in 1875, Annie allowed three years to elapse before filing. When Catherine Quigley divorced Daniel in 1876, she claimed he had deserted her five years earlier. She resumed her maiden name of Holland. Ten days later Catherine Holland and David Estey received a marriage license in Virginia City.[74]

Circumstances of divorce help us to understand issues involving gender, the law, and society, as well as views of women and the institution of marriage. In the case of the Comstock it is possible to see that early on women possessed legal advantages that their counterparts in other western states and territories achieved only through gradual transformation of the law. Utah Territory provided a female-oriented statute with a lenient residence requirement, liberal grounds, and an omnibus clause that allowed consensual divorce. Nevada legislators replaced Utah's statute with one based on that of California, but with slightly more liberal provisions. Judges in both Utah Territory and Nevada expanded the possibilities for divorce by interpreting the ground of extreme cruelty to include mental suffering. The ideology of companionate marriage and domesticity embodied in the laws contributed to the legal climate that permitted relatively easy divorce, even though such concepts had little relevance to the situation of most divorcing women.[75]

Although their post-divorce economic options were limited by Victorian expectations of women's proper place and work, the Comstock, like the mining communities of Montana or the transportation towns of Wyoming, offered sufficient opportunity for female employment to ensure survival. Only one divorced woman in this sample, a patient in the county hospital, appeared in the census as a pauper.

The American West provided a favorable environment for divorce, and the Comstock was no exception. Although divorce law in the West did not differ markedly from that in most eastern and southern states, western society's tolerance and acceptance of the practice meant that the West led the nation in the number of divorces per capita in the nineteenth century. Examination of women's experiences with divorce and its aftermath can help illuminate women's role in marriage and society on the Comstock, and, by example, it can assist in our understanding of a wide range of gender issues as they apply to the region.

❦ 5 ❧

The "Secret Friend"

Opium in Comstock Society, 1860–1887

Sharon Lowe

When Comstock prostitute Nellie Davis took her own life with an overdose of morphine in the spring of 1863, few were surprised. As the saying goes, "She had nothing to live for."[1] The manner of this woman's death illustrates the easy access to and frequent use of the "demon" drug, opium, and its derivatives by marginal groups in Comstock society.

The stereotypical image of drug use in nineteenth-century America is predominantly one of men and women in eastern cities, most often from the lower echelons of society, ensnared in a dark, downward spiral toward oblivion. Such was not always the case. Western cities, with their promise of boundless opportunity built on individual effort, were just as likely to have their share of problems stemming from opium. In addition, opium use in patent medicines was a socially acceptable form of addiction, where users were often women among the respected levels of nineteenth-century society. Conversely, recreational use of opium, usually by smoking it in a pipe (a practice first introduced in the West by Chinese labor in the 1840s)[2] was markedly less acceptable. Even so, this application of the drug was almost unrestricted because of its wide availability.

Virginia City, Nevada, was an inviting microcosm in which opium use flourished on the fringes of society. Miners and their camp followers, fresh from the California gold fields, rushed to the burgeoning Comstock towns that came into being almost overnight. Virginia City became the preeminent community, synonymous with the development of the great Comstock Lode.

Not only a prominent industrial center, Virginia City also had the façade of a Victorian urban center. As historian Rodman Paul succinctly describes the place, Virginia City had the "dimensions and character of a cosmopolitan city, with diversified interests and varied social features of a large and important center."[3] The underlying weakness of this cosmopolitan city was its boom-and-bust mining economy, which created the pervasive social disorder expressed in its haphazard infrastructure, inadequate health care, and high mortality rates from deadly disease and industrial accident. Virginia City thus presented ample opportunity for the introduction of drug use, especially of opium, in a society undergoing rapid industrialization.

Evidence suggests that opium had been used for therapeutic reasons in the eastern United States as early as the eighteenth century; recreational use, however, seems to have come much later and arrived in the West primarily via Chinese immigration. Chinese laborers came to California in 1849, during the Gold Rush, to work as laundrymen, cooks, and servants, and eventually to work on the railroads and in the mines. The Chinese were also an entrepreneurial class of merchants and contractors. Many came from Canton, which was closely associated with opium smoking and traffic.

The Chinese were often trapped in a hopeless system that reinforced the need for an escape outside the workplace. As Gunther Barth suggests in his *Bitter Strength: A History of the Chinese in the United States, 1850–1870*, release was to be found in the early Chinatowns of the West; there opportunities existed to indulge in gambling, prostitution, and smoking opium, often all in the same establishment. Opium became a form of escape in a hostile environment.[4]

The Chinese influx into Nevada was similar to that in California. Chinese had been in Nevada since 1851, when John Reese hired Chinese to build water ditches for irrigation and mining in the western Great Basin. When building of the Central Pacific Railroad began in 1863, thousands of Chinese workers came to labor in its construction.[5] There was considerable anti-Chinese resentment on the Comstock because of their employment in building the Virginia and Truckee Railroad from 1869 to 1872. Adding to anti-Chinese sentiment was the increased frequency of opium smoking on the Comstock and the beginnings of a pattern of recreational use of drugs. Though some citizens ignored the Chinese, others believed that they perpetuated vice and threatened community values. Opium dens exemplified this alleged lack of virtue and offended Victorian standards of propriety. As one upright citizen, widow Mary McNair Mathews, suggested, "Chinatown of Virginia City is like Chinatown of every other city of the Coast, a loathsome, filthy den. It is enough to breed cholera or any other pestilential disease."[6]

For many Euro-Americans, Chinatown and its opium dens were part of a culture that they considered impossible to assimilate because the Chinese engaged in vices and barbaric customs. The opium connection only solidified those beliefs. Mathews offers contemporary insight into the problem:

> Their opium dens, which seem to defy all police power to break up, are a nuisance, and are also ruining our people, for many have become slaves to this most destructive habit. But not only men and women visit the opium dens but I am informed, by good authority, that girls and boys visit them and often have to be helped home by their companions. Girls and boys, from twelve to twenty, are daily being ruined by this opium smoking. I never visited one of these dens, but have had them described to me. A table sets [sic] in the center of the room, a dish of opium upon that, and long pipes for each smoker is [sic] dipped in this, and they lie on bunks around the table and smoke till they become unconscious. After a person once smokes, he has created an appetite for a vice that he has no power or wish to refrain from. You who are so far from these scenes of vice have no idea of the baneful effects of this pernicious habit. It is utter ruin to smoke the first pipe, for there is but one way to keep them from it afterwards, and that is the walls of an asylum.[7]

Mathews may have exaggerated the connection between occasional smoking of opium and addiction. Some sources suggest that between 15 and 30 percent of the Chinese community in California was addicted.[8] Although there are no records to substantiate a statistic in Nevada, one would assume that the corresponding figure was similar. Statistics do suggest, however, that not all Chinese were addicted to opium and that the majority who smoked were simply social opium users, not the addicts that Mary

A nineteenth-century lithograph of an opium den. (Courtesy of Nevada Historical Society)

McNair Mathews described. The extent of non-Chinese addiction is not statistically ascertainable. Whether there were groups in society willing to integrate with the Chinese and other fringe elements such as prostitutes, pimps, and gamblers in frequenting the dens is unknown. Although the extent of their addiction, if they were addicted, is conjecture, one point is clear: Members of the underworld "would have the fewest scruples about associating with Orientals or experimenting with their vices."[9]

There seems to have been a symbiotic relationship between drugs and various marginal groups, including prostitutes. Recreational use of opium and the technique of smoking it in a social atmosphere became part of an alternative commercial zone that included gambling establishments, brothels, and dance halls.[10]

The use of opium as a recreational activity among marginal groups developed as part of the economic system that characterized the instant city or industrial urban center that Virginia City had become. It provided the prostitute with a synthetic euphoria that made the misery of the job less extreme. The social atmosphere of the Chinese opium den also influenced prostitutes. After working into the night they would often drink with their customers at various saloons or dance halls. Then they would go to the dens, which served the dual function of meeting place and sanctuary from the working world. Prostitutes stopped by during the early morning hours and relaxed with a pipe. One historian suggests that "within the den a rigid code of honor prevailed: smokers would not take advantage of other smokers, or tolerate those who did."[11]

Another characteristic of the opium den that made it attractive to prostitutes and others on the fringe of society was its ubiquity. Dens were common in almost every major city in the West. One nineteenth-century western opium smoker observed, "it's a poor town now-a-days, that has not a Chinese laundry, and nearly every one of these has its lay-out. You get the first ticket [letter of introduction written in Chinese] and you're booked straight through. I tell you it's a great system for the fiend that travels."[12]

Thus, on the Comstock, as in other mining camps, the accessibility of opium, the camaraderie, and the belief that drugs offered a refuge from worldly pressures promoted institutionalization of the opium den, especially at the margins of society. This was especially true for a place marked by rapid in-and-out migration and with a population unbalanced in terms of age and gender, conditions typical of the Comstock and other western industrial mining complexes. By the 1860s "nine tenths of the population were males and three fourths of the sector of the population were aged between twenty and forty."[13] This type of community welcomed the availability of opium. The expanded personal freedoms of a modern city encouraged experimentation and ensured that drug use would occur.[14]

Though hypodermic injection of drugs had been introduced in New York as early as 1856, it never became as popular as opium smoking for recreational use on the Comstock, even though it was less expensive, took less time, and was much stronger. One of the main attractions of opium smoking was that it was a social, almost ceremonial procedure. In his early study of opium smoking in America, Dr. Harry Hubbell Kane, writing in 1882, relates that he had "never seen a smoker who found pleasure in using the drug [smoking] at home and alone, no matter how complete his outfit, or how excellent his opium."[15] He further suggests that the allure of smoking opium was the fascination of a vice that was worth "moral ruin, the charring and obliteration of every honest impulse and honorable sentiment, the sweeping away of every vestment of modesty, by such associations and such surroundings."[16]

For many, opium smoking proved a vicious cycle. The addict's daily supply cost from fifty cents to three dollars, more than a day's work at unskilled labor. The loss of work from impaired ability and perhaps the loss of a job made it an even more expensive habit. The addictive nature of opium induced the user to consume more as the habit continued.[17] For the prostitute, the price was sometimes the highest of all—suicide or overdose.

One authority argues that Comstock prostitutes were major opium users as is shown by their frequent presence in the dens of Virginia City.[18] Storey County District Court indictment records indicate that use was quite extensive among prostitutes, and that many were prosecuted.[19] It is difficult to ascertain the extent of use in terms of ethnic groups. Clearly, however, prostitutes seem to have been the largest occupational group of users.

The Journals of Alfred Doten, diaries kept by a Comstock newspaperman from 1849 to 1903, refer several times to opium overdose among prostitutes. Doten's February 6, 1868, entry reports: "Little Ida, that I used to [erasure] some 2 years ago was found dead in her bed at the 'Bow Windows' (Jenny Tyler's) this morning. She had been rather dissipated for some time past and latterly had taken to opium—Ida Vernon was her name—about 32 years old—a man was sleeping with her and found her cold in the morning-rest in peace Ida-She was her worst enemy."[20]

It is difficult to determine an exact number of suicides caused by smoking opium, because the records often lump all opiate use together either as prescription medicine or as patent drugs, which often contained opium and its derivatives. The Storey County Hospital records between 1865 and 1880 do not list an instance of opium or suicide as a cause of death. Causes that are acknowledged include mania, mania-a-potu (defined as delirium from either alcohol or drugs), and other conditions whose direct cause could have been opium intake, implying that the drug was indeed a cause of death. In addition, the death category "unclassified" may well have included suicide or drug overdose.[21]

Another entry in Doten's diary, dated August 16, 1872, suggests the extent of opium overdose and suicide on the Comstock among prostitutes: "2 whores at Rose Benjamin's corner of D St. and Sutton Avenue, committed suicide by taking laudanum this PM—I went to see them, one not quite dead, but died about 11."[22] These are but a few of the instances of suicide among prostitutes. Between 1876 and 1880 coroner's records describe at least twenty deaths by suicide, which does not include attempted suicides using some form of opium. Unfortunately, the coroner did not always mention the cause of death.[23] As Goldman points out:

> Prostitutes of almost every age, nationality, race, and status within the irregular market place killed themselves. At age twenty-two, Laura Steele, a beautiful black-haired Scots-woman, took lethal laudanum at Rose's fancy brothel; Nellie Davis, down-and-out thirty-three year old prostitute, overdosed on morphine in her solitary room at Mrs. Gray's B Street lodging house; and an anonymous China-woman who had cost her owner $800 suicided by means of laudanum in the back of Stern's store in Gold Hill. She was one of the six Chinese women who chose death over a life of hopeless slavery.[24]

By the 1870s opium smoking had reached all levels of society and had become a social problem that could no longer be ignored. The Chinese migration, prostitution, and the ready availability of the drug added momentum and ushered other levels of society into recreational drug use. Furthermore, Virginia City was a male-dominated society, with few legal or social restraints. Because initially there were so few families, residents found their nights open to many forms of unrestricted recreation—drinking, card playing, prostitution, and opium dens.

Establishments catering to all of these tastes were conveniently located in close proximity to one another. Doten spoke of a typical night on the town during the mid-1860s in Virginia City: "The same-passed as usual-after I got through I went with Sam Glessner down to Chinatown-drank at Tom Poo's-went to Mary's house-we were in her room with her-she gave us each a cake left from the holiday of yesterday-filled with nuts & sweet meats-we laid on the bed with her and smoked opium with her-a little boy some 2 yrs. old sleeping there, belonging to one of her women . . . On my way home I stopped in at the Great Republic saloon-big whore ball going on."[25] By the 1870s many of Virginia City's proper citizens, both men and women, had also begun to smoke opium freely. According to Dan De Quille:

> Not a few men in Virginia City-and a few women-are opium smokers. They visit the Chinese opium dens two or three times a week. They say that the effect is exhilarating—that it is the same intoxication produced by drinking liquor except that under the influence of opium a man has all his senses, and his brain is almost supernaturally bright and clear. An American told me that he had been an opium-smoker for eighteen years, and said there were about fifty persons in Virginia City who were of the initiated. In San Francisco he says there are over five hundred white opium smokers, many women among them.[26]

It is difficult to determine whether all the women referred to were prostitutes, or to what ethnic group they belonged. Nevertheless, opium smoking was no longer restricted to the lower levels of society.

Though the monetary cost of smoking opium was not insurmountable, the cost in human suffering had to be addressed. Opium smoking was tacitly accepted until the sons and daughters of mainstream Comstock society began to engage in the practice. In the *Virginia Evening Chronicle,* dated July 9, 1877, an article entitled "In the Cradle of Hell" shocked citizens as to the extent of opium abuse in their community:

> An extraordinary scene was enacted on D street last night, the villainous aspect of which casts a shadow even over the contaminated locality. It appears that about four days ago a laboring man living near the C and C shaft sent his daughter to the grocery store for some soap and she did not return that night. . . . She was in the house of Rose Benjamin, a den which is ranked as the most notorious deadfall in the place and the father demanded admittance. Mrs. Benjamin, the harridan who runs the establishment, refused him admittance and slammed the door in his face. . . . A crowd of over 100 men collected in the street.
>
> When the gathering crowd took in the situation the excitement became intense and cries of "Pull down the House!" "Gut the den!" "Clean out the dead-fall!" and similar expressions were heard on all sides.[27]

The men finally rescued the young girl from the house, but it was rumored that another thirteen-year-old was held there as well. When interviewed with her mother present, the young girl explained how she entered into the den of inequity:

> I was four days in the house, and was induced to go there by a girl named Frankie Norton. . . . We smoked opium together. Reporter-How long have you smoked opium? Girl-About seven months. Here the mother threw up her hands in astonishment and burst into tears. . . . The reporter continued: I smoked in Chinatown and Gold Hill. The places are open every night.
>
> Reporter-How many pipes did you smoke in a day? About thirty-five, but sometimes not more than twelve. If I don't smoke I feel sick; I want some now.[28]

The reporter went on to question the girl and found out that the other girl was apparently still in the den and seemed "stupefied with the drug and utterly indifferent to her situation."[29] The reporter also discovered that many men went to this establishment to smoke opium and engaged in commercial sex there with young girls.[30]

After it was recognized that opium smoking was no longer restricted to the fringes and that it thus threatened middle-class propriety, action by local authorities was swift. Nineteenth-century author Harry Hubbell Kane quotes a letter he received when investigating opium smoking in America. Though the name of the doctor who wrote the letter is not known, he was apparently from Virginia City.

> Opium smoking had been entirely confined to the Chinese up to and before the autumn of 1876, when the practice was introduced by a sporting character who had lived in China, where he had contracted the habit. He spread the practice amongst his class, and his mistress, a woman of the town, introduced it among her demi-monde acquaintances, and it was not long before it had widely spread amongst the people mentioned and then amongst the younger class of boys and girls, many of the latter of the more respected class of families. The habit grew very rapidly until it reached young women of more mature age, when the necessity for stringent measures became apparent, and was met by the passing of a city ordinance.[31]

This source is mistaken as to the date that smoking by non-Chinese began, but the doctor does provide insight into the size of the problem and the evolution of reform in Virginia City. Reform was also beginning in other cities in the West, such as San Francisco. The *San Francisco Chronicle* of July 25, 1881, reported that the "habit in past years, so far as whites are concerned, was confined to hoodlums and prostitutes mostly. . . . Now that there are scores of places where the habit can be contracted in clean rooms in respectable portions of the city, the practice will gradually extend up in the social grade."[32] Other sources indicate that opium smoking was rapidly becoming a form of recreation in other cities in the West.[33]

By the mid 1870s the menace of opium had become a public controversy in Virginia City as well as in the West in general. Newspapers regularly described its dangers, reporting its growing spread outside the narrowly circumscribed lower classes. The reality evoked public outrage once it was known that young girls and middle-class women had begun to find their way into Chinatown. Where Euro-American smokers once came from the criminal class, now the rich frequented the dens as well. It was not, however, "just the prospect of upper class addiction that aroused concern, but also the fear that respectable women were being seduced in the dens."[34]

Inflamed public opinion prompted the passage of restrictive laws. In September 1876 the Board of Aldermen of Virginia City passed an ordinance that abolished opium dens and fined those who smoked opium.[35] By 1877 the Nevada legislature had also restricted opium to medical use by prescription only: "From and after the last day of March, A.D. eighteen hundred and seventy-seven, it shall be unlawful for any person or persons, as principals or agents, to sell, give away, or otherwise dispose of opium in this State except druggists and apothecaries; and druggists and apothecaries shall sell it only on the prescription of legally practicing physicians."[36]

Not only was it a crime to possess opium but also to possess pipes and related apparatus. The statute also made landlords responsible for anyone who used the drug on their premises. The problem, as always, was that these substance-abuse prevention laws were difficult to enforce and rarely solved the problem. A few weeks after passage of the state law prohibiting the sale of opium, the Virginia *Evening Chronicle* commented:

Some months ago the Chronicle made a thorough exposure of the prevalence of opium smoking in Chinatown. Public sentiment became so strong upon the subject that the Board of Aldermen passed an ordinance prohibiting the sale of the drug and imposing a heavy fine upon all persons having it in their possession. The passage of this ordinance resulted in a temporary scare to the opium smokers, and a few poor Chinamen were arrested and fined for keeping dens for smokers. In a few weeks however the resorts were again in full blast, and the traffic in the drug was carried on under the very noses of the police. . . . The place is a plague-spot, which should be eradicated at once. Young girls are introduced to the dens, seduced while under the influence of the drug, and then they drift rapidly down into some crib on D Street.[37]

The article went on to state that the police were not doing their job and that the dens were still in operation. The implication was that Chinese smoking among themselves were left alone. If Chinese and Euro-Americans were smoking together, however, sporadic enforcement of the laws would occur for a month or so.

This laxity may indicate collusion between law enforcement and "vice elements." There is no evidence to indicate that law enforcers were being paid off, but, curiously, many dens remained in full operation. Euro-American smokers simply transferred their habit to new quarters while continuing to obtain their opium from "a network of Chinese dealers." Later, once the heat was off a particular den, the Euro-Americans would "drift back to the Chinese den or to a den run by a Chinese but largely patronized by white customers. Then there would be another crackdown, and the white smokers would disperse again."[38]

Goldman suggests that on the Comstock the laws enacted over the opium issue during this period illustrate the anti-Chinese sentiment that was brewing in the West. She argues that "neither law stemmed the brisk trade on opiates. They merely validated Caucasian moral ascendancy and facilitated police harassment of the Chinese."[39] Euro-American addicts continued to get their opium from the Chinese. Their sources, however, seemed to relocate to other areas like Gold Hill, Sutro City, or even old "abandoned mine tunnels, cellars or back rooms of restaurants and stores," according to historian Phillip Earl.[40]

In 1909 the federal government enacted the Smoking Opium Exclusion Act, which prohibited importation of the drug and inadvertently created other problems. Smugglers increased the price of opium so that other derivatives such as heroin and morphine, which were more addicting and less expensive, were substituted illicitly. Furthermore, users switched from smoking opium to administering it hypodermically or ingesting it orally, because the latter methods were cheaper.[41] These changes, however, had little effect on the Comstock. By the end of the second decade of the twentieth century mining prosperity had declined, and the Queen of the Comstock was a ghost of her former self.

Medical addiction to opium followed a different pattern than did recreational use. The medical addict, unlike the recreational user, was not limited to a particular geographical locale. Medicinal use of opium was similar in both eastern and western states and was most common among middle- and upper-class women. Moreover, medicinal use did not carry the stigma attached to the recreational use of either opium or alcohol. Indeed, the growing temperance movement encouraged drug use by "respectable" women who would never consider alcohol as a means of alleviating distress. They easily succumbed to the calming effects of laudanum and its derivatives, replacing one drug genre for another. In addition, a small medicine bottle was much easier to conceal than a flask of bourbon was.

Another reason for the popularity of opium was that its derivatives, including laudanum and morphine, could treat a wide range of ailments, including tuberculosis, rheumatism, neuralgia, insomnia, diarrheal disorders, and everyday coughs. Doctors were not able to cure many of these illnesses, but with opium they could at least treat the symptoms and make the patient comfortable. As David Courtwright suggests: "In spite of the kindly doctor on horseback, riding for miles to tend the sick and injured, there was in reality, little the harassed nineteenth-century physician could do to cure his patients' ills. Only a few of the many drugs he prescribed were demonstrably effective (some, like calomel, were positively dangerous), and the causes of most diseases were still poorly understood."[42]

Opium and its derivatives became the universal pain remedy of the nineteenth century throughout the United States. It was often prescribed as a panacea by doctors in their treatment of disease. It had the added advantage of being inexpensive. "Happiness might now be bought for a penny, and carried in the waistcoat pocket; portable ecstacies might be corked up in a pint bottle," wrote the Englishman Thomas De Quincey, describing the relative ease with which the drug could be obtained.[43]

One medical historian suggests that because opium was considered the most effective painkiller, "physicians even invented a new disease for it to treat, called neuralgia (a term still in use today, but in a more restricted sense)."[44] Neuralgia was "a term which is frequently employed, both tech-nically and popularly in a somewhat loose manner, to describe pains the origin of which is not clearly traceable."[45] Opium then, was the perfect "cure-all" to treat this particular disease. Concern over addiction, however, became more vocal after the Civil War, when medical doctors were criti-cized for over-medicating patients with the narcotic, because they often had little practical experience and there were no legal or professional limi-tations on administering opiates. Also, few medical schools instructed stu-dents about the possibility of addiction as a result of freely prescribing morphine. As Charles Terry and Mildred Pellens point out in their survey on addiction, "the authors of a great majority of textbooks on the practice of medicine, materia medica, and therapeutics failed to issue any warning of the dangers involved in the use of morphine."[46] Thus doctors had a free hand in the dispensing of the addictive drug.

Doctors prescribed opium for many illnesses without identifying the cause. Some segments of the public were skeptical of doctors in general and relied on self-medication rather than paying high medical fees for an educated guess. In the case of Annie Alchoon, a Virginia City wife and mother, self-medication of laudanum and morphine became not only ad-dicting, but fatal. In the Virginia City Coroner's Report, dated January 9, 1879, the cause of her death is noted as an overdose of laudanum. Appar-ently Alchoon had been buying laudanum for years, with no restriction. The day before her death, her son bought a bottle for his mother because, as the pharmacist stated, "I have been acquainted with her and have sold her laudanum for years."[47] When her landlady was asked if "Mrs. Alchoon was in the habit of taking laudanum or morphine" she answered, "She was in the habit of taking four bits worth [4 ounces] in a day."[48] Her husband also stated that his thirty-nine-year-old wife had been "so used to taking laudanum that it became a necessity," and that she had been in the habit of taking laudanum for some twelve or thirteen years: "I am aware she took great quantities of it—I was in Bodie at the time of her death—The dis-patch announcing her death was sent down to me while in the Bodie mine."[49] With few restrictions, self-medication was a matter of going to the closest pharmacist and asking for one's drug of choice. In this case, as in many others, the result was death.

For some women, addiction did not lead to death, but often equally dis-tressing was institutionalization. In the Nevada State Hospital commit-ment records from 1873 to 1890, over half of those committed were women.

Reasons for institutionalization included "morphine and whisky," as in the case of Mrs. Sue Rothenbucher of 1879.[50] Another example was Mrs. Caroline Cassidy, whose third commitment for opium use, in 1873, apparently resulted in further court proceedings in which she lost custody of her two children, aged six and eight. The children were subsequently placed with an adoption agency in 1874.[51]

Indiscriminate drug use was furthered by the patent medicine industry. Women were the chief users of opiated tonics, which doctors freely prescribed for such ailments as "menstrual and menopausal discomforts." Sixty to seventy percent of women between the ages of twenty-five and fifty-five used these remedies.[52] Rebecca Hastings, a thirty-five-year-old native of Cornwall, England, died in Virginia City of hemorrhage during her menses. Apparently one Dr. Heath prescribed opium, iron, stimulants, ergot, and brandy, the usual remedies. The doctor stated that she died of exhaustion, but this prognosis was highly questionable in light of the amount of opium he had prescribed.[53]

Women in Victorian society frequently complained of nervous exhaustion or depression. Carroll Smith-Rosenberg, commenting on women in the late-nineteenth century, states:

> American girls seemed ill-prepared to assume the responsibilities and trials of marriage, motherhood and maturation. Frequently women, especially married women with children, complained of isolation, loneliness, and depression. Physicians reported a high incidence of nervous disease and hysteria among women who felt overwhelmed by the burdens of frequent pregnancies, the demands of children, the daily exertions of house-keeping and family management. The realities of adult life no longer permitted them to elaborate and exploit the role of fragile, sensitive and dependent child.[54]

Under these constraints, "hysterical" women were given laudanum, pills, paste, tinctures, camphors, and injections of opium to alleviate female ailments. The euphoric state of oblivion resulting from the tranquilizing and sedative qualities of opium numbed the nervous stress that women complained of and often "lured others into fatally increasing their dosage."[55] Easily obtained over-the-counter medical preparations containing either alcohol, opium, or both, caused addiction and often resulted in a self-administered overdose. In the Reno *Evening Gazette*, on March 7, 1881, an article entitled, "Who is to Blame" explained the problem at both the national and local levels:

> Physicians are no doubt largely to blame for the alarming increase of the opium habit in this country. An old physician recently said, "I'm sorry to say that some of our oldest physicians seem to be falling into its constant use and it appears to me out of laziness. . . . After having once been called in an emergency, to save the life of a patient who had taken an overdose of morphine . . . he finally succeeded in restoring the lady to consciousness, when just at the time, the family physician arrived and took charge of the case. The first thing he did was to whip out his little syringe and administered a dose of morphine hypodermically.[56]

An 1879 example from Virginia City provides further illustration of over-medication. In the case of Julia Hynes, who apparently was pregnant and in a great deal of pain, the physician, Dr. M. W. Heath, prescribed "an opiate to ease her pain; 1/2 of a grain of morphine and a 1/2 grain of syrup of poppies. The doses were intended to be 1/2 of a grain to be given every 3 hours if she was in pain."[57] When he came back the next day, she had apparently over-medicated herself and died. "I thought she was dying of what we call embolism. I did not see any medicines left—there was an empty bottle, with the client's name on it. I didn't know she had another doctor who had prescribed Ergot, which was used by women when they wanted to get rid of a child."[58] Presumably, the combination of drugs that she was receiving from more than one doctor was the cause of death. This was not an isolated case.

Not only were women more apt to be drug users, but as one historian suggests, the typical "addict appeared to be a respectable woman, whose dependence began in prescriptions from a physician for either real or fancied ailments."[59] Though research indicates that medicinal use among women clearly existed, the associated secrecy prompted alarm within the medical community. Without a realistic idea of which class and how many women were actually addicted, authorities remained ambiguous about addiction. Furthermore, many women took their medicine in private, letting few others, often not even their husbands, know the extent of their reliance on their "secret friend." By the latter part of the nineteenth century, the medical community itself was aware of the opium problem among women: "We have an army of women in America dying from the opium habit—

larger than our standing army . . . The profession is wholly responsible for the loose and indiscriminate use of the drugs."[60]

Perhaps the fact that women were under the care of doctors to a greater extent than were male patients explains the belief that women were more prone to excessive use of the drug. As drug historian Wayne Morgan points out, "reported generalized ailments . . . increased the tendency to prescribe a sedative."[61] The medical establishment also theorized that because they dealt with gender "illnesses" such as reproduction and menopause, women were sick more often than men and were thus apt to rely on medications such as laudanum and opiate derivatives.[62] Contemporary observers suggested that industry had molded the entire generation, "enhancing a sense of nervousness to new pressures stemming from the industrial order. The impact was especially hard on women . . . for a woman is more nervous, has a finer organization than man, [and] is accordingly more susceptible to most of the stimulants."[63] This popular view of drug use supported the belief that women indeed were "finer than men, had special natures and roles and needed unusual care."[64]

Another reason for drug use among middle- and upper-class women was social. On the Comstock, respectable women were stigmatized for seeking release in the many saloons dotting the city's landscape, particularly when many of their peers were joining temperance leagues such as the Sons and Daughters of Temperance and others that were established on the Comstock throughout the 1870s and 1880s.[65] As one doctor commented, "As a rule, women take opiates and men alcohol, both seeking the same pleasurable results. A women is very degraded before she will consent to display drunkenness to mankind; whereas, she can obtain equally if not more pleasurable feelings with opiates and not disgrace herself before the world."[66]

For whatever reasons, opium became problematic at both the national level and on the Comstock. While its use was not restricted to women, addiction through medical means was most prevalent among them. Opium was also used to treat other aliments, however, including diarrhea and hangovers, for which a morning-after injection of morphine was the preferred treatment. Some doctors even prescribed opiates for alcohol addiction.

Thus many westerners, as one author suggests, were dependent through "direct and repeated administration of an opiate by a physician; others simply continued purchasing the drug after the doctor ceased prescribing it."[67] The prescription records of Warner and Munkton, druggists on the Comstock during the early 1860s, reveal large quantities of opium administered

Virginia City's C Street was the heart of the community's commercial district. (Courtesy of California Historical Society)

to patients. A prescription dated January 1, 1863, calls for "a teaspoon of opium camphor every hour until gone." Approximately 50 percent of the prescriptions contained opium (either tincture, camphor, sulphur or pills).[68] Even when physicians refused to prescribe further opiates for patients, they could resort to using patent medicines, which were sold without prescription in drugstores, mail-order catalogues, and traveling medicine shows. Such medicines were likely also available in Virginia City at drugstores on C Street,[69] or from the local vendors that De Quille described on the streets of Virginia City:

> During the summer, men who have for sale all manner of quack nostrums, men with all kinds of notions for sale, street shows, beggars, singers, men with electrical machines, apparatus for testing the strength of the lungs, and a thousand other similar things, flock to Virginia City. Of evenings, when the torches of these parties of peddlers, showmen, and quack doctors are all lighted and all are in full cry, a great fair seems to be under headway in the principal street of the town-there is a perfect Babel of cries and harangues.[70]

He went on to describe the different "wonder cures" available on the Comstock, like "soap root" toothpaste, which "makes the old man feel young, and the young man feel strong."[71] Because there was no law that required listing ingredients on packages or bottles, opium was often the active, secret ingredient.[72] Even temperance workers were known to sip "Old Sachem Bitters" that contained opium and large quantities of alcohol. Housewives, affluent women, and prostitutes[73] alike could easily take "Wyeth's New Enteritis" pills, or "Mrs. Winslow's Soothing Syrup," which contained large amounts of morphia.

Though medical use of opium on the Comstock followed national patterns, recreational use was almost exclusively western, propelled and made possible by the rapid development of a phenomenal industrial boomtown. Consumption of narcotics for medical use in Virginia City was facilitated by readily available patent medicines and freely dispensed prescription drugs. For many women, opium provided a temporary escape from the restraints of the Victorian world, without the stigma associated with alcohol. Ultimately, however, its use on the nineteenth-century Comstock became not only an individual matter but also a social problem. Women's "secret friend" was no longer a secret, and the pipe dreams that obscured reality prompted public measures to curb its use.

PART THREE

❧ Occupations and Pursuits ❧

The gold fields of California gave women remarkable opportunities to make money by offering scarce services at inflated prices. Eventually supply caught up with demand, and the halcyon days of extravagant profits ended. On the Comstock, born in a West with an already established, if distant, infrastructure, the period of price gouging was brief. For most who settled there, opportunity meant carving out a niche in a burgeoning, diverse economy.

Women pursued dozens of occupations, each warranting a full analysis. Prostitution will, perhaps unfairly, always be a cliché of the western mining camp. Although many women were involved in the Comstock brothels and cribs, they were consistently a minority. Fortunately, Marion Goldman's *Gold Miners and Silver Diggers* provides a comprehensive study of the subject. Other occupations have not received this level of research. The needle trades, for example, were the occupations Comstock women most often engaged in to earn wages, and yet these trades have not received the same level of attention as has prostitution. Spiritualists were on the other end of the demographic scale. Few in number and poorly documented, they nonetheless played an important role in helping the populace deal with concerns through mystic avenues. Fortunetellers and needleworkers are only two of many occupations that deserve full treatment.

Motivated by spiritual concerns at once similar to and distinct from those of the fortunetellers, the Daughters of Charity were an important cornerstone of Virginia City society. They founded the local orphanage, an important private school, and the best hospital in the mining district. Like fortunetellers, they were few in number, but they, too, had a profound effect on the community. The larger number of women involved in general charitable and public causes also shaped and improved society. Reminiscent of the Daughters of Charity, altruism was an important factor directing their choice of pursuits.

The four kinds of endeavors discussed in the following essays are merely a sampling of the possibilities. They serve to underscore, however, the diversity of choices available to women on the Comstock while also emphasizing that not all women's undertakings were for profit.

❦ 6 ❦

Creating a Fashionable Society

Comstock Needleworkers from 1860 to 1880

JANET I. LOVERIN & ROBERT A. NYLEN

Introduction

T he Comstock Lode offers a unique opportunity to examine the production of women's fashionable clothing in a nineteenth-century western mining town. Virginia City and neighboring Gold Hill were sophisticated mining communities that established a pattern of social
and industrial development for the mining West. Fashionable clothing
played a crucial role in developing and displaying the cultural sophistication of the Comstock. A variety of needleworkers who created complex
garments and contributed to the rich diversification of this western mining
community provided the means by which to achieve stylish attire.

The inhabitants of Virginia City and Gold Hill were aware both that
they were at the heart of an international phenomenon and that they were
far removed from places like New York, London, and Paris, where the cul-
tured elite set standards for proper nineteenth-century society. Yet the
Comstock was not culturally isolated. Indeed, residents carried Victorian
value systems to the mining districts of Mount Davidson, where they transformed their houses into fashionable homes of elegance and style.[1] Virginia
City and Gold Hill were not unusual in this respect among nineteenth-
century frontier communities. What makes them exceptional is their early
development and diversity as well as the importance of the mining district.

Just as Comstock residents incorporated nineteenth-century mores and norms into their frontier lifestyle, they also paid attention to fashionable styles of dress. Still, achieving the fashionable look for most mid-nineteenth-century women was a new experience. Before this period, only women of upper-class status had the resources to purchase fabric and have a dressmaker create a stylish garment.[2]

The production and consumption of clothing provides insight into regional social behaviors—in this case, into of a prestigious western mining community. Lucrative wages and life-threatening employment gave the people of Virginia City and Gold Hill both the means and the desire to achieve a pronounced display of fashion. Advertisements emphasizing Parisian and New York goods testify to the fact that fashionable attire was meaningful to Comstock residents. Newspaper discussions of civic and social events reported in detail the costumes of the women who attended. These events provided an opportunity for women to exhibit the current vogue.

The West, as an extension of an increasingly worldwide Victorian culture and society, generally adhered to the fashionable standards of the day. Still, there is some question as to whether the region afforded women more flexibility than did other regions of the United States. Examples of strict conformity to formal custom on the Comstock occasionally occurred side by side with expressions of laxness. Whether nonconformity was a personal response to social pressures or an expression of regional variation is a subject for future discussion. What is evident is that the sewing trades on the Comstock reveal a complex social system that catered to the needs and ambitions of a diverse clientele.[3]

Clothing was gender, class, and age specific in the nineteenth century. Men wore trousers; proper women did not. In 1865 the editor of the Gold Hill *News* ran several articles about a rebellious local schoolteacher who wanted to wear breeches. Responding to her request, the editor replied, "she may assume to be a man, as soon as it pleases her convenience with every appendage there unto belonging."[4] Although some women did try wearing the new bloomer costume during their overland travels, it was not considered acceptable fashionable attire. Clothing was also dictated by the age of the wearer. Boys wore dresses until their "breeching" at age four or five, and age determined the length of girls' dresses. Short skirt lengths were appropriate for young girls and longer ones for young women aged fourteen and older. In addition to these restrictions, fashionable clothing, especially for women, was tied to the activity of the moment. There were clothes for morning, for visiting, for afternoons at home, for dinners, for walking, for balls and receptions, and for mourning.[5]

The Industrial Revolution, a period of technological advances, resulted in an increased production of cloth. The mass production of clothing, however, evolved slowly. It began early in the nineteenth century with men's wear, which quickly became popular, with many ready-made garments available by the 1850s. Virginia City dry goods merchants stocked a wide variety of shirts, cuffs, collar bands, waistcoats, underwear, trousers and coats.[6] Women's ready-to-wear was not widely available. Only undergarments and outerwear, such as capes and jackets, were sold off the rack in the 1860s. Most women's garments were custom-made, either by the women themselves or by dressmakers, milliners, or seamstresses.[7] Costume historian Claudia Kidwell observed that fashions in the second half of the nineteenth century were complex,[8] and yet they were manufactured by needlewomen with little formal education. Sewing skills involved a tradition of "women's work" that was taught to daughters by mothers and was, and still is, notably perishable.[9] It is the impermanence of the product that has customarily distinguished men's work from women's work. Possibly because of this difference, women's work has traditionally held less status. Yet before the creation of mass-produced, ready-to-wear garments, the manufacture of women's clothing was an achievement that required skill, expertise, and time.

In the United States, women's fashion innovations emanated from Paris, primarily through the wide distribution of women's magazines and detailed newspaper accounts of international social events. From Paris, images of the new styles appeared in New York, where they were filtered through to the lesser fashion centers, such as Boston in the East and San Francisco in the West. Business owners were quick to point out that their inventory of stock had arrived from these trendsetting communities. Advertisements for dressmakers and milliners in Virginia City continually referred to "Paris and New York styles." Mrs. Loryea, the first milliner to advertise on the Comstock, stated in 1861 that she had French merino, French flowers, and the latest styles of bonnets.[10]

Table 6.1 illustrates the concentration of personal clothiers from the urban states to the more remote states in 1880. It compares the number of

Bonnets from Godey's Lady's Book and Magazine, *January 1878.* (Courtesy of Frances Humphreys Collection, Nevada State Museum)

men and women tailors, tailoresses, milliners, and dressmakers to the general population in Massachusetts, California, Nevada, and Colorado. Massachusetts had the highest concentration of personal clothiers, which is not surprising, considering that it was a major fashion center and textile-producing state. California had a slightly lower proportion of personal clothiers, even though San Francisco was the fashion center of the West Coast. The mining states of Nevada and Colorado show markedly lower numbers of personal clothiers, with Nevada's numbers being higher than Colorado's.[11]

Although there was a lower concentration of personal clothiers in the western mining states than elsewhere, this does not suggest that these areas were less fashionable than were other places in the country.[12] Elizabeth Bliss of Gold Hill wrote to her aunt in June 1864: "The houses are furnished well and there is much more dress displayed here than on Beacon Street when the ladies make calls. . . . And such a display of dress I never expected to see here. They never think of wearing anything but point and thread lace."[13]

In fact, Virginia City residents were keenly aware of fashion and frequently tried to outdo each other, as Louise M. Palmer stated in 1869:

It is curious that we all protest against lunch parties, yet continue to give them and attend them. It is stupid to dress in one's newest silk and handsomest corals to partake of chicken, creams, ices and Champagne, with a dozen of one's own sex. Who can help be painfully conscious that each and every one of them have priced the silk at Roseners' and the corals at Nye's before they came to one's wearing. It is a trying thing for one's dress to be subjected to the test of its value, and not of its adaptability and becomingness.[14]

Fashionable expectations in Virginia City were a reflection of regional nativity. Easterners typically brought the latest fashion ideals to the frontier. Yet the Comstock was similar to other frontier communities, where some people relaxed their Victorian formality. Mary McNair Mathews, a widow from New York, arrived in Virginia City in 1869. As she began her search for work, she was surprised to find Mrs. Beck answering the front door, "wearing a black and white calico wrapper." Mathews explained, "Being dressed so much more plainly than the rich and gaudily-dressed ladies I had met with at the other houses, I took her for hired help, instead of the lady herself."[15] Although Beck was a merchant's wife and a member of upper-class society, she chose to answer the door in a casual manner.

During the last half of the nineteenth century, Virginia City emerged as a major community of the mining West. It boasted international entertainment, exotic foods, and a sophisticated lifestyle, and it was tied to the economy, politics, and fashions of San Francisco. San Francisco milliner Mrs. Cowles advertised in the *Territorial Enterprise* in 1862, "orders from Nevada promptly attended to."[16] She was apparently exploring business opportunities in Virginia City before migrating eastward to Nevada. Response must

TABLE 6.1 Percentage of Personal Clothiers, 1880

State	Total Population	Personal Clothiers	Percentage (%)
Nevada	62,266	434	.7
Colorado	194,327	1,168	.6
California	864,964	9,516	1.1
Massachusetts	1,783,085	26,462	1.5

Source: *Compendium of the Tenth Census* (June 1, 1880) part 2, revised ed. (Washington D.C.: Government Printing Office, 1888).

have been positive, as Mrs. Cowles arrived on the Comstock less than a year after her first advertisement appeared in the *Territorial Enterprise*.[17]

The first women to solicit sewing work on the frontier were undoubtedly the pioneer women themselves. Rarely identifying themselves as seamstresses, these women used their domestic skills to provide a much-needed service to the mining community. In 1861 a woman in Washoe Valley advertised that she was willing to do "machine sewing at Two Cents Per Yard, Canvass Bagging and Tents Stitched at the low price of two cents per yard on the Best Machine in the Territory."[18]

The sewing machine was patented in 1846 and although popularly accepted was not necessarily essential. Mary McNair Mathews, who became a seamstress in Virginia City, sewed by hand until she could save the thirty dollars she needed to purchase a machine. In some cases hand stitching was preferable. Mrs. Loryea offered her customers a choice between hand and machine stitching in 1861.[19] Farrell-Beck, in her study, "Nineteenth Century Construction Techniques," explains: "Until well into the 1860s, machine stitches were second-rate substitutes for fine hand work."[20] As late as 1877 in Virginia City, Madame Algo provided her customers with the option of machine or hand stitching.[21]

The producers of women's clothing appear in the census under four occupations: milliners, dressmakers, seamstresses, and sewing women. Milliners were the high-status needleworkers, designing and making women's headgear. Dressmaking, the cutting and constructing of dresses, had a lower status than did millinery. The dressmaker usually designed and cut dresses that were then sewn by apprentice dressmakers or seamstresses. Seamstresses and those who referred to their jobs as "sewing" or called themselves "sewing woman" constituted the lowest level of the needlework trades.[22]

Table 6.2 illustrates the frequency with which women's custom clothiers of the Comstock appeared from 1860–1880. It also illustrates their distribution. The largest jump is from 1860 to 1870, when the number of dressmakers increased from none to forty, and the milliners increased from two to fifteen. This dramatic increase parallels the boom years of this mining community.[23]

Some areas of women's clothing production were not evident on the Comstock. There were no hoop makers, lace makers, or galloon makers, for example.[24] In contrast, needlework represented the third most popular occupation for women on the Comstock in 1880. According to three de-

TABLE 6.2   Frequency of Women's Custom Clothiers in Storey County
by Occupation, 1860–1880

Year	Milliners	Dressmakers	Seamstresses	Sewing Women	Total
1860	2	0	0	1	3
1870	15	40	15	1	71
1880	22	97	21	5	145
Total	39	137	36	7	219

Source: Eighth, ninth, and tenth U.S. manuscript censuses of 1860, 1870, and 1880, respectively.

cades of census material, there were 219 people, only one of whom was male, manufacturing women's clothing in the area.[25] As a whole, these people were young and primarily single, and the majority were born in the United States. Most stayed on the Comstock no more than two or three years. The majority worked as independent proprietors; a select few were employers, and some were employees.[26] Examining these various custom clothiers of women's dress during the twenty-year boom cycle of Virginia City and Gold Hill provides additional insight into this important type of women's work on the Comstock.

Milliners

Milliners were the first group of professional needleworkers to establish themselves on the Comstock Lode.[27] The *Encyclopedia of Victorian Needlework* defines millinery as "the composition of headdress, whether bonnet, cap, veil or other decorative or useful headcovering."[28] Of the total fashion silhouette, millinery was the forerunner of style. A beautifully trimmed hat or bonnet was essential to the Victorian woman. It was the portion of nineteenth-century vogue which allowed for personal taste. Adorning the hats was not just the finishing touch but was the expression of artistry and design. Fashionable millinery was reserved for the upper classes of Comstock society. The hard-working widow, Mathews, never mentioned going to a milliner, yet headgear was an essential part of a Victorian woman's toilette. Mathews tartly judged the high-society Comstock women as "wearing a whole flower garden on top of their heads."[29]

The production of millinery required "good taste, an eye for colour, and

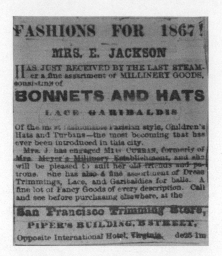

An advertisement for bonnets and hats appearing in the 1 January 1867 Territorial Enterprise. (Courtesy of Nevada State Museum)

a light hand."[30] There was a traditional hierarchy in the millinery profession. After finishing an apprenticeship, which could last one or two years, a girl then became a "maker," the worker who fashioned hat shapes from foundations. Then, if she was fortunate, she could be a "trimmer," whose job it was to adorn the hat with various forms and combinations of lace, feathers, flowers, and ribbons. According to Mrs. E. Jackson of Virginia City, a trimmed hat in 1868 sold for $2.50, whereas straw hats were $1.00, and shakers were $.50.[31]

There does not appear to have been a formal system of apprenticing for the millinery trade on the Comstock. No one advertised for millinery apprentices, and there is no evidence of a segregated system of assembly. The census enumerator indicated only "milliner" or "worked in millinery store." The only millinery training that is evident is that which was passed down from mothers to daughters. Jackson, Mayer, and Vincent all worked with their daughters in their millinery businesses.[32]

In many instances the millinery and dressmaking trades overlapped. Mrs. Loryea had a fashionable millinery and dressmaking shop in Carson City in 1861, where she made "ball Dresses of every description. . . . [and stocked] kid gloves, mantillas, etc." She even advertised, "ladies in the interior wishing bonnets sent to them, will have their orders promptly attended by stating their age, complexion and the color desired."[33] Loryea opened another shop on C Street in Virginia City when her husband became a partner in Almack's saloon.[34] Here she advertised "Ladies Fine Cloaks [probably ready-made], Dry Goods, Fancy Goods, Perfumery, Stationery, Fine Crockery, Glassware, etc.," as well as "stamping and embroidery," which were women's recreational sewing.[35] This expanded inventory indicates that Loryea was adapting to the wide variety of needs and desires of the few women on the Comstock in the early 1860s. At the height of Loryea's career, she had two single women, Agnes Maas and Eliza McAvoy, in her employ.[36]

Overall, milliners on the Comstock were rarer than dressmakers, except in 1860, when there were two milliners and no dressmakers. The 1870s and 1880s show at least two to three times as many dressmakers as milliners (see table 6.2). The milliners on the Comstock Lode tended to be younger than their dressmaking counterparts (see table 6.3). The exceptions were the proprietors, who tended to be older and married.

There are few studies of needlework available for comparative purposes, but a detailed analysis comes from Boston. Although far removed from the Comstock, the case of Boston's needleworkers provides at least one means by which to evaluate Virginia City in a broader context. The profile of nineteenth-century milliners in Storey County is consistent with that of milliners in Boston. There were fewer milliners than dressmakers in both communities. Milliners in both places were primarily single and native born. Bostonian and Comstock milliners were younger than their dressmaking counterparts. The only exceptions were the employers, who tended to be older.[37]

The most prominent millinery establishment on the Lode was owned by Mr. and Mrs. E. Jackson. They arrived with their family on the Comstock in the early 1860s.[38] Mrs. Jackson began as a dressmaker, opening her shop on A Street between Union and Sutton. Her husband joined her in business, and they moved into the Pipers Building, opposite the International Hotel—a prestigious location with high visibility. Originally called the San Francisco Trimming Store, the Jacksons's establishment stocked a wide assortment of "bonnets, hats, lace garibaldis, dress trimmings and lace" in addition to "fancies" and "curls."[39] The Jacksons worked together in their shop, a unique business arrangement in the needle trades of the Comstock.[40]

TABLE 6.3 Age Distribution of Women's Custom Clothiers in Storey
County by Occupation, 1860–1880

Age	Milliners $N=39$ (%)	Dressmakers $N=137$ (%)	Seamstresses $N=36$ (%)	Sewing Women $N=7$ (%)
55–59	0.0	2.2	0.0	0.0
50–54	2.6	.7	0.0	0.0
45–49	2.6	2.9	5.6	14.3
40–44	10.3	5.1	5.6	0.0
35–39	5.1	12.4	19.5	14.3
30–34	10.3	12.4	11.1	14.3
25–29	17.9	25.6	22.2	28.6
20–24	28.2	24.1	19.5	0.0
14–19	23.1	14.6	16.7	28.6
Total	100.0	100.0	100.0	100.0

Percentages are based on cumulative totals as expressed during the federal census years,
1860–1880.
Source: Eighth, ninth, and tenth U.S. manuscript censuses of 1860, 1870, and 1880, respectively.

The other side of the International Hotel boasted another fashionable
millinery shop, where Fanny Mayer operated at No. 6 North C Street.
Mayer also shared a business arrangement of sorts with her husband, who
owned the saloon below her millinery shop. Both businesses advertised as
being across from the International Hotel.[41] Mayer, a Prussian, first began advertising in the city directories in 1863, a year after the birth of her
daughter. She combined the skills of millinery and dressmaking. Mayer
and Jackson each hired employees; in fact there must have been some competition between the two firms, because Jackson advertised that she had
just hired Miss Mary E. Curran, who formerly had worked for Mayer.[42]

All of the milliners on the Comstock were Euro-American, except for
forty-year-old Elizabeth Vincent and her twenty-four-year-old daughter
Annie, who were African American.[43] The Vincents employed two other
Euro-American milliners, nineteen-year-old Kate Foley and fifteen-year-
old Rosey Gallagher. The Vincents did little advertising, yet they made an
important contribution to Virginia City's history. On April 7, 1870, the
African American people of Virginia City staged a celebration in honor
of the passage of the thirteenth amendment ending slavery. As part of the
festivities, Vincent and her milliners made a silk flag, which led the long
procession of dignitaries, marchers, bands and saddle horses. After the

procession, Vincent presented the flag to the newly formed Lincoln Union Club.[44]

Aside from the owners of millinery shops and those employees who worked in them, the majority of the milliners listed in the census and the city directories worked alone out of their residences. In some cases a milliner and a dressmaker teamed up, and occasionally two milliners worked together. After Mary Curran worked for Mayer's and Jackson's millinery establishments, the city directory listed her with her own address and did not mention any other employment relationship.[45] None of these independent milliners remained on the Comstock long. Most of them are listed in either the 1860, the 1870, or the 1880 census and maybe in a city directory, but then they disappear from the records.

Mrs. Lash, a fashionable New York milliner, set up shop in 1868. She advertised the usual inventory of hats, bonnets, flowers, and straw trimmings. But she did not remain in business long; by October, twenty-eight-year-old fellow New Yorker, Mary Wolter, succeeded her. Wolter worked as a milliner underneath the International Hotel on C Street for almost two years and then moved down C Street to a location opposite the Presbyterian church, where she sold off her inventory and retired in 1870.[46]

As with many aspects of "women's work," employment could and did fluctuate. It is unclear, for example, whether Wolter retired at age thirty or married. Perhaps she reappeared with her married name, or possibly the economic fluctuations of the mining society caused her to withdraw from business and move to another community. Whatever the reasons, Comstock milliners enjoyed a sense of business independence. Fanny Mayer, who advertised as a milliner in 1863, does not appear as a milliner or dressmaker in the 1870 census but reappears in the 1880 census, still married and now working with her daughter. The same was true of Jackson, who advertised in 1863 and then ceased to advertise again until 1867, at which point she continued to advertise through the mid-1870s.

Independence and women's rights were emerging in the nineteenth century, and successful women engaged in the needle trades often played an active role. Oregon milliner Abigail Duniway became a West Coast suffrage leader, and Bostonian children's clothier Maria Hollander was an ardent advocate of woman suffrage. In Virginia City, Mrs. E. Jackson advertised for the "Women's Rights Convention" in 1868.[47]

Gamber's study of Bostonian needleworkers notes the smaller number of milliners compared to the larger number of dressmakers. She proposes that the small number of milliners and the fact that milliners stocked inventory indicated exclusivity.[48] This appears to have been true on the Comstock, as well. In addition to their stock of headgear, most of the milliners who advertised listed various items of dress trimmings. In a mining community with constant ups and downs, the millinery business enjoyed some stability. Two milliners, Jackson and Mayer, remained on the Comstock for at least ten years. Millinery shops were located in a prestigious area of town. Both Mayer and Jackson eventually trained and employed their daughters in the art of millinery. In analyzing the advertisements of people who were both dressmakers and milliners, millinery was always, with no exception, listed first. Opportunity, prestige, and status appeared to be more attainable to the milliner than they were to the dressmaker.

Dressmakers

Women's clothing on the Comstock was a reflection of status, values, and amount of disposable income. Within this complex urban mining community there were varying expressions of fashion. From the Paiute wearing the cast-off clothing of an employer to the ladies attending the Inaugural Ball, the power of fashion to direct behavior was apparent.[49] Mathews was teased for not succumbing to the latest fashionable apparel: "Mrs. Calvin often laughed at me, and said: If you would go down in that old stocking, and get out some of the gold you have hoarded up, and put it on your back in fine clothes, you would stand some show to get a rich husband, for they would know then that you did have something; and now they don't know you are worth anything."[50]

A wide variety of dress choices were available to the fashionable and the not-so-fashionable women of the Comstock. Most women strove for upper-class comfort, hiring a dressmaker or seamstress to make a garment. Some women's ready-made clothes were also available from dry goods merchants by the 1870s, while other women received hand-me-downs or made their own clothes.[51]

Whether a woman made her own dresses or hired a dressmaker, achieving the fashionable look required skill. Before the advent of ready-to-wear or the popular acceptance of the paper pattern industry, fashionable women's attire was cut and fit for the individual. Gowns of the 1860s, the 1870s, and the 1880s were complex, with an extremely tight-fitting bodice and sleeves and elaborately draped skirts. Whereas the production of millinery was an art, the manufacture of fitted dresses was a precise science.

Dressmaking consisted of two major skills: cutting, and sewing seams. Cutting, by far the more difficult of the two skills, involved "the mental process of determining the shape of the pieces as well as the physical act of cutting."[52] Sewing the seams of the garment was more time consuming but required considerably less skill.

Originally called mantua makers, early-day producers of women's clothing traditionally used the pin-to-form method of clothing construction.[53] With simple, loose-fitting fashions, such as those of the Empire period, it was relatively easy to cut the garment pieces and fit them directly to the wearer. But as the fashions changed and became tighter fitting and more complex, cutting the cloth required greater skill.

Whereas there was no formal apprenticing program for milliners on the Comstock, there do appear to have been apprentice opportunities in dressmaking. Miss Diamond advertised in May 1872 that she was looking for apprentices.[54] And in the 1880 census Emma Buteau, a sixteen year old from Saxony, appears as a dressmaker's apprentice. Whether this is a reflection of her age or her skill level is not known. She is the only dressmaker's apprentice listed in either the 1860, the 1870, or the 1880 census, and Diamond's advertisement is the only record of a solicitation for apprentices.

Aside from apprenticing, there were alternative methods to learn the science of dressmaking.[55] Drafting systems evolved into "charts" or "rules" that could be used to produce women's fashionable ensembles. They were devices used to design the garment and, at the same time, size it to the wearer.

Three different drafting system methods were available throughout the latter half of the nineteenth century: proportional, hybrid, and direct measure. The proportional system was based on the theory that *one* body measurement correlated to other idealized measurements, which could be used to draft and fit a garment. The hybrid system combined idealized proportional measurements and several direct measurements to create a garment. And the direct measure system measured all the critical body circumferences and created a customized pattern.[56] Advertisements from Aaron Tentler's proportional system stated: "The learning of the trade [dressmaking] requires a long time and is . . . expensive and difficult." This was true, especially in rural areas, where there were limited or nonexistent apprentice

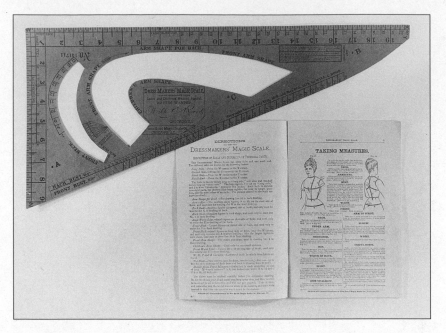

Drafting rule and instruction booklet for Will C. Rood's Dressmakers' Magic Scale.
(Courtesy of Nevada State Museum)

opportunities.[57] With his system however, "every lady may learn to make any kind of dress herself, in a short time."[58]

So attractive were the advertisements that many dressmakers bought these systems to learn the science of cutting. Once familiar with the systems, dressmakers used them to enhance their advertising. Miss Leighton, a Virginia City dressmaker, mentioned in June 1878 that she used "The Peerless Drafting Rule." And Mrs. Paul, another fashionable dressmaker of the Comstock, stated that, "all garments [were] made and cut by actual and correct measurement." Both methods were successful in creating a fitted garment; the direct measure method, however, provided a more precise fit, which was critical in the nineteenth century. The proportional method evolved into the standardized paper pattern industry of today.[59]

Along with these drafting systems, drawings and diagrams of the latest fashionable attire were available through monthly women's magazines. In 1853 *Godey's Lady's Book* began supplementing fashion illustrations with corresponding diagrams. Two years later *Peterson's Magazine* began including inch measurements on the miniature pattern drawings. Separate pattern pieces then became available, but only in one size. *Godey's* informed its readers that this proportionally sized pattern would fit "a lady of middle height and youthful proportion."[60] Individual fitting required implementing the traditional "pin-to-the-form" method. Paper patterns, although available, did not become readily accepted until the beginning of the twentieth century. Yet paper patterns were available in 1878 for the Comstock woman through Mary J. Cromwell, agent for Demorest's Pattern Co.[61]

Dry goods houses and millinery shops stocked vast arrays of fabrics and trims. Sam Rosener, an 1868 dry goods merchant, advertised that he was importing fabrics specifically from Paris and Lyons. Also mentioned in Rosener's ad were inexpensive black silks, some ready-made ladies' cloaks, and cashmere and paisley shawls. At this time there is no mention of ready-made dresses for women. But in 1874 dry goods merchant Ira Berck listed ladies' underwear, dresses, and cloaks as available. These early ready-made garments would more than likely have been calico wash dresses—casual, comfortable everyday dresses. Mathews explains that her son, Charlie, always bought her a Christmas present, and sometimes it was a new dress.[62] More than likely, it would have been a ready-made wash dress of the type Berck sold.

Dressmakers tended to create custom-made silk ensembles, which were costly. Mathews reported that many Virginia City women paid twenty to forty dollars for a silk to be made, and half that for fabric and materials.[63] Some women spent even more. Julia Bulette, a prostitute who was later murdered, had a large inventory of fashionable clothing.[64] In fact, the conviction of her murderer, John Millian, was due to his sale of a "dress pattern," which was verified as having been purchased by Bulette at Rosener's dry goods store. This dress pattern later sold at auction for forty-five dollars.[65] High prices and exquisite dress patterns were not uncommon on the western frontier.[66]

Of the 137 dressmakers on the Comstock between 1860 and 1880, 135 were Euro-American women, and 2 were African American. Dressmakers tended to be slightly older than their millinery counterparts. The majority were between the ages of twenty-five and twenty-nine; milliners tended to be in the twenty to twenty-four-year-old range (see table 6.3). Dressmaking could be performed at home and required little capital, so it appealed to married women more than millinery did. Widows constituted 13.6 percent of the milliners, 14.3 percent of the seamstresses, and 13.4 percent of the dressmakers.

The profile of nineteenth-century dressmakers in Storey County, Nevada, is consistent with that of dressmakers in Boston, a major fashion center. There were at least twice as many dressmakers as there were milliners in both communities. Bostonian and Comstock dressmakers were older than milliners. Bostonian and Comstock dressmakers were predominately single and native born. The exceptions for both communities were the dressmakers who hired employees; they tended to be between the ages of thirty and thirty-nine.[67]

Trautman's study of Colorado needleworkers provides additional insight. In 1880 Denver, another western mining community, had a population of 25,000 and listed forty-eight dressmakers in the city directory. In contrast, Virginia City's population in the same year was 16,000, and there were ninety-seven dressmakers. Although Trautman based her study solely on directories, Virginia City appears to have offered a greater opportunity for fashionable dressing than Denver did (see table 6.1). In addition, more Denver dressmakers (23.7 percent) than Comstock dressmakers (43.7 percent) were married.[68]

With the emphasis on Parisian goods and styles, it is surprising to find that there were so few French dressmakers on the Comstock. Two dressmakers referred to themselves as "Madame," but neither of them appear in the census, so verification of their French nativity is problematic. More than likely, the preface of Madame was meant to attract customers rather than to indicate French origin.[69] Madame Aglo advertised that she received the latest French novelties every ten days directly from Paris. And Mrs. Haynes mentioned that she used "The Great French System of Cutting."[70]

Although there were more dressmakers than milliners in Virginia City in the 1870s and 1880s, the former were also more transient. Not one dressmaker stayed from 1870 to 1880 (see table 6.2). Mrs. Gray, who advertised in the 1871 city directory and then again in the 1878–1879 directory, was gone by 1880.[71] Even Miss Diamond, who is listed in the 1880 census, arrived in 1872. More common were dressmakers such as Maria Berry, Nancy Baldwin, and E. Chamberlain, all of whom listed in the 1870 census and in the 1871 city directory, and who then vanished from the record.

Using the Sanborn Perris map to track the locations of dressmakers demonstrates that most tended to work out of their dwellings, with intense concentrations in the Flowery District, C Street, and the northeast sections of Virginia City. In 1880 six dressmakers lived in the "red light district" and possibly engaged in prostitution. Three of these lived on the corner of D Street and Union, the only instance of a dressmaker, milliner and seamstress living and working together in Virginia City. Twenty-seven-year-old, divorced Jessica Klunter was the listed seamstress; Alice May, a twenty-six-year-old single woman, was the dressmaker; and twenty-two-year-old, married Blanche Lebo was the milliner. This was an interesting combination of women, each of whom was consistent with her occupational profile: the younger milliner, a slightly older dressmaker, and an older divorced seamstress, all lodging at 1 D Street in Virginia City. Whether they catered to nearby streetwalkers or exaggerated to the census enumerator and instead sold sexual favors is impossible to determine.

Given the transient nature of dressmaking, there were few formal employer-employee business relationships among dressmakers in Virginia City. The foremost dressmakers appear to be thirty-year-old Miss Diamond, who advertised for apprentices, and Mrs. Gray, age forty-six, who employed Miss C. H. Johnson and Mrs. M. E. Thompson. Occasionally a younger dressmaker hired a seamstress. For example, Miss Annie Macguigan, age thirty-one, hired Irish seamstress Miss Ellen Reudenbough, thirty-seven. Dressmakers were sometimes employed by milliners to provide the total fashion silhouette. For example, Mrs. Elton worked for Fanny Mayer.[72]

Most dressmakers appeared to work independently, developing their own clientele and emphasizing the "latest styles" and "fit" in their advertising campaigns. Miss McCarrick, who was both milliner and dressmaker, mentioned that "she is prepared to exhibit the Very Latest Eastern and European Styles and Fashions."[73] Where fit of the garment was of the utmost importance, working with repeat customers was helpful. When their business locations changed, both Miss Diamond and Madame Moinet solicited business from "[their] old friends again."[74]

Dressmakers did not necessarily stock an inventory of fabrics or trims, as did milliners. Most patrons purchased their fabric at either the dry goods merchant or the milliner and contracted with the dressmaker to make up the garments. Dressmakers were sometimes hired by the woman of the house to make a custom pattern out of a piece of lining fabric. To save money, the woman herself could sew up the garment. This may have been what Mrs. Haynes is referring to in her advertisement, "Cutting and Basting a specialty."[75]

Another option for dressmakers and seamstresses was "going out by the day." This refers to working in someone else's home and doing personal sewing, as Miss Bailey did for Mrs. Wagner.[76] In 1878 Miss Leighton advertised that she also would be willing to "go out by the day."[77]

More common than a formal business partnership was an informal, mutually beneficial association with other dressmakers. Most frequently, sisters worked together. The 1880 census and the business directories gave evidence of three pairs of sister dressmakers present on the Comstock.[78] In addition, there were three combinations of widows and married dressmakers; of these, two were young widows working with older married women.[79] In the third instance, the widow was the older woman, and the younger dressmaker was married.[80] Such combinations provide insight into the survival skills of the frontier women.

Ideally, dressmakers tended to make the latest fashionable toilettes for the Victorian woman, yet there were also exceptions. Possibly in response to slow times, several Virginia City dressmakers made simpler, more functional clothing. Miss Leighton advertised as a fashionable dressmaker and then two months later teamed up with Miss Gibson. Together they advertised that they could sew "calico suits" cheaper than anyone in the city. These "wash goods" sold for $3.50 each and would not have been considered high-fashion attire.[81]

Another exception was Ellen Duignan, who listed herself as a dressmaker in the 1871–1872 city directory and yet on the same page advertised "Plain Sewing." Plain sewing, according to *The Dictionary of Needlework*, included "sewing, stitching, hemstitching, running, whipping, tacking, herringboning, finedrawing, darning, quilting, overcasting, etc." This type of needlework was not usually done by skilled dressmakers but rather by seamstresses.

Duignan, Gibson, and Leighton illustrate the occupational overlap between dressmaker and seamstress that was evident on the Comstock. In order to survive on the frontier, people had to be flexible, finding work wherever they could. Charles Nathan, a twenty-eight-year-old Prussian tailor in the 1870 census, advertised as a "Fashionable Ladies' Dress and Cloakmaker" in May of the same year. This switch from tailor to ladies' dressmaker probably reflects the "dullness of the times"; his listed prices suggest that he was supplementing his tailoring business by manufacturing mid- to low-end women's wear.[82] Male dressmakers were rare; fashion centers such as Boston reported that only 5 percent of women's custom clothiers were men.[83]

TABLE 6.4 Marital Distribution of Women's Custom Clothiers in
Storey County by Occupation, 1880

Marital Status	Milliners N = 22 (%)	Dressmakers N = 97 (%)	Seamstresses N = 21 (%)	Sewing Women N = 5 (%)
Single	63.6	55.7	42.9	40.0
Married	18.2	23.7	23.8	40.0
Divorced	4.5	7.2	19.1	20.0
Widowed	13.6	13.4	14.3	0.0
Total	100.0	100.0	100.0	100.0

Marital status was not recorded in previous U.S. census manuscripts.
Source: Tenth U.S. manuscript census, 1880.

Three dressmakers expanded their needlework activities and opened employment offices for women. Two, Mary Conway and Susan Carroll, opened a "Female Employment Office and Dressmaking" in 1866. A few years later, Miss Dowd also opened a "Female Employment Office."[84]

Dressmaking was the most popular needleworking occupation on the Comstock. Dominated by single American-born women, dressmaking provided opportunity. Both millinery and dressmaking involved apprenticing to learn technique, but millinery relied on "intrinsic artistry," whereas dressmaking could be learned by "every lady."[85] Dressmakers made women's fashionable toilettes, yet they did not stock an inventory of dry or fancy goods. Primarily independent workers, they usually worked out of their dwellings, or sometimes went out by the day or worked for private families.

Seamstresses

Seamstress is a generic term for anyone who sews clothing, but in the nineteenth century it referred to a specific task of sewing seams. The *Encyclopedia of Victorian Needlework* defines a seamstress as "a needlewoman whose department in her particular art is to perform Plain Sewing, as distinguished from dress or mantle making, and from decorative Embroidery."[86]

The sewing of seams, although not as technical as dressmaking, does involve learning different stitches, such as hemstitching, running, whipping, tacking, finedrawing, overcasting, buttonholing, gathering, and ruching. But the most common use for plain sewing was hemming sheets, towels,

TABLE 6.5 U.S. Nativity of Women's Custom Clothiers in Storey County by Occupation, 1860–1880

State	Milliners N= 24 (%)	Dressmakers N= 83 (%)	Seamstresses N= 22 (%)	Sewing Women N= 4 (%)
California	12.5	15.6	22.7	50.0
Connecticut	—	2.4	—	—
Delaware	—	—	4.5	—
Florida	—	1.2	—	—
Georgia	—	—	4.5	—
Illinois	4.2	4.8	—	—
Indiana	—	2.4	4.5	—
Iowa	—	3.6	—	—
Kansas	4.2	—	—	—
Kentucky	—	1.2	—	—
Louisiana	—	3.6	—	—
Maine	12.5	6.0	9.1	—
Maryland	12.5	3.6	—	—
Massachusetts	8.3	4.8	4.5	—
Michigan	4.2	1.2	—	25.0
Minnesota	4.2	—	—	—
Missouri	8.3	2.4	—	—
Nevada	8.3	3.6	4.5	—
New Hampshire	—	3.6	4.5	—
New York	16.6	18.1	—	25.0
Ohio	4.2	3.6	4.5	—
Pennsylvania	—	7.2	13.6	—
Tennessee	—	1.2	—	—
Vermont	—	1.2	13.6	—
Virginia	—	3.6	9.1	—
Wisconsin	—	4.8	—	—
Total	100.0	100.0	100.0	100.0

— means no individuals borns in that state.

Source: Eighth, ninth, and tenth U.S. manuscript censuses of 1860, 1870, and 1880, respectively.

tablecloths, and napkins, and sewing the straight seams of undergarments and ordinary work clothing.[87]

Whereas there were no formal millinery apprenticeship opportunities and only a limited number of dressmaking apprenticeships on the Comstock, the art of plain sewing was taught at the Virginia City School of the Daughters of Charity in 1871. The school's advertisement specifically stated: "Board and Tuition [shall include] English, French, Spanish, Ornamental

Needlework, Tapestry, Embroidery, Plain Sewing."[88] Interestingly, the sisters did not offer dressmaking or millinery, yet recreational sewing courses including ornamental needlework, embroidery, and tapestry were listed.

In Christina Walkley's study of British Victorian seamstresses, she states, "No matter what their social standing, little girls were taught to sew."[89] Sewing was an essential aspect of female education in the United States as well. Plain sewing was a critical part of "women's work," which is probably why the Daughters of Charity included it in their curriculum. In addition, sewing was genteel and could provide a girl with a modest means by which to make a living.[90]

Seamstresses, the least skilled of the needleworkers, were able to construct simple garments and do repairs and alterations.[91] A seamstress ranked lowest on the needlework ladder of status; since basic sewing was a skill every girl should have possessed, a seamstress, unlike a dressmaker or a milliner, did not require any additional training. Almost any woman could resort to working as a seamstress, if need be. This seems to have been the case with Mary McNair Mathews. Arriving on the Comstock in the 1870s and finding herself out of money, she immediately began soliciting sewing. An experienced seamstress who informed her readers that she owned a

TABLE 6.6   Foreign Nativity of Women's Custom Clothiers by Occupation, 1860–1880

Country	Milliners $N=15$ (%)	Dressmakers $N=54$ (%)	Seamstresses $N=14$ (%)	Sewing Women $N=3$ (%)
Britain	—	22.2	14.3	33.3
Canada	6.7	29.6	14.3	—
France	13.3	1.9	28.6	—
Germany	13.3	5.5	7.1	33.3
Ireland	33.3	20.4	28.6	33.3
Mexico	6.7	3.7	—	—
Prussia	13.3	5.5	—	—
Russia	13.3	—	—	—
Scotland	—	9.3	—	—
Switzerland	—	1.9	—	—
Other	—	—	7.1	—
Total	100.0	100.0	100.0	100.0

— means no individuals born in that country.
Sources: Eighth, ninth, and tenth U.S. manuscript censuses, 1860, 1870, and 1880, respectively.

"hoop skirt factory" before starting her adventures on the Pacific Coast,[92] she was a hard-working widow, and her memoir provides considerable insight into needlework on the Comstock in the 1870s.

Most seamstresses did not advertise for work but rather found employment by word of mouth. Mathews's first job was working for Mrs. Hungerford, mother-in-law of wealthy mine owner John Mackay. Many such private employers became long-term clients, but in this case Mathews was offended at having to eat her meals in a cold, damp room, removed from the family, so she remained only a few days. During her stay, Mrs. Hungerford paid Mathews the equivalent of three dollars per day.[93]

Working for families could either be "going-out-by-day," as Mathews did for Hungerford, or it could include room and board. One job of the latter type for Mathews included sewing a wedding outfit for a family in Dutch Flat, California, and then, upon referral, going to a neighboring family to sew children's clothes in preparation for school. Mathews seems to have had some knowledge of constructing garments. While it is doubtful that she had technical dressmaking skills, she may have had experience in the traditional pin-to-form method of garment construction.[94] In the case of the wedding outfit, she did not refer to using a "chart" or to creating the latest fashionable toilette for the bride; she may have been assigned simply to make the outfit, which probably had been cut from an existing garment and was then stitched by Mathews. In another example, when working for her friend, Mrs. Beck, she waited until Beck had "cut out some underclothes" before she could sew.[95] Here, sewing appears to be just that—stitching the seams together and possibly finishing the raw edges.

Whereas dressmaking and millinery advertisements put emphasis on the "latest fashionable" styles, seamstress work did not emphasize fashion, eastern or Parisian styles, or glamorous fabrics. Mathews appears to have maintained the fashion image of a proper and conservative widow, never "[buying] needless finery." Her views on fashion were strictly utilitarian, as she sharply stated: "The ladies dress so rich and gaudy, and use so much paint and powder, that they are really not themselves when dressed for church or ball, or for the street, but only painted dolls, dressed in silks and satins."[96] This response is not necessarily surprising; seamstresses were not the forerunners of fashion. For the most part, they had neither the time nor the excess disposable income necessary to enjoy such indulgences. But they, along with the dressmakers, were the makers of fashion.

Walkley, in her discussion of English needleworkers, goes into detail regarding the sweatshop conditions and long hours of the Victorian seamstress.[97] There do not appear to have been any industrialized sweatshops in Nevada, but Mathews did mention her long day: "I often sewed till twelve and one o'clock at night. After all was quiet, I could do a great deal of sewing."[98] She never discussed how long it took her to construct a dress, but since she earned three dollars a day and sold a calico suit for eight dollars, one can ascertain that it would have taken her over two days to make the garment.[99]

There were few unemployed seamstresses on the Comstock. Mathews did mention having run onto hard times when "business became dull and everybody tried to do their own sewing."[100] During these periods Mathews did laundry, took in boarders, or taught school. She also mentioned the difficulties she encountered when customers did not pay her. At one point, she carried over $700 on her books. This certainly was not all from sewing, but Mathews' problems with nonpayment were perhaps typical of other needleworkers.[101]

Louise Hungerford Bryant was another widow on the Comstock who worked as a seamstress. With her mother's help Bryant began sewing as a young child. When her husband died, she depended upon her needlework skills for survival. Almost destitute, she went "out-by-day," sewing for some of the Comstock's wealthiest families, and did piecework for Rosener's dry goods store. Her seamstress work also included trimming a bonnet and repairing a dress for her landlord.[102] Bryant's story was a rags-to-riches dream for many struggling seamstresses, for she married John Mackay, one of the wealthiest men on the Comstock.

Seamstresses tended to perform the tedious and mundane tasks of sewing a fashionable woman's toilette. Nineteenth-century fashionable women's gowns had detailed inside seam finishes. Sometimes the seam edges were pinked and cast with an overcast stitch; and sometimes they were encased in binding; seamstresses performed both techniques.[103] Perhaps this is the type of work Ellen Reudenbough did when she worked for the dressmaker Annie Macguigan.

The arrival of seamstresses on the Comstock came later than one might expect, since they were the least skilled and presumably the most common of the needleworkers. Nevertheless, there were no seamstresses recorded in the 1860 census in Virginia City. The stratification of the needleworkers becomes evident as the mining camp increased in population. The appearance of seamstresses on the Comstock between 1870 and 1880 is similar to that of milliners; the number of milliners went from fifteen to twenty-two, and the number of seamstresses increased from fifteen to twenty-one in the 1880 census (see table 6.2).

Trautman concludes that working as a seamstress "was an occupation for single females," a statement consistent with the current Storey County findings.[104] She further states that in 1880 Denver city directories list only five seamstresses among a population of 25,000. Four of these five worked for department stores, and only one worked out of her residence. This low number of seamstresses, compared to the general population, is consistent with the information in table 6.2, which shows that Colorado had a lower ratio of personal clothiers to the general population than occurred in Nevada.[105]

Most women used sewing skills at home, as part of their household routine, but for many who were in need, working as a seamstress provided a way to make a living. Mathews and Bryant obtained work whenever and wherever they could. Some needleworkers "went-out-by-day" or boarded with families, worked with dressmakers, or did piecework at dry goods stores. A sewing machine saved the seamstress time but was not a necessity.[106] With nothing invested, it is not surprising that seamstresses were the most transient of the needleworkers. Perhaps they moved on, or possibly, like Mary McNair Mathews, they worked at other occupations, such as that of laundress, nurse, schoolteacher, letter writer, or boardinghouse operator.[107] The rags-to-riches story of Marie Hungerford Bryant Mackay was unusual. Lucrative seamstress work was rare, but Mathews, with her frugal lifestyle and diverse occupations, did manage to enjoy a respectable living.

In addition to the others in the needle trades, seven women in Virginia City listed themselves as a "sewing woman" or as just "sewing" in the census. None of them were listed in the city directories, none of them advertised, and no additional information about them is available. Neither Trautman nor Gamber have identified this lowest level of needleworking. Perhaps the one "sewer" in the 1860 census sewed "tents and ore bags" as did the women, discussed above, who advertised in Washoe Valley in 1861. These women may have only sewed with a straight stitch, not having the stitching skill and diversity of the seamstress. Or maybe they did not sew clothing and so did not consider themselves seamstresses.

Conclusions

The producers of women's custom clothing on the Comstock were diverse in their skills but markedly similar in their profiles. Most were young, single, native-born Euro-Americans, coming primarily from California and New York. Of the foreign born, many came from Ireland, Canada, and England.

Needleworkers on the western frontier and eastern seaboard are consistent in their profile. Bostonian and Comstock milliners were younger than their dressmaking counterparts (except for employers, who were older). There were more dressmakers than milliners, suggesting a more select status for the latter. Both were primarily native born, although Virginia City did have more foreign-born dressmakers and milliners than did the East.[108]

Trautman's study of Colorado dressmakers and seamstresses illustrates the low frequency of these needleworkers compared to Denver's larger population. This study is problematic, however, since Trautman uses only listings in city directories and not census material. In spite of its limitations, however, it is consistent with the census figures used in table 6.1, and it illustrates the fact that Nevada had a higher proportion of personal clothiers than did Colorado in 1880.

The first needleworkers on the Comstock, the milliners, were the most fashionable. Arriving one year after the discovery of silver in 1859, the area's two milliners provided fashion and gentility for the frontier woman.[109] Contrary to popular belief, unskilled women were not the first to arrive; rather, those who came first were skilled and stylish. Some milliners worked out of their homes, while others hired employees and set up shop. Milliners showed the most stability of the Comstock needleworkers, presumably because they invested capital to purchase their inventory of dress trimmings and fancy goods.

Dressmaking was by far the most popular needleworking endeavor on the Comstock. Drafting systems enabled many women to pursue this trade.[110] Dressmaking could supplement a married woman's income or provide a modest means of income for a single independent woman. It was a mobile business; dressmakers either worked out of their homes or "went-out-by-day." Some worked for milliners to provide the complete fashion silhouette, while others were employers and hired seamstresses. Seamstresses and "sewing women" were the least skilled of the needleworkers. Using their sewing skills to get by, seamstresses tended to get work whenever and wherever they could.

These occupational titles can be misleading, since many of the needle-workers performed several tasks. Louise Bryant trimmed a bonnet, yet she did not consider herself a milliner; Mary McNair Mathews sewed a wedding dress and was not a dressmaker; Ellen Duigan was a dressmaker, yet she advertised for "Plain Sewing." Comstock needleworkers were flexible, adjusting to the diversity of the community and the economic fluctuations of the times.

Change dominated life on the western mining frontier. It is not surprising to see that the needlework trades mirror the economic fluctuations of the mining industry. None of the 219 needleworkers reviewed in this study, except Jackson and Mayer, stayed in Virginia City for longer than ten years. Most stayed a year or two, then possibly married and changed their last names or moved on. The case of the Reno tailors who opened for business on December 14, 1880, is typical. In January 1881, the Nevada *State Journal* reported, "Epstine and Jacobs, who have been doing business in Reno for some time past, as merchant tailors, closed up their business place last Tuesday and have left for Tombstone, Arizona."[111]

The clothing styles of the fashionable nineteenth-century woman were restrictive, but the custom production of these goods encouraged free enterprise for many women.[112] While there were few other opportunities open to women, the milliner and dressmaker found a demand for their skills in stylish Comstock society. Virginia City was a fashionable western frontier community, providing independence and a chance for opportunity and freedom. Mrs. Cowles was looking for business potential when she solicited for Nevada customers in 1862; Louise Bryant found a husband; and African American Elizabeth Vincent was able to own her own business. Virginia City milliners and dressmakers (but not seamstresses or sewing women) did have opportunity and the prospect of modest financial independence.

Virginia City's population peaked in 1875 with over 20,000 people. From the late 1870s through the 1880s the town began its slow decline in population, mining production, and stature. The 1890s witnessed further decline; nonetheless, even with a population of 3,673 in 1900, there were still three milliners and thirty-two dressmakers on the Comstock.[113] Clearly, these few needleworkers were in jeopardy, but this was not only because of the decline of Virginia City and the Comstock. By the end of the century, new technologies in the garment and textile trades, combined with new, more relaxed, less fitted clothing styles for women, increased the mass production of women's apparel.[114] This in turn improved availability, lowered prices, and heralded the success of the department store. But it also brought about the decline of the custom needleworker.[115]

The development of the industrialized garment industry included the incorporation of men in what had previously been a female occupation. Men dominated the large-scale factories and retail department stores.[116] Thus, while one freedom was achieved, another was lost. The independent dressmaker did not fit into the mechanization of mass-produced clothing. Seamstresses continued to do unskilled sewing and found work in the factories of the garment industry. Milliners survived for several decades but stopped carrying an inventory of fancy goods and became retailers.[117] In their day, however, the women involved in the needle trades had a profound effect on the Comstock. They fashioned its look, enhanced its diversity, and found their own means by which to share in the big bonanza.

❦ 7 ❦

Mission in the Mountains

The Daughters of Charity in Virginia City

Anne M. Butler

The discovery of gold in these hills gathered our people here. From distant land they came, from Ireland and their Priests came with them, from Italy and Portugal and every nation, across the desert plains, in hardships, in trials, in thirst and famine, . . . they came and they built a new West and brought their Faith with them. Their bodies' strength was spent . . . little did they receive for their sacrifices.[1]

These evocative words came from the Most Reverend Robert J. Armstrong, bishop of Sacramento, at Virginia City, Nevada, on the occasion of the diamond jubilee celebration of the Roman Catholic Church in Nevada. A committee of at least 125 men, clergy and laity, had organized a festive commemoration that reverberated across the Silver State.[2] On that date, September 8, 1935, an impressive collection of luminaries gathered at St. Mary in the Mountains church for joyful reminiscences of the pioneer struggles of frontier Catholicism.

Yet, amidst the celebrations, of one group might it be said especially, "little did they receive for their sacrifices." The Daughters of Charity of Saint Vincent de Paul of San Francisco, who served the parish and the town from 1864 to 1897, seemed to take no place in the day's orations or published accounts of the festivities. The bishop of Reno, Reverend Thomas

K. Gorman, mentions the sisters briefly in his diamond jubilee publication but gives no indication that representatives of the women's religious order attended the 1935 celebration. No Daughter of Charity, with her distinctive blue habit, guimpe, and soaring starched white cornette stands among the dignitaries whose photographs grace Bishop Gorman's text.[3] Nor are the Daughters of Charity central to the work of Sister Ann Frederick Hehr, O. P. in her "History of the Catholic Church in Virginia City, Nevada: St. Mary in the Mountains, 1860–1967." This account of the frontier parish makes only three or four thin references to the more than thirty years the Charity sisters lived and worked at Virginia City.[4]

This historical invisibility of women religious meshes with the traditions of church history and with that of the American West.[5] Under each of these rubrics, the actions and experiences of men dominate the written record. Yet, the most recent trends in women's history push back the earlier boundaries of scholarship to encompass a rich documentation of the nature of womanhood and its impact on the nation's heritage. Nowhere has this research been more dramatic than within the nineteenth-century American West, where the import of women's history currently assumes new and interesting patterns.[6]

Within this context, the time has come to reevaluate the meaning of pioneer living for congregations of religious women. By 1890 several thousand Roman Catholic sisters lived west of the Mississippi River. They represented a mix of religious congregations, some with European roots, others formed on American soil. All wrestled with the conflicts between formal community rule and the exigencies of western life. Ultimately, they all contributed to a transformation of religious life that saw the European monastery change to the American convent.

The Daughters of Charity of Virginia City, Nevada, provide an excellent example of the ways in which Catholic religious women responded to western life, while they retained the integrity of the community rule, a highly codified mix of religious discipline and social service directives. The Daughters of Charity maintained this delicate balance, revealing themselves as thoughtful, active players, women who responded to Comstock life in many different ways, understood the economic dynamics of the mining world, navigated through the local political realities, introduced accessible social services to the citizenry, enhanced the quality of life for people in need of educational and medical care, and added to the history of religious women in the American West.[7] They accomplished all this under long-standing gender constraints imposed both by their church and by the masculine society around them. Accordingly, they deserve historical reassessment from religious and western historians.

The epic of the American West, especially those brief, chaotic years associated with the mining industry bonanzas, took shape as largely a male venture. In the early 1860s Virginia City exploded into life as one of the most dramatic and productive of these mining towns. Miners, engineers, merchants, and adventurers, eager for quick and easy wealth, streamed into the new city, precariously perched along the eastern face of Mount Davidson. Masculine community dominated, even though a few women and children came to inhabit the mountain town. Overall, Virginia City became a man's world of sweat and toil, speculation and violence.[8]

In 1864, after repeated requests from the clergy in Nevada, three young women, members of the Daughters of Charity of St. Vincent de Paul, journeyed to this bustling urban center to establish a school and hospital.[9] Sister Frederica led the small band and emerged as the guiding force among the sisters for the next twenty-two years.[10] By their very arrival, Sister Frederica and her companions, who at first glance seem alien among the frenzied argonauts, mirrored the experiences of many frontier newcomers.

Commonly, travelers to the West followed a circuitous route, which, by force of weather, transportation, and finances, they broke into installments. Each segment brought novel conditions and sharpened the regional savvy of the migrant. This pattern marked the frontier trek of Sister Frederica, born in Ireland and raised in Philadelphia. As a young sister in the community, she worked at various assignments in the Maryland and Washington, D. C. area. However, in 1855 her superiors selected the thirty-year-old Frederica to lead four other sisters to a recently established mission in California.

Their trip cut across locales increasingly lashed by the bitter acrimony of a growing slavery debate and took the sisters from Emmitsburg, Maryland, to Baltimore, to Philadelphia, to New York. From New York, they sailed for Jamaica and made their way across the Isthmus of Panama, completing the trip to San Francisco by water. They stayed in public hotels—elegant and humble—negotiated with all manner of guides, witnessed political tensions in Jamaica, noted racial disharmonies, traversed remote jungle areas in a primitive train, resolved a commercial dispute for a male traveler, took fragile shuttle boats to reach their final transport, and successfully interacted with strangers of diverse race, culture, and class.[11]

In California, Sister Frederica expanded her administrative skills by organizing the Children of Mary Society and serving as the principal of St. Vincent's School. In 1864 she again answered her superiors' directive that she set forth, this time for the mining regions of Nevada.[12] These collective experiences, accumulated through nearly a decade of leadership assignments, prepared Frederica well for the sights and sounds of a swirling, raucous mining camp.

Her companions, Sisters Xavier and Mary Elizabeth, equally versed in the realities of the West, were drawn from those in the California community considered best suited for the arduous rigors of establishing a remote mission house. The three turned their faces to the Nevada mountains as veterans of the westward movement—seasoned travelers and cultural observers, equipped with an adaptable nature and a sense of innovation. Convent life, far from protective, had put Sisters Frederica, Xavier, and Mary Elizabeth directly in touch with the secular world's means and manners. Mature in their religious vocations and focused on their mission goals, these three migrant single women arrived in Virginia City with more western knowledge and sense of personal purpose than many of the gold-crazed miners and entrepreneurs of the Comstock possessed.

Besides experience as nineteenth-century western travelers, the Daughters of Charity paralleled Virginia City's population in other ways. According to the 1870 census for the convent, whose number had more than doubled, six of the eight sisters had been born in Ireland, one in Australia, and one in the United States.[13] That trend continued up to 1880, when the number of sisters stationed in Virginia City peaked at sixteen. Among this group, six had been born in Ireland, while another seven were born of Irish parents who had already emigrated to the United States. The 1870 group ranged in age from twenty-two to forty-five, although Sister Frederica was fifteen years older than any other woman in the house. The mission retained its youthful cast in 1880, with four sisters in their twenties, eight in their thirties, one in her forties, and the remaining three in their fifties.[14]

In addition to nativity and age, gender also played a role in the demographics of Virginia City. The 1860 census indicated that the nearly 2,500 men had arrived without wives, as the record shows fewer than 120 women.[15] Within a few years, more women arrived, but the gender ratio hovered around two to one, in favor of males. Still, some families did live in the town from the first. Wives followed husbands into the district and gave birth to children, who grew up amidst the noise and grit of a mining rush.[16] Despite its few families, which either inhabited the small shacks along the mountainside or stood apart with great wealth and social status, Virginia City remained largely a bachelor community. By 1875 the combined population of Virginia City and its neighbor, Gold Hill, tilted toward 20,000 residents. Men made up almost 70 percent of the total population, and most of them attached themselves in some manner to the mining operation.[17] Overall, the city took shape as a strange mix of young, single migrants and far fewer family units. These elements helped to shape its urban persona, remembered as a blend of the coarse and cosmopolitan, the ribald and genteel. Newspaper editor Wells Drury described Virginia City as a place where, despite the growing "marks of civilization," "the rough element was never entirely dispersed," and the "community gloried in a crude opulence."[18]

In this swirling boomtown, the Daughters of Charity fit the general population profile. Either immigrants or first-generation Americans, they were young adults, supporting themselves in work created by and dependent on the mining boom. Also, unalterably committed to remaining unwed, the sisters had neither spouses nor the desire to seek any and lived outside of usual family arrangements. Thus, socially uninterested in the available bachelor pool but cut off from traditional avenues of financial support for women—most typically from the wages provided by a husband—the sisters had to carve out a living, constrained by the dictates of their profession as religious women.

That profession represented the major element that distinguished the sisters from other Comstock residents. The Daughters of Charity grounded their formation in service to the poor and the sick outside the religious grille. Based on the seventeenth-century charitable plan of St. Vincent de Paul, the governance design for the Daughters of Charity bypassed the European notion of strict enclosure for women religious and outlined an apostolate carried to the public arena.[19]

From its outset, the Vincentian secular, non-cloistered philosophy, underwritten with a deeply spiritual component, caused no small amount of consternation among church officials, who defined a single strictly enforced cloistered life for religious women. From the 1630s on, the Catholic hierarchy from Paris to Rome watched the new community of women with a wary and restrictive eye. The popular response to the willingness of the sisters to undertake social missions that other enclosed orders rejected, the rapid growth of the Daughters, and the skill with which the founders, Vincent de Paul and Louise de Marillac, navigated the turbulent sea of church politics, overtook official resistance. By 1660, when Vincent de Paul and de Marillac both died within six months of each other, the Daughters of Charity had established an increasingly strong position within the Roman Catholic church.

At first glance, the sharply defined public service component of the Daughters of Charity seemed to make their migration to the understaffed mission region of the early United States a natural. Yet, between 1810 and 1812, frustration and misunderstanding marked the first efforts, by Mother Elizabeth Seton of Emmitsburg, Maryland, to organize an American community around the principles of Vincent de Paul. U.S. church and social conditions did not align with those of France, and great distances thwarted communication between Seton and various clerics. Despite these problems, the American Daughters of Charity flourished, although disputes over jurisdiction and organization led to splinter communities that broke from the Emmitsburg foundation.[20] Known interchangeably as the Daughters of Charity or the Sisters of Charity, the Emmitsburg community involved its missions in a surrounding society and showed a willingness to accommodate existing social conditions. These characteristics fueled its purpose and ultimately made it a popular choice among American bishops who needed mission workers, especially in remote western regions.[21]

These historical and philosophical underpinnings, rather than only the pleading of bishop and priest, account for the decision of the Daughters to live in the chaotic mining district of nineteenth-century Nevada. While other migrants came for personal enterprise or a chance at good luck and found themselves absorbed into a maelstrom of social and economic upheaval, the Daughters of Charity traveled to Virginia City precisely because of that unsettled environment. Although house rule guided their personal decorum, the sisters expected to confront the social needs created by the bonanza atmosphere of a mining town.

Beyond the "glamorous" accounts of stage robberies, high stakes gambling, wide-open saloons, disorderly houses, murders, and mayhem, lurked the real world of men and women struggling to cope with a precarious life in an uncertain economy.[22] For those trying to mark out an existence, whether in the saloons or the mines, the Virginia City ambiance may have seemed at times overwhelming, or at least personally taxing. However, for the Daughters of Charity, Virginia City, Queen of the Comstock, was a place to be embraced, as the needs of its population so closely meshed with the stated directive of their religious organization—carrying physical and spiritual comfort to the sick and afflicted of a community. An explosion in the Consolidated Virginia mine meant injured workers, shootings at the Delta Saloon left the dead and the maimed, orphans lacked shelter, working parents sought care for children, who in turn needed to read and write. Throughout the town, ordinary people suffered from the exigencies created by spontaneous urban development. The Daughters of Charity came to the Comstock to address, in many ways, the subsequent difficulties produced by the unpredictable environment.

They observed the results of boomtown building on a daily basis. They saw the thrills and excitement as well as the human confusion and need. They also recognized that ever-present physical hazards in Virginia City gave rise to a variety of social problems.

For example, in a report to the San Francisco motherhouse, Sister Frederica singled out the unstable character of the town's buildings as a particular hardship. Homes and businesses had been hastily thrown into place, with little regard for structural planning. Frederica thought the system of shoring up the excavated mines with large planks, designed to prevent the overhead dwellings from collapsing underground, particularly dangerous. Often these makeshift arrangements broke down without warning, and a house tumbled into the mines. The ever-present threat of fire, above or below ground, also troubled her. She wrote about the "unfortunate men," unable to escape the "gulf of fire," and consumed by the flames before water could be secured. Another risk came from embankments that collapsed onto the miners, so that "their lamentable cries are heart-rending—no one can reach them."[23]

The Daughters of Charity, far from retreating from the world's realities or expressing shock at social unrest, understood the complexities of Virginia City and took up a public ministry that responded on several levels. The nature of their community allowed them, unlike many religious women, to provide social services to the poor by actually leaving the convent and taking solace into the homes and workplaces of the people. The sisters expected, in accord with the recommendations of Vincent de Paul, to minister directly to the corporal needs of the sick or the suffering, while refraining from spiritual judgments in matters they might personally categorize as "sinful."[24] This mandate brought the Sisters of Charity into close contact with the Virginia City miners in their daily lives. One sister, writing in 1874, commented: "[Virginia City] being a mining district, it is composed of all classes of people. The poor miners, who live in the bowels of the earth, frequently suffer from ill health and sometimes encounter serious accidents, . . . never fail to be consoled by the care and sympathies of the Daughters of Charity who carry to their poor 'shanties' little delicacies, and words of consolation, which are so valued by these Men of Faith."[25]

Later in her remarks, the sister added, "The Sisters visit the jail, and much good for the poor convicts' souls and bodies is done there."[26] The Daughters' commitment to their own religious conduct in no way clashed with their perception of humanity, in both its negative and positive aspects, or dampened their enthusiasm for moving into the world of those in need. Based on the direct, regular interaction with both Virginia City's residential communities and law enforcement agencies that this reminiscence suggests, it appears fallacious to view Catholic sisters as uninformed, unsophisticated recluses. Such common stereotypes of women religious simply do not apply to the Daughters of Charity in the Washoe mining district.

Given their perspective as social activists, the sisters, charged with operating a school and a hospital, set to the task of fulfilling their religious mission and establishing their fiscal stability. The desire to maintain cordial relationships with bishops and pastors, while retaining self-governance, created no small conundrum for women's congregations. Not surprisingly, the church hierarchy resisted the idea that communities hold the deeds of ownership on mission property. Congregational ownership, although couched in language about humility and poverty, actually involved matters of power and control, especially for frontier bishops. Too often, they wrestled with the management of vast territories, far-flung parishioners, and few dollars. They did not want to add to their burden religious houses fueled with an independent spirit. This proved a futile wish.

Individual congregations, perfectly willing to defer to the church on matters of faith, often sought as much autonomous fiscal management as possible. This was especially true for the Daughters of Charity, which Vincent de Paul had designed to free sisters from diocesan and papal control. Concerned about the questionable merit of investing their funds in a lot and a house owned by the diocese, a move that would have violated basic organizational precepts of the community, the Daughters of Charity delayed opening the Virginia City mission.[27] They eventually won the acquiescence of Bishop Eugene O'Connell, who explained away his earlier objections as merely a desire for Vatican clarification on property title rights of individual congregations. O'Connell, faced with limited funds, vast

Virginia City in 1878 with the Daughters of Charity hospital standing prominently in the middle foreground. (Photo by Carleton E. Watkins; courtesy of Nevada State Historic Preservation Office)

spaces populated by about 14,000 widely dispersed Catholics, and few missionary nuns and priests, conceded that the sisters had secured and improved the property. He suggested, for the sake of peace in the diocese, that they should be given the deed.[28]

In addition to keeping a watchful eye over these larger fiscal interests of the Daughters of Charity, Sister Frederica needed money in hand for the Virginia City operation. She did not find it waiting when she reached the new mission. As frequently occurred for pioneer missionaries, upon their arrival in Virginia City, the sisters discovered the promised convent and school to be uninhabitable.[29] Adjusting to the inconvenience, they moved into a room at St. Mary's Church. A week later they shifted to their own house, originally a single room to accommodate the sisters and twelve children. In these crowded quarters, they opened their doors to Virginia City on October 15, 1864.

A combination orphanage and boarding school, this institution grew faster than the space and the funds it needed. Little by little, the sisters expanded their physical plant, but the demands always seemed greater than

St. Mary Louise Hospital of the Daughters of Charity. (Courtesy of California Historical Society)

The music class of the school and orphanage of the Daughters of Charity. (Courtesy of Comstock Historic District Commission)

the available room. Within the first month the number of children attending the school reached ninety. By January 1865, twenty occupants lived at the small convent; by July, twenty-five crowded into the small building, and 112 students attended the classes.[30] The boarders paid what their families could manage, and the schoolchildren contributed from two to four dollars a month. The frugal Sister Frederica had saved $500 by summer but lacked the money to launch a construction program for the school and hospital.[31]

A partial solution emerged in an early arrangement with the state of Nevada for a much-needed orphanage.[32] In 1867, using the state's requirement that an agency be incorporated as a condition of funding, Sister Frederica further legitimatized the status of the Virginia City mission. This cooperation with the state brought the sisters $2,500 and advanced both their secular and church legal stature.[33]

The arrangement, however, threw the sisters into a fierce debate over the use of public funds. Funding opponents objected that the legislation explicitly stated that the asylum receive only white children for care. The controversy, played out on the pages of the *Territorial Enterprise*, charged that Sister Frederica had informed the legislature that the Daughters of Charity would not accept black orphans with the white children. Pointing to the establishment of impartial suffrage in every state and territory, the newspaper argued that funds drawn from a common taxation of citizens could not be used for a public agency with exclusionary policies. The Daughters of Charity, the article continued, could run their orphanage any way they chose, but not with state money.[34]

In response to this strong editorial, an answering letter, which denounced the proposal to house the black and white orphans together as an "experiment," appeared in the Gold Hill *News*. It accused the *Territorial Enterprise* of inflaming anti-Catholic prejudices. The latter paper considered this letter, published by a competitor, over the signature "Sisters of Charity," a personal affront. The *Enterprise* writer shot back another volley with heightened acidity. He pointed to the solid record of the *Enterprise* in supporting the orphans' asylum, insisting it had done so to such an extent that other denominations grumbled about favoritism bestowed on the Daughters. However, avoiding a direct attack on the Daughters, the editorial asserted that the letter to the Gold Hill *News* had obviously been penned by someone other than the sisters and their name had been illegally attached to it.

While in the heat of the moment it seemed that the *Territorial Enterprise* had assumed a more egalitarian stance than the Daughters of Charity in a matter of social justice, the high tone of the second article ended on two weakening points. The writer insisted that the *Territorial Enterprise* did not suggest that "the children should sit side by side." Indeed, he continued, "It is no injustice to colored children to educate them separately, and we do not maintain the propriety of placing them in the white classes, but it is unjust that they should be deprived of the full benefits of the public school or of any other institution sustained by common taxation."[35]

Although Sister Frederica, a product of the bitter border regions of pre–Civil War days, apparently assumed a position at odds with concepts of political justice, her perception of this as an attack by anti-Catholic forces in Nevada influenced her responses for the next several years. Additionally, if she sent the alleged telegram to the state legislature and the letter to the Gold Hill *News,* her position seems to have paralleled the ultimate resolution of "separate but equal" promoted by the *Territorial Enterprise,* so she must have been somewhat confused by the uproar. The nineteenth-century Daughters of Charity, who worked with African Americans in many locations, may have defined themselves exclusively in terms of social ministers, rather than as political reformers.[36] If that was the context, Sister Frederica felt the Vincentian call to offer care for African American orphans but remained unwilling to adopt the political stand of social integration.

This paradoxical position flowed naturally out of the Daughters' early days in a tumultuous France, where Vincent de Paul had vigorously tried to secure congregational approval, not only from church leaders but from secular ones as well. In those delicate years, during which one cleric could banish another, the Daughters had learned to work diligently for the poor but to avoid even the appearance of political involvement. In any event, the *Territorial Enterprise* ended the discussion on a final negative gender note, as it dismissed the sisters' ability to reason out these changing political issues by asserting, "They are not expected to understand either constitutional law or party politics."[37]

Despite this controversy, state funding continued, and another allocation of $5,000 over a two-year period came to the Nevada Orphan Asylum. The situation refused to go away, however. In 1873, despite a recommendation from the Committee on State Affairs that the sixty-five orphans and "twenty-five half-orphans" continue to receive the annual award of $2,500, Sister Frederica withdrew the sisters' request for support and Nevada opened its own asylum.[38]

In an unusual political move, Frederica sent the legislature a long letter, which revealed her wounded feelings in the dispute. After effusive thanks to the "members of the present legislature who have shown a kindly feeling [to the sisters] and a due appreciation of their labors," she reviewed the history of the asylum, noting, "The sisters established the asylum in 1867, when the State allowed but a heartless indifference for parentless children." Warming to her subject, she continued:

> All know, though some reluctantly admit, they were the Pioneers in the state for the orphans. When they threw their doors open to the public, no questions were asked, there was no distinction of persons; no discrimination of creeds—distress and poverty directed their

[Daughters of Charity] actions. The State . . . not from charity, but a sense of justice, rewarded them. But of late, a hostile feeling has risen against them. . . . If we are not entitled to the appropriation in justice, we do not look for it. . . . in charity.[39]

This loss of funding represented a financial setback for the Daughters of Charity but demonstrated their intense commitment to the maintenance of community identity and the refusal to allow civil or ecclesiastical encroachments on self-governance.[40] Sister Frederica did charge that candidates for election to the Nevada legislature had been approached by anonymous persons and informed that a qualification for election included opposition to the Daughters of Charity and the orphans. Whether Frederica knew this based on valid information or hearsay remains unclear, but the accusation speaks to her ever-present sense that anti-Catholic feeling permeated her negotiations with the state.

In part, Sister Frederica could feel confident about her decision to step away from state funding, for, as she stated in her letter, "The asylum has friends and these friends are good enough to see that neither ourselves or the orphans under our charge can want for food or clothing."[41] Her description of the sisters' lives may actually have defined the hostility she believed emanated from part of the legislature. Her explanation that, "We have no salary or wages; we have consecrated our life, our time, our attention, our care and all that this world could afford us, to help the distressed, the afflicted and especially the poor orphans," perhaps touched some as offensive, and the subsequent state monies expended on the sisters appeared as a tangible violation of the separation of church and state.[42] Nowhere in the records are there indications of how many African American children actually lived at the orphanage, nor is it clear whether the problem centered more on the constraint, as defined in the sisters' rule, against care of boys over a certain age.

Despite the conflict that unraveled over several years, the overwhelming acceptance of the sisters' school/orphanage indicated the demand within Virginia City for services to supplement family functions. Families in differing circumstances quickly turned to the sisters for care and education. The services offered by the Daughters of Charity transcended the narrow definitions of "school" or "orphanage." Certainly, children without any parents came to live at the facility. In addition, single parents—those widowed or separated—boarded children with the sisters. While mothers and fathers routinely sent youngsters for the traditional secular education and religious training, the sisters increasingly answered a much more complex demand from the community. The sisters' facility, long before the professionalization of social work, combined aspects of school, orphanage, boardinghouse, and day-care center.

Overall, the Daughters of Charity provided a series of social services to Virginia City before the young state of Nevada had such agencies in place. Every indication suggested that the demands from the Virginia City populace would only increase as the mining industry continued to mushroom in Nevada. Such issues as a congregation's stated mission, its agreements with the bishop, or the sisters' personal and professional resources rarely concerned needy people who typically expected nuns to answer any social call. The Virginia City residents proved no exception, as can be seen by the rapid increase in the sisters' charges. One sister called the appeals "constant," with petitioners, "asking the kindness lost when a parent dies, others to receive care they never knew before, because of their poverty and distress." [43]

Hampered by their small numbers, but enjoined by their religious purpose and their own employment needs, the sisters seemed inclined to enlarge their mission. However, tuition payments from the pupils and uncertain state support barely kept the sisters operative. To move forward in the expansion of their work, they needed a firmer financial grounding in Virginia City.

Calling on their many years of experience as administrators, the Daughters of Charity moved to an important aspect of their mission work. They forged close, important relationships with those Virginia City residents most able to assist in broadening their vital service institutions. In this, they benefited from their association with St. Mary's well-known and highly popular pastor, Father Patrick Manogue.

Father Manogue, a vigorous, almost myth-like pioneer priest, laid the groundwork for Virginia City's Catholic institutions with his arrival in 1862. A former miner, this large, earthy priest carved out a highly public role for himself on the Comstock. [44] Anxious to convince the Daughters of Charity to accept the Virginia City mission, Manogue had in place, before their arrival, solid contacts with local residents who would be interested in the work of the sisters. [45]

The sisters quickly took advantage of these contacts and of the vigorous Irish Catholic spirit in Virginia City. Little more than two months after their arrival, they capitalized on that religious spirit with a fair to benefit the building fund.[46] Such fairs became a regular feature of the Virginia City social scene. On December 22, 1864, the editor of the Gold Hill *News*, Alfred Doten, reported that he visited the "Ladies Fair for the benefit of the Sisters of Charity," and left "minus four bits" after only ten minutes. For his money, Doten, who moved at the center of the action on the Comstock, received a raffle on a saddle and a flirtatious letter.[47]

Doten typified the sort of support the sisters needed. It was imperative that they draw on all available monetary sources in Virginia City. The interaction the sisters enjoyed with a core group of lay women who organized charity functions over the next three decades assisted them in accomplishing this goal. These women—Catholic and non-Catholic—moved among the Virginia City elite soliciting donations, selling raffles, and drumming up attendance for various benefits. They mingled with the wealthy of the Comstock in such environs as the private boxes and orchestra seats of the Piper's Opera House and attended late night after-theater dinner parties.[48] They had recreational access to social circles where an Alfred Doten might be persuaded to endorse the community projects of a small group of nuns. The constant efforts of the women of Virginia City could yield as much as $16,000 for the sisters from a single fair.[49]

From August through October 1868, with the nastiness of the orphans' asylum dispute apparently set aside, the *Territorial Enterprise* ran a series of articles promoting various social activities to benefit the sisters. Lavish in its praise of the asylum, the newspaper made special mention of all the women who organized and ran these events, complimenting their appearance, their handiwork, and their organizational skills. All appeared forgiven between the *Territorial Enterprise* and the Daughters of Charity.

At these fairs, which typically ran for a week, men and women mingled together, enjoying festivities planned to encourage mixed company. After wandering through the booths and games, chancing the raffles, and overindulging in elaborate food and drink, the men helped to clear away the floor for a lively midnight dance each evening.[50] Although the sisters did not attend the evening galas, their unspoken presence motivated the festive crowds to loosen purse strings. These benefits represented the combined planning and action of religious and lay women. As they worked with their secular backers, the Daughters of Charity forged personal friendships with these women, bonds that reflected the importance of female networks across secular and cloistered worlds in the growth of the Catholic church.[51]

In a further effort to connect with benefactors and to make the asylum's efforts more visible, the sisters used the local newspapers to advertise their activities and their appreciation for favors. Donors and patrons received generous thanks. On occasion, the sisters published exact lists of donations, a way to highlight their needs and to prod the less giving to match the spirit of neighbors and business competitors. Fourteen sacks of potatoes, a dress pattern and some calico, two heavy cloaks, a little pig—the presenters of each found an eloquently phrased thank you in the *Territorial Enterprise.*

These messages carried an explicit reminder to all citizens of the sisters' most powerful weapon for the public, that the "prayers of the orphans will ascend on high for him and his that they may ever enjoy those blessings ... beneficial to them in this life and assist in procuring for them above all the blessings and happiness of the world to come."[52] Who among Virginia City's unsteady populace, where life often came to a sudden and violent end, would not want to feel comforted by the sisters' promise of some assistance in the afterlife?[53] Which business would want to see a competitor act more generously toward the orphans? Use of the newspaper as a forum for appreciation helped to keep the work of the Daughters of Charity before the public eye.

A large immigrant population gave a decidedly Catholic flavor to the Comstock, where Germans, French, Italians, Hispanic, Cornish, and Irish mingled in the mines. Not all these people adhered to the Catholic religion, and other denominations were represented in the mountain town. Methodists, Episcopalians, Presbyterians, Baptists, and an African American Baptist group all moved into the area by the 1860s, although not all assumed a vibrant, long-standing role in the community.[54] Several of these groups, especially through their ethnic benevolent associations, felt inclined to support the sisters' work.

Despite the assistance, the sisters did not always mesh entirely comfortably with the non-Catholic groups and sometimes believed themselves the focus of religious prejudice.[55] Obviously, Sister Frederica felt that anti-Catholic sentiment accounted for the withdrawal of funding for the orphanage by the Nevada State Legislature.[56] In the 1870s a spate of concern over the dangers of Catholicism surfaced in the organization of a local branch of the virulently nativist group, the American Protestant Association, which found grist for its mill in the Vatican's pronouncement of papal infallibility.[57] Despite the discomfort the sisters expressed about "Protestant pressures," such tensions were entirely predictable in a community of such diverse composition and in a nation where notions of nativism intensified with each surge of immigration.

These controversies paled beside the solid record the sisters built with many non-Catholic associates. In fact, one dimension of the Daughters' relationships in Virginia City included an emerging ecumenical tone, long before such notions achieved common popularity. In the 1870s, during the height of the American Protestant Association's nationwide anti-Catholic campaign, Mrs. Sunderland—an Episcopalian woman—and another non-Catholic, Mrs. Theall, undertook regular "begging" errands for the sisters. According to the Annals, these women, "left none of the mines unvisited" and trekked through the "most dangerous places," to "get the last dollar."[58] For many years, another Episcopalian, Dr. Wake Bryarly, gave free medical care to the children at the orphanage.[59]

Additionally, the sisters enjoyed cordial relations with the Jewish community in Virginia City. Jewish girls attended the St. Mary's School, some as boarders. At the holiday season the sisters received cash, staples, and confections from these families. Even within a religious community accustomed to secular contact, the sisters sensed that their close interaction with the Virginia City Jews made the St. Mary's mission unique.[60]

Central among those concerned with the well-being of St. Mary's and the work of the sisters was John Mackay, Irish-born mining baron, who contributed generously to the parish across a number of years. His marriage in the 1860s to a widow, Marie Louise Bryant, who taught French at the sisters' school, reinforced his connections to the convent. Mackay, a close associate of Father Patrick Manogue, underwrote a variety of St. Mary's functions, especially the rebuilding of the church after a disastrous town fire in 1875.[61]

The generosities of the Mackays, although important and regular, may have assumed unrealistic legendary proportions. At least one sister stationed at Virginia City remembered Mackay as a devoted friend to the institution, always ready to send wood for the orphans, "not lavish in his bounties, but generous. . . ."[62] Yet, his donations in 1867 seemed lavish when on one occasion they included 200 pounds of flour, a gallon of brandy, a case of wine, 115 pounds of beef, and assorted other goods.[63] As for his wife, rescued from a life of genteel poverty through her marriage to the business scion, the sister recalled, "Mrs. Mackay gave not so freely; perhaps she thought her husband's donations sufficed for both."[64]

Nonetheless, it was Marie Louise Bryant Mackay who donated the six acres of land on which the Daughters of Charity built the town's first medical facility, graciously and diplomatically named St. Mary Louise Hospital. Known to the locals as St. Mary's, it opened in 1876, twelve years after the sisters arrived in Virginia City.[65] This four-story, state-of-the-art building accommodated sixty to seventy patients and included gifts from all the well-known local benefactors, such as Mrs. James G. Fair, who provided a massive kitchen range and a special ironing apparatus, and John Mackay, who started things off with ten tons of coal and twenty-five cords of wood.[66]

The issue rests not so much with whether one or two wealthy couples single-handedly responded to every financial need of the Daughters of Charity. Of greater significance are the indications of steady charitable support to the convent, school, and hospital. This regular auxiliary backing allowed the sisters to plan on a more even supply of funding. It granted them the grace of knowing others valued their presence. It reinforced the efforts of single women to survive as immigrant workers on the frontier. The manner in which they built these relationships with lay people bonded ordinary citizens to the sisters and the sisters to the larger community.

Most importantly, through this system of fund-raising, the Daughters of Charity formed a bridge from the mining industry rich to the working poor of Virginia City. They forced those who traveled in the cosmopolitan section of town, enjoying a life of opulence, to acknowledge the existence of struggling worker families. If the wealthy did not themselves choose to mingle with the poor, then the Daughters of Charity did so for them. The efforts of a Mrs. Sunderland and a Mrs. Theall underscore the direct impact the Daughters of Charity exerted on the surrounding secular community. If that served as an impersonal way for some of the fortunate to dismiss their guilt about social conditions, it nonetheless meshed with nineteenth-century notions of social responsibility and advanced the sisters' endeavors.

For workers in Virginia City, fluctuating economic conditions and a severe landscape combined to make for a harsh life. The added intensity sparked by the competition among the diverse immigrant groups made Virginia City an even more unpredictable environment. The firm control wielded by the Miners' Union did much to keep matters balanced, as did the various foreign language benevolent societies, newspapers, and military guards. The presence of the Daughters of Charity accounted for an additional component in maintaining community interaction. Although they cannot be credited with negotiating labor relations or initiating political referenda, they helped to direct attention to social needs. By their distinct appearance and their personal behavior, the Daughters of Charity reminded differing immigrant peoples of common institutions and corresponding values. Regardless of national or religious persuasion, everyone understood on sight the meaning of the blue habit and white cornette, giving visual affirmation to people's hopes for stability and a better life in both practical and spiritual terms. Whether one approved of Catholicism or not, the sisters' social service transcended national, linguistic, and religious boundaries.

Unfortunately, the impact of that service became lost as both clergy and laity assigned a self-effacing, anonymous role for religious women that made it unseemly to point to their accomplishments. That attitude translated into invisibility for the Daughters of Charity in Virginia City's secular and religious record. The Daughters of Charity at Virginia City demonstrated through their community action, financial planning, and personal interactions the historical loss generated by this neglect of religious women.

The sisters' place within their own church has also been disregarded. Barred from the most sacred actions of Catholicism—saying the Mass and administering the sacraments—women religious, described as sweetly passive in church history accounts, seem to have played no active part in the development of their church. Strictly enjoined from the priesthood, or, in the nineteenth century, from any contact other than as communicant with the sacred host, women religious nonetheless served when called. In times of religious crisis, expectations shifted for sisters, but their contributions remained unacknowledged.

For example, in August 1871 a fire in the business portion of Virginia City threatened the Catholic church. For safety, a priest removed the sacrament from the church and brought it to the sisters' chapel. As the fire swept on through the town, the sisters' establishment appeared destined for destruction. Father John Nulty then gave into one sister's care the sacrament and directed the startled woman to carry it, late at night, over several miles to the Gold Hill church.[67] Within Catholic tradition, this sister could hardly have been charged with a more serious and sacred task. At the very least, the priest's directive violated convent regulations about sisters' night travel. Yet, Father Nulty apparently expressed neither reservation nor hesitation about his order. Under "frontier" circumstances, prohibited religious behaviors for women took on temporary acceptability. In other words, circumstance negated gender restrictions concerning women's religious functions.

By the 1890s the fortunes of Virginia City had declined. The usual demons of mismanagement and depleted ore veins wiped out the once dynamic town. The Daughters of Charity faced the reality that they no longer had a substantial clientele in the Washoe District. The legendary Sister Frederica left in 1886, assigned to the Daughters' hospital in Los Angeles and later as principal at St. Patrick's School in San Francisco. Twenty-two years on the Comstock had drained her health, but apparently she departed with great sadness. Her replacement, Sister Baptista Lynch, stayed in Virginia City until the Daughters of Charity withdrew, giving the little mountain mission rather amazing continuity for a nineteenth-century bonanza town.[68] In 1897, after much consideration and resistance from many quarters, the Daughters closed the Virginia City school, hospital, and convent. An unfortunate tension with the pastor, dwindling numbers in the school and hospital, and the press of new calls from other mission fields all combined to bring about the decision.

Despite practical reasons for closing the mission in the mountains, the Daughters' leaving-taking was not easy for the sisters or their Virginia City friends. Mamie McCarthy, a recent graduate, wrote to her brother in California of the event, capturing the sadness felt by the Daughters and their students. Mamie told her brother Joe: "The Sisters went away, a week last Sunday. Sister Josephine felt worse than any because she was here the longest, ten years in April. A great many of the girls do not intend to go to the Public School and I do not blame them. I am so glad I finished before they left."[69]

Clearly, the Daughters of Charity had done more than staff a school and a hospital. They had made a home and built personal relationships in the little mountain town. Their departure created a permanent void in the social institutions of Virginia City, as well as in the emotions of those comforted by the presence of the sisters with the blue gowns and white cornettes.

For over thirty years, the Daughters of Charity fulfilled a vibrant social activist role on the Comstock, one in which they reached across religious, gender, class, and perhaps racial lines to provide care for the Virginia City poor and needy. They did so as single women working to support themselves and those under their protection. Many of their endeavors brought them praise and gratitude, but resistance greeted certain of their actions. Regardless, the sisters worked tirelessly to support themselves and to make social assistance available to a broad range of Virginia City residents.

From their own perspective, the Daughters simply lived out the apostolate of Vincent de Paul. However, the public nature of that apostolate made these religious women uniquely suited to deal with the extreme social demands of a western bonanza town such as Virginia City. The Daughters of Charity came to the West unhampered by the traditional European model of enclosure for religious women, a system that kept sisters virtually withdrawn from the public sphere and limited the nature of their missions. The Daughters of Charity, by turning aside the enclosure rule, avoided struggling with the vexing problem of adjusting European monastic expectations to American social and cultural realities, a process that proved wrenching for many other religious congregations with western missions. This allowed the Daughters to move into Virginia City as activists and immediately launch their caregiver programs on behalf of Comstock residents.

The Daughters of Charity did not seek to cloister themselves from the raucous, earthy world around them: rather they aggressively sought to carve a space for themselves in it, even as they vigorously maintained their commitment to the religious rule of their congregation. In the process, these efforts of the Daughters mixed with a larger initiative that gradually brought a transformation to religious life for American Catholic women. From a secular viewpoint, the Daughters of Charity on the Comstock participated in the growth and development of Euro-American institutions and community. Further, these women added to the mounting evidence that, in the American West, bold and meaningful womanhood assumed many roles, undertook many challenges, and left many legacies.

❧ 8 ❧

Divination on Mount Davidson

An Overview of Women Spiritualists and Fortunetellers on the Comstock

BERNADETTE S. FRANCKE

She may be consulted in regards to events only the shadows of which have yet been projected into the planes of our lives. She sees and feels the presence of these shadows and in her mind they take apprehensible shape.[1]

The mystic figures of the Comstock and their cryptic messages were part of everyday life for many residents of that famed mining district. The above newspaper ad for the woman known as the Washoe Seeress highlights the pervasive influence of the supernatural in the area. Her prognostications were widely publicized, as were advertisements for and accounts of other local soothsayers. While the search for wealth in the Comstock mines increased the market for seers, people also sought the services of those with second sight for their day-to-day concerns.

At its heart the mining district was tolerant. Virginia City, known for its well-attended, traditional churches, also had practitioners of divination. These practitioners added diversity to the population and freely developed their skills. In a place noted for being on the scientific and technological vanguard, it was not unusual for residents to be drawn to nineteenth-century spiritualism, itself a cutting-edge phenomenon.

Although the role of spiritualism on the Comstock deserves to be underscored, this kind of interest is age old. Prescient abilities hark back to the beginnings of recorded history, and permutations are evident in all known cultures of the world. Throughout history people have been curious about the unknown and have searched for answers concerning their futures and destinies. For example, in ancient Egypt fortunetellers stared into a pool of ink, while shamans of the sub-Sahara used clear pools of water to predict the future. The oracle of Delphi, a priestess in ancient Greece through whom a deity was believed to communicate, was renowned for trance speaking. Astrology, another well-known form of divination, was popular in China, Mesopotamia, India, Mexico, and Central and South America.[2]

While taking various forms, these ancient traditions and the evolving Spiritualist Movement flourished throughout nineteenth-century America as well as in Virginia City. The Spiritualist Movement was primarily centered on communication with the spirits of the dead. While other cultures had participated in various forms of spirit communications, this philosophy was new to most Euro-Americans.

There are several accepted reasons for why the Spiritualist Movement became so popular. It is possible, for example, that the nineteenth-century technological expansions and accompanying quest for knowledge overflowed into the mystifying realm of the spiritual world. Nothing was beyond explanation, and the Comstock residents did not shy away from questioning. The Spiritualist Movement can also be seen as an extension of the numerous U.S. religious reform movements of the nineteenth century. The Spiritualist Movement, born during the late 1840s in Upstate New York, first thrived in an area known as the "Burned Over-District."[3] As historian Slater Brown notes, this area had seen such "successive waves of revivalism that few souls remained combustible."[4] In these smoldering embers the spiritualists found their following.

As the Spiritualist Movement spread across the country, consultation with the spirits through a medium was employed for every imaginable concern, foreseeing the future, looking for lost or stolen items, or discussing life after death. Throughout the country, people formed spiritual circles, conducted seances, demonstrated automatic writing and levitations, and spoke with spirits through mediums. In individual sessions and using the more traditional forms of fortunetelling, seers interpreted astrological charts, gazed into crystal balls, and studied palms and tea leaves. All of these forms of divination were known in the Comstock mining district from the time of the lode discoveries in the early 1860s through the great bonanza of the 1870s.

Various authors wrote books interpreting the phenomenon, and the movement provided material to many well-known writers of the day. Mark

Twain called Spiritualism the new "wild cat religion,"[5] and included the phenomenon in his writings. Twain attended seances with various mediums, including Ada Hoyt Foye, and wrote about them in the San Francisco newspapers, often with tongue in cheek. Olivia Langdon, before marrying Twain, had some experiences with a faith healer and developed friendships with mediums. She continued to consult them at several points during her lifetime.

As participants in the Spiritualist Movement or as part of their own cultural traditions, men also worked with divination on the Comstock. The focus of this investigation, however, is directed toward women, whom census records and primary literature indicate constituted the greater percentage of practitioners. The Comstock was a diverse place that included women of many different backgrounds. Information on African American, Chinese and Northern Paiute women's association with divination or spiritualism on the Comstock is sparse, however. Tremendous prejudice existed against these cultures during the nineteenth century, which may account for the lack of information in the newspapers about their occult activities. It also raises questions about the forthrightness of those accounts that were printed. Further complicating the matter is the difficulty of distinguishing between traditional religious practices and occult activities. For all these reasons, this account concentrates on the activities of Euro-American women, who were by far the best-documented group involved in the various forms of divination.

Reviews of lectures and forecasts were patronizing at times but did give credence to women's spiritual abilities. Victorian society viewed women as innately qualified to be mediums, a role that required one to be intuitive, able to set aside one's self in order to receive the spirits. In the nineteenth-century United States these characteristics were commonly attributed to women more than they were to men. Women were also responsible for the moral and spiritual education of children, and society encouraged women to conform to rigid standards of social acceptability.[6] It was considered inappropriate for women to be vocal about social or political issues. However, with the Spiritualist Movement as a vehicle, women entered a previously unavailable public forum. Mediums could philosophize publicly on matters that might until then have been reserved for the judgments of men.[7] They could offer advice based on a spirit's interpretation, even if it conflicted with the conventions of society. Many mediums became popular and influential lecturers. Spiritualist philosophy became associated with women's

*Emma Hardinge Britten, a spiritualist who practiced
on the Comstock during the nineteenth century.* (Courtesy
of Special Collections, Getchell Library, University of Nevada, Reno)

rights, abolitionism, and other progressive movements. The core of the
spiritualist philosophy, however, remained focused on communication with
the dead.

In 1864 Emma Hardinge Britten lectured on Spiritualism in Como,
near the Comstock. By the time she appeared in Como, Britten was already a noted speaking medium and lecturer and had left her theatrical career.[8] She became a national historian of the Spiritualist Movement, publishing works such as *The Place and Mission of Woman, an Inspirational
Discourse delivered at the Melodeon, Boston, Massachusetts February in 1859*
and *Modern American Spiritualism: A Twenty Year Record of Communion Be-*

tween the Earth and the World of Spirits; 1870.[9] Her early presence in the vicinity underscores the fact that spiritualism prospered on the Comstock.

Alfred Doten, editor of the Gold Hill *Evening News,* frequently mentions events involving spiritualists on the Comstock in his journals, perhaps because of personal interest. Doten's sister Lizzie was a spiritualist lecturer and trance speaker residing on the East Coast. Lizzie Doten supported the social reforms of the day and championed the cause of equal pay for women to diminish the necessity of marriage as a means of survival. A friend of Emma Hardinge Britten, Lizzie Doten also spoke at the Melodeon in Boston, Massachusetts in 1859 and sent her brother copies of her poetry and her lectures. In her early years, Doten supported herself as a seamstress and teacher.[10] At the age of seventy, after she had retired from the lecture circuit, Lizzie Doten married her companion, Z. Adams Willard.

The Comstock newspapers provide an important local source of information on the supernatural. Reporters for the Gold Hill *Evening News* covered a spiritualist meeting at the courthouse in Virginia City in February 1866, focusing more on the discussion that took place while waiting for the medium to arrive from Silver City than on the contents of her presentation. Residents such as General Jacob Van Bokkelen gave a rousing testimony to his belief in the spirit world. Apparently the reporters were not convinced of the validity of Van Bokkelen's declaration, because they thought he described himself as a "free drinker" before realizing he had stammered and had really said "free thinker."[11]

Of course, there were charlatans who duped the public with fabricated messages and earned large sums of money for their efforts, but legitimate spiritualists strove to alleviate the mystery and fear associated with death. They considered the end of life as a transition into the next world, a place not to be feared but one where the deceased could speak with the living.[12] Grief experienced over the demise of a loved one played an important role in the development of spiritualism. The need for reassurance on the spiritual whereabouts of the deceased led many to the seance table.

Mrs. Ada Hoyt Foye, the noted spiritualist writing medium, exhibited her powers in Virginia City in 1867. Alfred Doten was fascinated with what he saw and wrote, "It is wonderful to me—1st of any manifestations of that kind I ever saw in my life—I don't know what to think."[13] Doten had several sittings with Mrs. Foye. At one particular gathering, he received communication from his deceased father, Samuel. The message detailed his son's background. It is interesting to note that Foye was reportedly also a

Laura DeForce Gordon. (Courtesy of American Antiquarian Society)

friend of Lizzie Doten and could have known about Alfred Doten's background. Another session included Joe Goodman, proprietor of the *Territorial Enterprise,* and Rollin M. Daggett, who later became one of Nevada's representatives in Congress.

During this same time other spiritualists arrived on the scene. Laura DeForce Gordon lectured as a speaking medium to a crowd at the District Court Building. Doten described the hall as "jammed closer with people than I ever saw it before even at John Millian's sentence."[14] (Millian was hanged for the murder of Julia Bulette, a Virginia City prostitute.) Reverend James E. Wickes of the Methodist Episcopal Church attended one of the many lectures given by Gordon. On the evening of Saturday, December 6, 1867, Wickes interrupted her lecture, arguing biblical philosophy. Organizers of the lecture decided that a special debate would be held at the Methodist Episcopal Church on the following evening. Typically, Spiritualists interpreted the Bible as supportive of contacting spirits of the dead, in contrast to those belonging to traditional religions who viewed "holding traffic with the spirits as sinful."[15] Neither the trustees of the church nor the congregation allowed the debate to be held in their building, and on the following night a packed crowd of five hundred to six hundred attended the lecture, again at the District Court. The Reverend Wickes did not attend, and according to Doten the audience felt it was "tacit acknowledgment on his part of his inability to prove the inspiration and infallibility of said book."[16]

Doten apparently had a soft spot in his heart for Gordon. Gordon's hair became disheveled when she removed her hat prior to beginning her lecture. He passed a note to her that read: "Mrs. G. your hair sticks up where your hat caught it.—Yours Truly, Alf Doten."[17] Gordon was well received during her visits to Virginia City. She would return four months later for additional speaking engagements and private appointments. At the conclusion of her lecture on Sunday, December 8, some of Gordon's friends presented her with a silver bullion brick valued at $20.63. Doten commented in his diary that "an association was formed for the purpose of carrying out Spiritualist Humanitarian and investigating ideas; over 100 people signed the roll."[18] This was the beginning of the formal Virginia City Spiritualist Society. Laura De Force Gordon went on to speak on the cause of women's suffrage in the Northwest, was nominated for the California State Senate, and became one of the first two women admitted to the California bar.[19]

The Virginia City Spiritualist Society Hall was located under the dental office of Dr. Powers at 34 South B Street between Taylor and Union Streets. The Society sponsored lectures such as the one on October 4, 1868, in which a woman named R. N. Gore gave a lecture on Jesus Christ and provided psychometric readings. The term psychometry refers to divination of facts concerning an object or its owner through contact with or proximity to the object. On the same day that the newspaper advertised Gore's lecture, it also carried an ad for books on psychomancy, or soul charming, which promoted prosperity in love or business.[20]

Spiritual circles were also held at the house of Lavinia Lanning (which also appears as Lannen) for at least nine years. It is unknown if her husband, who operated a furniture shop on B Street, participated in the gatherings.[21] Lannen would hold on to her belief in spiritualism throughout her lifetime. Her marble marker in the Virginia City cemetery commemorating her death in 1888 carries the following inscription: *Death is but a kindly frost that cracks the shell and leaves the kernel room to germinate.*

Proponents of the Spiritualist Movement continued to suffer criticism from established Euro-American churches of the time. While certain types of spiritualism were essential in the doctrines of organized religions, this new movement was a deviation from what most religions traditionally had accepted. For example, the Catholic clergy acknowledged the occurrence of spirit manifestations, although these were never on demand. The clergy perceived the timed spirit manifestations among the spiritualists as suspect.[22] Nonetheless, according to the *Book on Mediums,* the spiritualist doctrine "respects all beliefs" and encourages questioning of the spiritualistic philosophy. The *Book* further states, "One should guard not only against recitals that may be more or less exaggerated, but against his own impression, and not attribute everything he cannot understand to an occult origin."[23]

In November 1872 a phenomenon occurred in Virginia City that would merit publication in the *Catholic Guardian,* a noted national religious publication. Father Manogue, the revered pastor of St. Mary in the Mountains Catholic Church and Vicar General of Nevada, recalled his experiences with Agnes McDonough, fourteen years old, who apparently had been visited by her father six years after his death. Father Manogue was summoned to 163 North B Street, where McDonough was living with the Masel family, who were relatives. John Masel was the well-known proprietor of the Empire Market. Also residing in the home on B Street were the Masels's two young sons and Agnes's brother. Wherever Agnes McDonough was in this particular house, a series of rappings could be heard, and she reported having had conversations with her father, though only when in a room by herself. Father Manogue positioned himself near the doorway of the room to witness the event and supplied McDonough with a series of questions for the spirit. The questions concerned where the spirit came from, if Christ was a stern and severe judge, and where the spirit was going from here. They produced responses concerning purgatory, the kindness of Jesus Christ, and the spirit's ascension to heaven.

Father Manogue was opposed to the new spiritualist philosophy. In an interview with the press, however, he said that he believed in Agnes McDonough's sincerity. She herself "took out a card" in the newspaper saying she was sincere. Father Manogue, who did not want to see this event become public, could not stop the enthusiasm of the press. The occurrence received detailed press coverage, complete with a floor plan of the Masel home in the *Territorial Enterprise*.[24] Newspapers as far away as the San Francisco *Chronicle* mentioned the incident: "There is considerable excitement in the spiritualist circles in regards to recent singular manifestations. They regard the appearance of McDonough to his daughter as the most striking manifestation they have had in years. Miss McDonough is regarded as a very powerful medium and one who will yet do greater things."[25]

Everyone had an opinion on the matter, and it seemed that the newspaper interviewed almost anyone with any association with the event. The editors of the *Virginia Evening Chronicle* soundly chastised Joe Goodman of the *Territorial Enterprise* for the extensive coverage he gave the story. They claimed that "insinuations muttered through the teeth of this silly if not artful child are recorded in detail by this historian of the new revelation as conclusive evidence."[26] The *Virginia Evening Chronicle* quoted a *Catholic Guardian* editorial that suggested it was possible for these events to occur but that "in this, as in all things else, all we, as Catholics have to do is to put on the brakes, go slowly and listen to the never-erring voice of the Church."[27]

The press continued to cover other local spiritualist-related events. In 1875 the *Virginia Evening Chronicle* concluded that it was the influence of a medium, Lou Finch, and a spirit named Starlight that caused Euginia Turner to seek a divorce from her husband.[28] Several months before the divorce trial, Turner had fled Virginia City with their son but was intercepted by her husband, George E. Turner. Apparently an argument ensued and George Turner brought their son back to Virginia City. Euginia Turner sought a divorce from her husband on the grounds of cruelty and testified that he had "violently and angrily assaulted her and twice threw her down."[29] A trial began in the First District Court with a jury of his peers consisting of notable residents such as A. M. Cole, druggist; Fred Boegle, stationery and bookstore owner; A. M. Kruttschnitt, former County Treasurer; and E. M. Chappin. Five people were subpoenaed for the case, and in the end the jury decided that "Mr. George E. Turner treated his wife with great kindness."[30]

The court divorce records do not discuss the financial controversy surrounding George Turner that had begun a year earlier with bankruptcy proceedings. Another court case that involved George Turner and was pending at the time of the divorce proceedings concerned the repayment of a four hundred dollar note to Timothy Dwyer. The court records indicate that Turner may have drawn on his wife's account for repayment of his debts and that he signed her name to receive these funds. Eugenia Turner had purchased sixty shares in the Belcher and twenty-five shares in the Crown Point Mining Companies in July of 1874. She had had this transaction recorded in the County Recorder's office specifically under Inventory of Separate Property.[31]

The dates on which Eugenia Turner sought the advice of Lou Finch are lost to history. The above circumstances indicate that perhaps it was more than just a whimsical visit to a spiritualist that led Eugenia Turner to leave her husband. Finch may have provided support that bolstered her confidence.

Finch's advertisements listed her as a celebrated clairvoyant, physician, optician, aurist, and a graduate of the Female Medical College of Philadelphia, Pennsylvania. The advertisement also stated that "She charges nothing for examining the sick." The Female Medical College, established in 1850, was the first school of medicine for women in the world. The college triumphed in the face of much opposition to the education of women as doctors. The name was changed in 1869 to the Women's Medical College. The institution still exists today.

Throughout the reign of the Spiritualist Movement on the Comstock, the traditional forms of fortunetelling were always present. Gertrude Remington, E. F. Thorton, and H. A. Perrigo were three women living in New York State who promoted their spiritual talents in the Gold Hill *Evening News* in 1867. Their advertisements proclaimed that they had abilities to "reveal secrets no mortal ever knew" so that clients could "know thy destiny."[32] They required the seekers to send information on their place of birth, age, disposition, and complexion. In the case of Perrigo and Thorton a lock of hair had to be included. The women promised to reveal insights using their astrological and clairvoyant expertise. Both Remington and Thorton charged fifty cents, while Perrigo's fee was one dollar. Doten reacted to a similar ad and sent Mr. and Mrs. A. B. Severance of Wisconsin a lock of his hair and one of a woman named Lizzie Landell. For the price of two dollars he was assured that he would receive information on each of their characters.[33] The results are lost to the historical record.

The most popular fortuneteller on the Comstock was the "Washoe Seeress," Eilley Bowers. A native of Scotland, she followed her homeland tradition of developing second sight and had been consulting her peep stone long before arriving on the Comstock. Second sight was a term recognized in the nineteenth century as one describing those with clairvoyant

Eilley Orrum Bowers. (Courtesy of Nevada Historical Society)

abilities. The peep stone that Bowers used was a "ball of glass shaped like an egg."[34] By gazing into the glass, Eilley Bowers could see images. She was able to "find out all manner of things," such as the location of missing persons, stolen articles, ore bodies, and spirits of the dead.[35]

In 1858 Bowers appealed to Snowshoe Thompson, the famous Sierra mail carrier, to secure a new stone for her on his next trip to Sacramento. Bowers had requested the stone because her existing one had become cloudy and she wished to view a mine that she had seen partially through her old stone.[36] Thompson was unable to fulfill her request. He did, however, have a firm belief in her abilities and viewed her as responsible for great lode discoveries.[37] Bowers continued to rely on her powers for direction after the death of her husband, Lemual "Sandy" Bowers. Her friend, Mrs. Dettenreider, who provided furnished rooms for boarders, often offered her home for Bowers's sittings. Bowers advertised her seeing abilities, and the local newspaper followed her predictions more closely than it did those of other fortunetellers in the area at the time.

Individuals vouched for Bowers's abilities by recounting for the newspaper her assistance in locating missing items. For example, G. L. Whitney located a lost gold watch chain after consulting the Washoe Seeress in February 1878. In another instance, Bowers correctly predicted that a stolen ring worth $450 would be returned and that the reward payment would not need to be made. During a seance with Bowers, a woman who had been looking for a certain individual requested information. Through Bowers, the spirit of the missing person informed the woman of his death and burial in the Virginia City Cemetery without headstone or grave marker. The spirit asked the woman to have undertakers Wilson and Brown show her the location of his grave. He requested that on Decoration Day she place flowers on this grave. He said, "If . . . you could but see the thousands of disembodied spirits hovering over the cemetery on the occasion, you would feel amply repaid for your trouble."[38]

Poor financial management that had begun before the death of her husband made it impossible for Bowers to purchase mining stocks. At the precise time she had foreseen as fortuitous, Bowers had little money. She was concerned that others might suffer financial constraints and stated that she did not "wish any of her friends to purchase stock upon the strength of her predictions."[39] Eilley Bowers continued to share her gift until her hearing made it impossible for her to decipher the questions of those who would employ her services. She died on October 27, 1903.

There were many other fortunetellers on the Comstock about whom little is known. During the Comstock heyday of the 1870s, advertisements for fortunetellers appeared frequently in the local newspapers, featuring women with surnames such as Martelle, Hoffman, Smith, LaRue, Solama, Baumann, Lewis, and Eckhard. The one for Madame Solama, for example, described her as "The Celebrated Moorish Spiritual Medium" and suggested that she "may be consulted for a few days in any language desired, upon all matters in relations to business or trouble of any kind."[40] The *Virginia Evening Chronicle* Business Directory heading for fortunetellers suggests that, "Those who wish to explore the mysteries of the future are promised assistance by the following named Astrologers."[41]

Maria Smith advertised in the 1878–79 Bishop's Directory as a clairvoyant. Madame Smith, not the same person as Maria Smith, advertised in the *Virginia Evening Chronicle* as the "Wonderful Gifted Astrologer and

Advertisment placed by Madame Solama
in the Virginia Evening Chronicle, *1876.*
(Courtesy of Nevada State Library and Archives)

Fortune Teller." Minnie Hoffman, a native of Germany, advertised frequently as astrologer and fortuneteller in the *Virginia Evening Chronicle* during 1876. In 1880, at age fifty-nine, she was still commercially advertising her occult abilities.[42] According to the tenth U.S. manuscript census of 1880, Minnie Hoffman listed her occupation as that of keeping house. For whatever reason, she chose not to promote her role as a fortuneteller to the enumerator. On the other hand, Rosa Grose did list her profession as fortuneteller in the 1880 census. A native of Hungary, Grose was divorced and thirty years old. She evidently did not use advertisements.

Mary McNair Mathews, known for her book, *Ten Years in Nevada*, practiced fortunetelling by reading "the cup." Mathews realized that she could make more money as a fortuneteller than as the manager of her lodging house at the corner of C and Silver Streets, but she decided not to pursue fortunetelling as an occupation. She did not want to be labeled a fortuneteller, and to make this clear she would not accept payment for her spiritual services.[43] This stance may have reflected evolving standards about women and employment. A dichotomy existed for fortunetelling: although using mystic means to divine the future was widely accepted among social acquaintances, and the skill of the seer was worthy of recognition, fortunetelling was not always sanctioned as suitable employment.

Whether they were lecturing on the limited view of the Bible or advertising their abilities to tell past, present, and future events by studying cards and the movement of planets, fortunetellers on the Comstock offered alternatives to traditional spiritual beliefs. They fulfilled people's need to delve into the unknown, a need that has been prevalent throughout history. As the *Territorial Enterprise* described them, these women were able to "look into the seeds of time, and say which grain will grow and which will not."[44]

We can only speculate about the psychic abilities of these work-a-day prophets. The overall validity of their predictions is impossible to ascertain. While many seekers saw and heard what they wanted, an article that appeared in the Gold Hill *Evening News* on September 15, 1875, is worth noting. It said that, "Some of the Virginia Firemen say that numbers of people in that city are firm in the belief that the whole or at least the greater part of the place will shortly be destroyed by fire. They say that the disaster has been predicted by certain Spiritualists." A little more than a month later Virginia City was devastated by the greatest fire in its history.

9

"The Advantages of Ladies' Society"
The Public Sphere of Women on the Comstock

Anita Ernst Watson, Jean E. Ford,
& Linda White

It is a singularly remarkable fact that the public journals of the Territory have had so little to say of the rapid increase of women, children and homes here, that beyond our borders . . . [we] are not conclusively known as possessing the advantages of ladies' society.

This oversight regarding the image of the Comstock disturbed Minerva Morris, a woman living in Gold Hill in the 1860s. She believed that "women are the most necessary and indispensable ingredient in human society," and openly expressed her opinions about the important, albeit under-emphasized, role of women. Appearing in the second issue of the Gold Hill *News,* Morris's letter to the editor rejoiced "at the thought of having a village newspaper" but implored the publication not to limit coverage to "questionable stock reports and police items." Although Gold Hill was a mining community and as such was overwhelmingly male, Morris stressed the presence and importance of female interests and influences.

In Mrs. Morris's opinion, life on the Comstock had not been depicted accurately: "People elsewhere now only know that this is some sort of a place more natural than artificial, more rough than elegant, more rich than refined." In fact, the mining communities did "abound in wives, maidens, and children," and Morris asserted that "the growing moral excellence of our famous Territory is very fully due to family and female influence, an influence which is increasing every day." Women's contributions to the community, too worthwhile to ignore, should be properly acknowledged: "In [the] midst of all the rush and noise and illusions of wealth, womanly influence will, while it solaces the careworn, still further innoble itself by nourishing the fine sentiments of nature, education and refined intercourse . . . and I for one am resolved to urge you to transcend the usual gallantry of our newspapers which never allude to the ladies except under the merely utilitarian captions of marriages and births."[1]

Minerva Morris's letter is one of relatively few public statements regarding the civic role of women on the Comstock, and it highlights several important aspects of life for middle- and upper-class women there. Although their numbers were small,[2] women were involved and effective in social and charitable activities in the mining localities. They entered and were acknowledged in the public sphere[3] when they dispensed charity and also when they campaigned for temperance and worked for enfranchisement.

Morris was correct in her charge that women were only rarely acknowledged publicly; recognition of their contributions to community life and its improvement was infrequent and understated. When credited, women were generally not mentioned by name but were referred to as a group; they became the "ladies." The "ladies" were faceless, shadowy figures, a dimly reflected image of their meaningful reality in the mining communities.

Despite their anonymity, as Minerva Morris so clearly stated, those women perceived their role and talents in social and civic activities as vital to the quality of life on the Comstock and as unique to their gender.[4] The exclusion of women from the pages of the newspapers, an omission which Morris referred to as gallantry, reflected the prevailing perception that a woman's proper place in mid-nineteenth-century, middle-class society was in the home, not on the front page.

The lives and limited public activities of women on the Comstock are representative of both constancy and change in the roles and expectations of many middle- and upper-class women in the post–Civil War decades, not just in the West, but nationally. The mining communities of the Comstock, while isolated, were not unaffected by national trends. They materialized as sophisticated, industrialized urban centers clinging to stark hillsides and squatting in the rocky gullies and canyons, oases of culture and grace in the "howling wilderness" of the West. The residents of Virginia City, Gold Hill, and Silver City had more in common with the urban

The finer houses of Virginia City were located above C Street, the commercial corridor of town. (Courtesy of California Historical Society)

dwellers of Philadelphia or San Francisco than they did with an isolated Nebraska homesteader, and the Comstock's communities embodied urban culture and expectations.

One shared aspect of culture was a public concept of the proper role for "ladies." Historian Barbara Welter has identified a middle-class cult of domesticity that was promulgated by many magazines and advice manuals of the day. In them a concept emerged of "true womanhood" that was based on an assumption of women's natural morality. Each sex worked within the domain most suited to its talents and natural abilities, women separately from men.[5]

The prevalence of such an ideal and the limitations imposed by, or the possibilities for control created within, a domestic sphere are debatable.

The residents of Virginia City have always enjoyed a parade. (Courtesy of Nevada Historical Society)

Not all women of the middle and upper classes accepted the ideal or lived in a separate female sphere, and it was certainly never a reality for poor or working-class women. Even for those women who operated within a domestic sphere, the boundaries appear to have been fluid. Many men and women did, however, accept a prescriptive ideal of true womanhood, and they brought that concept to the West.[6]

Women's civilizing mission[7] was predicated on the assumption that women had a specific function in American society, one that they were naturally suited to assume. The frequently articulated view, espoused both publicly and privately by the middle class, presumed that women were innately more moral than men. Women embodied traditional Christian virtues of humility, modesty, submission, and piety, qualities that were necessary underpinnings for the success and expansion of the American republic.[8] As the United States modernized and industrialized over the course of the nineteenth century, the concept of the family as an economic unit shifted to an ideal of the family as a refuge from an increasingly busy, complex public life. The home was the calm in the eye of the storm; the woman created and controlled the tranquillity, while the man embarked upon daily forays into the tempest to provide for his family.[9]

According to this model of domesticity, women's presumed natural sphere lay within the home. Incursions into the public sphere, where the fragile female had no natural skills for survival, were constrained by an understanding of a woman's limits. Women who wanted to move beyond the boundaries of the home were most successful when they ventured in directions that were seen as a natural extension of the domestic sphere: charity work and reform projects. Such women enlarged the domestic realm into public life and created a feminine sphere, where women exercised their inherent talents on acceptable female activities.

Upper- and middle-class women on the Comstock worked well within these boundaries, as is evidenced by their charitable activities and reform efforts. Female benevolent efforts were both secular and religious. When working outside the church, women were involved with private philanthropy or were members of auxiliary groups of fraternal organizations. The soup kitchen operated by Mary McNair Mathews and Rachel Beck, discussed below, seems to have been an anomaly in Virginia City, where the bulk of the eleemosynary work appears to have been done through church-affiliated groups.

One secular charity organization, the Sanitary Commission, involved the women of the Comstock in philanthropy with national connections. When decades of compromise between northern and southern interests failed, and Civil War erupted, the Union was unprepared for the carnage that modern technology facilitated. The Sanitary Commission emerged as the benevolent support of northern hospital and medical services. Between 1862 and 1865 over 7,000 local societies were organized, and together they contributed $4,800,000 to the Commission's funds. Branches in Nevada raised $163,581.07; the bulk of that came from Storey County's contribution of $109,760.07. Men were the Commission's figureheads, but it was primarily administered and staffed by women.[10]

Much of the money raised on the Comstock for the Commission came from small bazaars, fairs, and balls. Typical of such efforts was a fund-raiser held at La Platte Hall on B Street in Virginia City in January 1863. The women organized a dinner and dance and were publicly commended for their efforts. The supper they made was labeled "excellent." Guests at the ball tried several new dances, including the Virginia Reel and a plain quadrille called a medley. In spite of clumsy feet and ripped hems, everyone had a good time; the Sanitary Ball raised $400 for the Commission. There was also a fancy ball held at Gold Hill and one at Steamboat Springs, both attended by people from Virginia City, Gold Hill, and Silver City.[11]

Most occasions for fund-raising were locally oriented, however. The Comstock communities were located "on a slope of a mountain speckled with snow, sagebrushes, and mounds of upturned earth, without any apparent beginning or end, congruity or regard for the eternal fitness of things," as famous reporter and traveler, J. Ross Browne, described Virginia City in 1860. With a growing population, "frame shanties . . . tents of canvas, or blankets, of brush, of potato-sacks and old shirts . . . smokey hovels of mud and stone [and] coyote holes in the mountain side"[12] were replaced by fine wood and brick homes, elegant hotels, and sturdy boardinghouses. But the towns in the shadows of Mount Davidson were still intruders in an often inhospitable landscape. Their citizens endured fires, floods, blizzards, and other environmental calamities.

When an emergency developed, relief committees emerged to cope. Public recognition of such activity indicates that ladies' societies were most often the source of aid and support. In March 1867 newspaper commentary revealed that the Comstock was still in the grip of a harsh winter: "Storming—Storming—At 1 o'clock last night it was still storming. It does nothing but storm. It still blows and snows as though never going to let up. Good bye, world! we no longer delight in thee."[13] Three days later there was a "rare view" to the east, "five or six miles of snow-clad mountains, then fifty or sixty miles of valleys and low hills free from snow." There had been problems with the roads, but Geiger Grade was reported "passable for sleighs"; with the drifts shoveled, "we hope to see the road fully open in two or three days at the furthest."[14]

The closing of roads by heavy snow adversely affected the residents of the Comstock, and a number of groups attempted to help. A notice for a meeting to plan relief, to be held in the sheriff's office, noted that without the storm and the resulting mill and mine closures, most families would have been able to provide for themselves. In hard times, however, it was up to "the wealthy and those in easy circumstances" to help a little.

A Calico Ball, given by the "charitable ladies of this city," was part of the benevolence that occurred during the weather crisis. "The party was one of

the most sociable and agreeable of the season, for all felt that they were doing something in aid of a most excellent cause and naturally were in good
humor with themselves and everybody else." That good cheer raised $1,000
"for the relief of the poor." The day after the ball the leftover food was distributed "to those in need."[15]

Women of more modest means were also involved in charity work on
the Comstock. Mary McNair Mathews and a friend, Rachel Beck, opened
a soup kitchen to feed the destitute and out of work when business stag-
nated in Virginia City in 1877. With the mines shut down, the women
feared the hungry might riot and destroy the town. They collected donated
supplies from restaurants, butchers, and markets; bakers sold them leftover
bread. Soup that was unused by the end of the day was given to "Chinese
and Piutes . . . daily hanging about the door at mealtimes." With only two
paid employees, McNair and Beck prepared meals and fed 400 to 500 people three times a day for about a month, noting that the "bonanza ring
never gave us a dime."[16]

Mining was a hazardous occupation, and the newspapers of the Comstock reported death and injury in the mines daily. One page of an 1870
issue of the *Territorial Enterprise* included an announcement of a benefit
performance to assist a widow, Mrs. Curran, and her children, and a brief
article about the crushed pelvis and other severe injuries of a workman.[17] A
fire a year earlier in the Yellow Jacket Mine in Gold Hill had resulted in the
deaths of as many as forty-five men. Within days the Gold Hill Relief
Committee had organized to receive and disburse the money being sent
from around Nevada and California.[18]

There were a number of lodges and fraternal organizations on the Comstock. As a result of the high percentage of foreign-born residents, some
groups had ethnic affiliations. There were several Irish military companies,
German Turnverein societies, a Scots Caledonian Club, the Italian Benevolent Society, and gatherings of Mexican and Chinese. Many of these associations had charitable goals.[19]

The Odd Fellows organized in Virginia City in 1861, with the expressed
intent of "visiting the sick, relieving the distressed, burying the dead, or
caring for the orphan." Women were not members of these lodges, but
most had auxiliary organizations for them. A Rebekah Lodge affiliated
with the Odd Fellows, for example, was established on the Comstock in
1868. The influence of women was noted by observers of the fraternal
orders: "There is no enterprise that is not elevated and inspirited by the

co-operation of woman. She purifies whatever she mingles with and refines wherever she associates."[20]

One of the bulwarks of Comstock social life and charitable endeavor was entertainment, both professional and amateur. Professional entertainment to rival the best that cosmopolitan eastern cities offered was notable on the Comstock. Public diversions started with the presentation of "The Farces of Toodles and Swiss Swains" at the Howard Street Theater in 1860, and by 1863 Maguire's Opera House had opened. Adah Issacs Menken, a notorious courtesan-actress, presented her most famous play, *Mazeppa,* to enthusiastic Virginia City audiences. Major theatrical troupes entertained in the elegance of Piper's Opera House, which superseded Maguire's in 1868. Theater patrons enjoyed a wide variety of offerings; Edwin Booth did "Hamlet," Henry Ward Beecher lectured, and "The Montgomery Queen's great show, with an African Eland, and Abyssinian Ibex, Cassowaries, and the Only Female Somersault Sider [sic] in the World" thrilled audiences with the wonders of the world.[21]

Amateur theatricals competed with the professional troupes as charitable fund-raisers on the Comstock and provided another acceptable public forum for women. In 1875 the Virginia Benevolent Association gave a "grand concert" at Piper's Opera House, raising "several hundred dollars, which will be judiciously expended for the relief of poor and destitute persons during the present winter." Women provided much of the evening's entertainment. Mrs. L. B. Moore played the sailor's hornpipe; Miss Felicia Genesey, dressed as an Indian maiden, recited from Hiawatha; Miss Katie Beck sang "Good Night My Sweet"; and Ettie Adams "showed a high order of dramatic talent" when she sang "Good-bye, Susan Jane." Although several men performed, the women were reported to be the favored attractions, gathering the money that was tossed on the stage.[22]

There was more to the social round than entertaining and charitable evenings at the theater, however. Louise Palmer, wife of a mining superintendent, wrote of her life in Virginia City for the *Overland Monthly* in 1869. Like Minerva Morris, she perceived that the image of life on the Comstock was not representative of reality and declared with pride, "We move in the best society." Palmer occupied her days with lunch parties and shopping but reported plenty of evening entertainment available: "We have club parties and public balls, interspersed with private card and dinner parties. And we are all very gay and fashionable—exhibiting our diamonds and laces to the eyes of rival mine and millmen's wives and daughters with as much eagerness as would a New York or Parisian belle."[23]

Lest she be considered shallow and flighty, Palmer noted that there was more to social activities on the Comstock than frivolous parties and glittering balls. Churches were an important part of social activity: "We Virginians are a church-loving people, too . . . our churches wax rich and strong." She thought it curious that religion was "performed by proxy," by the wife who "appears both on her own behalf and that of the husband," the men too busy with business to attend church.[24] In contrast to Palmer's observations about women attending church in lieu of their husbands, an 1876 history of the First Presbyterian Church decried the lack of women. The early difficulties of church building and growth were blamed on the dearth of women to make up the congregation and do church work, and it was noted that "the prayer meetings are mainly supported by men, few women attending."[25] Despite an initial perception of scarcity, there were eventually many women active in church life in the mining communities, which was another part of women's public sphere on the Comstock.

Palmer labeled women's participation without their husbands in church activities "curious," but it was typical for the West and for the United States in general. Religious activities were an accepted public interest for women, and church congregations were often predominantly female.[26] As part of the domestic ideal, religion was considered a natural outlet for women. One physician, addressing medical school graduates in the East, informed the young men that women were naturally religious, with a "pious mind"; a woman's "confiding nature leads her more readily than men to accept the proffered grace of the Gospel."[27]

From the outset there were numerous church congregations on the Comstock, and thus the opportunity for social activity through religious endeavor dated from the beginning of community building in the district. The Catholic church was present in Virginia City by 1860; an early history of Nevada noted the charitable work of this church and that the altar and rosary societies of Virginia City were united and "chiefly composed of ladies."[28] A Methodist Episcopal congregation had been organized by 1861. The Episcopal church established St. Paul's in Virginia City in 1861 and St. John's in Gold Hill the next year. Fairs and festivals put on by the "ladies" furnished the church, purchased the rectory, and raised money to retire debt for St. John's.[29] The Presbyterians followed the same pattern, with a congregation in Virginia City in 1862 and one in Gold Hill by 1863. Jewish miners and merchants were among the earliest settlers on the Comstock, and by 1865 there were two Jewish benevolent associations. The B'nai B'rith Lodge of Virginia City funded space for patients and nurses during an 1869 smallpox epidemic.[30] The Baptists had a chapel by 1863, with a small, primarily African American congregation. That church property was sold in 1867, and the congregation was not reorganized. A second congregation, the Tabernacle Baptist Church, was formed in 1864. Services were conducted in the courthouse, the Miner's Union Hall, the Washington Guard's Hall, and private homes until a church was constructed in 1873.[31]

The first two Sunday school teachers at St. John's Episcopal Church in Gold Hill were Mr. Hale and Mrs. Pfoutz, which indicates that both men and women were involved with Christian education. Women's work in churches, however, went beyond teaching; a number of church voluntary associations were formed, among them "mite" societies. The Ladies' Mite Society of Gold Hill and the Episcopal Mite Society held "mite meetings" to organize benefits for church and charity. News of the Carson City Mite Society also appeared in Comstock newspapers.[32] The name "mite" reflects the expectation that those invited to the events would give a mite, a small amount of money, for the worthy cause.[33]

The deserving recipient of the generosity of the Ladies' Mite Society of Gold Hill in early 1871 was William J. Evans. The ladies sponsored a "social hop" to raise money for Evans, "a gentleman who is paralyzed and unable to work." Evans's injuries were illustrative of the hazards of mining. The evening at the Miners' Union Hall netted $370 and "enables this needy gentleman to go home to his family and friends in England in good style. All honor to the ladies of Gold Hill."[34]

Descriptions of women's church activities are commonly found in newspaper announcements or accounts of charity auctions or balls. In October 1873 a fair, "under the management of the ladies of the Catholic church," was held at Armory Hall to benefit the Nevada Orphan Asylum, run by the Daughters of Charity. Newspaper reports of the event noted that the "receipts are most flattering to the ladies who have interested themselves in its success," in spite of the fact that the fair was competing with Chiarini's Circus. Reports mentioned the women participating in the fair by name, praising their handiwork effusively and including detailed description of the "fancy articles such as are unusually found upon a table of the kind at a ladies' fair." The individuals involved, the articles raffled, and the food available for the "most fastidious epicurean" at the fair clearly point to the participation of the upper and middle classes.[35]

In addition to being refined enough to appreciate gourmet foods and fine furniture, the citizens of the Comstock were considered openhanded, almost to a fault. As Mathews observed: "Nevada men and women are very generous; they will divide their last potato with you, or give their last 'bit' to a charitable cause; and not many even stop to inquire whether it is for charity or not."[36] An 1870 account of life in Virginia City judged the residents to be "utterly incapable of doing things slowly or by parts." This attitude carried over from business and personal life to charity. As the commentary continued, "The poor we have with us to be sure . . . But, all honor to Virginia's big-hearted citizens, the poor are cared for, and generous bounty is dispensed day by day without fail or stint, to the unfortunate but worthy poor."[37] The people of the Comstock gave extravagantly and often; reports of charity fund-raisers frequently mentioned large profits.

The specific extent of women's participation in church affairs, as in many public activities, could be difficult to ascertain. As an example, a newspaper announcement for an 1867 charity ball indicated that "the ladies of Virginia will hold a St. Patrick's Ball" at the Athletic Hall. The ball was connected to St. Mary's church, and members of the Arrangements and Invitation Committees were listed by name. Only men appeared with both first and last names, however, so it is possible that some of the names appearing only with initials were women's.[38]

Most commonly, newspapers or church records mentioned a woman's first name only if she was unmarried. The records of St. Mary's Asylum and School referred to four women, Mrs. Thomas Sunderlund, Mrs. Thrall, Mrs. Lynch, and Mrs. Bain,[39] as "the most distinguished ladies of Virginia" who "worked with an energy never to be forgotten" at the school's Christmas Fair. The customary omission of a married woman's given name is evident in this reference, and in most contemporary accounts of the women's activities. The most important aspect of a woman's identity was her marital status, which was carefully indicated by her title and used in most public references to women of respectable standing in the community.

Even major female benefactors suffered this fate. Mary Louise Mackay, the wife of Bonanza King John Mackay, was a substantial donor to St. Vincent's School for Boys and St. Mary's Asylum for Girls. Mrs. Mackay also provided a parcel of land for the Saint Mary Louise Hospital, built in 1876 and operated by the Daughters of Charity until 1897.[40] In newspaper accounts of charitable activities, however, she was referred to as Mrs. John Mackay.

Even more difficult to determine than first names is information about the race and ethnicity of women in the public sphere. Early city directories refer to some Comstock residents as "colored," and newspaper accounts of accidents or criminal activity often referred to race or ethnic background in descriptions of individuals involved. Such was not the case, however, for accounts of charitable balls and fairs, where the "ladies" tended to be featureless as well as nameless. The public involvement of women of color and varied ethnicity provides a rich area for further research.

Charitable endeavor and Sunday school teaching were the common arenas for women's church work, but other public activity was not unknown. In 1873 the Methodist Episcopal Church had a religious revival meeting with a visiting female preacher, Maggie Van Cott. Van Cott, a widow, had traveled and preached widely; she was particularly interested in the rights of women as clergy and became the first woman licensed to preach by the Methodist Episcopal Church.[41] The *Territorial Enterprise* referred to her "powerful preaching and exhortations," but seemed more interested in using Van Cott's sermon as background for a comical story about one of the men in the audience who made light of the preacher's efforts: "Last Sunday evening, as we are told, Mrs. Van Cott approached a Cornish miner, who, from the serious face he wore, seemed to have been a good deal impressed, and laying a hand on his shoulder said: 'My friend, are you a laborer in the vineyard of the Lord?' 'No, mum,' said the man, scratching his head and rollin up the whites of his eyes at the face of his fair questioner, 'No, mum, I be workin' in 'ee Savage lower level.'"[42]

In 1875 Fanny Stenhouse's lecture about the inner life of the Mormons was held at the National Guard Hall and drew a "large audience, composed of the most substantial people of the city." A year earlier, she had written an exposé, *Tell It All*, of Mormon life and plural marriage; the newspaper account stressed her reliability and truthfulness, drawn "from her own full and bitter experience." Stenhouse was described as a lucid and powerful speaker: "There is nothing of the woman's rights or pedantic blue-stocking about [her]; she is a modest, quiet and retiring lady who steps from the retirement and quiet of a cheerful fireside onto the public platform to picture the actualities of polygamic life."[43] The description of Fanny Stenhouse as a "quiet and retiring lady" signalled that, although she was stepping beyond the domestic sphere into the public realm generally occupied by men, she did so with a "ladylike" demeanor. Stenhouse's modesty and restrained approach made her nontraditional role acceptable to the audience.

Newspaper reports on women's church and charitable activities were generally straightforward and laudatory accounts of the events. Occasionally, however, there was a condescending tone to the articles. The affairs became the background for jokes and ridicule, as with Maggie Van Cott's revival meeting. The Gold Hill *News* poked fun at a fancy dress ball sponsored by the Carson Mite Society, speculating about the program's note that "dress is optional."[44]

The *Territorial Enterprise* account of the Orphan's Fair in 1868 is an example of the patronizing undertones sometimes taken by newspapers on women's activities. After a detailed description of the articles for sale, the reporter launched into a muddled tale of ice cream consumption. Praising the abundance of frozen treats, which were enough to make a common man "groan aloud," the observer then portrayed himself as a helpless victim of female good intentions. He was led into overindulgence "quietly and pleasantly." Then, after receiving a proposal by mail delivered by the "postmistress," he required more ice cream to regain "his usual sanity." When the doll chosen by him from the grab-box cries the name of his would-be fiancée, he "breaks for the door." He was stopped "in his flight" to buy a raffle ticket but managed to "bolt from the hall in confusion" and board a train headed for Reno. He found himself "upon the Summit of the Sierras" before he recovered from the ice cream bacchanalia.[45] Obviously, the story was written as an entertaining account of the Orphan's Fair, but beneath the banter was a tale of a man who found himself helpless against the importunities of the women and was finally forced to flee for his physical and mental health. Only superficially admiring, the article sympathetically depicted the stereotypical befuddled male, out of his element among feminine fripperies and beset by manipulative women.

Mark Twain's mockery of women's charitable work was less subtle and resulted in serious consequences for his career. Working as a reporter for the *Territorial Enterprise,* Twain wrote a satirical piece about the diverting of charitable funds for a "Miscegenation Society somewhere in the East." Realizing that the commentary would be offensive, Twain tossed it aside, but it was inadvertently published in May 1864. A public apology was issued to the ladies involved in the charitable ball, held in Carson City, when they took umbrage at the story. One incensed husband challenged Twain to a duel, and the uproar escalated. For two weeks Twain and the editor of the *Virginia Union* squabbled via letter, calling each other "fools, cowards, poltroons, liars, and puppies." The excitement died down when Twain left Virginia City for San Francisco, or as one Comstock newspaper noted, "*vamosed,* cut stick, absquatulated."[46]

Charitable and church-related activities were an important element of women's public work on the Comstock, but social reform efforts were another acceptable forum for women. Reform was, like religion, closely tied to the perceived moral superiority of women and an arena that they had dominated since the early nineteenth century.[47] The women of the Comstock who attempted to improve the physical and moral quality of life in their communities were fulfilling traditional and approved roles. Among the causes that provided a fertile field of opportunity for betterment in the mining communities was temperance. Faced with male-dominated communities where the saloon was the center of social life, women in the West and on the Comstock attempted to impose their vision of morality, propriety, and sobriety.

Alcohol has always been a part of American life, and efforts to restrain the vice of intemperance date to the eighteenth century. Over the course of the nineteenth century, as society altered and expanded, a growing number of people began to perceive drunkenness as a problem. The customary restraints that community and family exerted over drunken behavior were less apparent and less effective in an urbanized, industrialized society. In the West those controls were believed to be even less manifest in a population dominated by transient, single males.[48]

Temperance activities were widespread in the decades before the Civil War, but in the post-war years the emphasis changed. Rather than trying to reform the individual, temperance groups entered the political arena and focused on changing the law to control the sale and distribution of liquor. Men and women organized to harass saloons out of business and to support local option laws.

The citizens of Nevada and the Comstock were similar to those in other western regions in their efforts and ultimate failures. Since saloons and drinking were so pervasive in mining society, it was difficult to make people on the Comstock take the temperance mission seriously. One Nevada newspaper offered a tongue-in-cheek caution for potential reformers: "Nevada affords a splendid field for missionary enterprise in temperance matters, but a reasonable allowance should be made for those who reside in localities where the waters are more or less alkaline."[49]

There was support for a temperance movement in Nevada, however, and groups were organized early to combat the menace of whiskey and saloons, desiring that "the time [may] soon come when the greatest curse humanity ever knew—intemperance—shall be driven from our land."[50] Some of the citizens of the Comstock shared this aspiration, and by 1863 the Sons and Daughters of Temperance were meeting in Virginia City every Monday evening at 7:30. The Gold Hill group, Pioneer Division No. 1, met at the Odd Fellows Hall on Wednesday evenings.[51]

The temperance organizations toiled to rid their community of alcohol-related vice. It was a daunting task: By 1876 there were 137 retail liquor dealers in Virginia City and Gold Hill, along with 10 wholesalers and 5 breweries.[52] The Sons and Daughters of Temperance persevered, working to change the law. Mary McNair Mathews recalled: "The temperance people worked very hard to get laws passed at the assembly by sending mammoth petitions. Several of the ladies of the lodge generally went around with a petition, each trying to get the most names. I used to take my paper, and go to the post-office, and there take names as the people came in for their mail. I got nearly a thousand names there. I also stood in front of my house, twice a day, for two weeks, and took names. I got fourteen hundred names in all."[53] One temperance bill that she supported proposed to limit the dollar amount of liquor that a dealer could sell to an individual. The credit extended for liquor could consume an entire paycheck, and Mathews disapprovingly noted that the men made it a point to pay those bills first. There was general support for the bill among those approached with a petition, because according to Mathews "it not only benefited the man himself, but his family, and the community at large."[54]

Supporters of temperance attempted to spread the word about both temperance and the accomplishments of women through educational and entertaining lectures. In 1875 the National Guard Hall was the site of a lecture given by Miss Sallie Hart, "the little Temperance heroine." Hart was engaged to speak about the life of Margaret Fuller, labeled by the *Virginia Evening Chronicle* "the most brilliant woman of her time." Hart was described with typical condescension, as "young, pretty and pert . . . [amusing] her audiences with her freshness as much as she instructs them with her historical facts."[55]

The Women's Christian Temperance Union (WTCU), organized in 1873, came West by the late 1870s. By the 1880s the leader of the WTCU, Frances Willard, was promoting a widening of the women's sphere and indicated through her activities that "women intended to move from the private, domestic sphere into the public world."[56] The women on the Comstock anticipated Willard's bold proposal with their foray into political petitioning. The public activity of the women involved in temperance went beyond the traditional participation in fund-raising but was still within the female sphere of moral guidance. Other public activity engaged in by women on the Comstock pushed the limits of the public sphere just a bit further, when some women supported more direct involvement in the political process.

A newspaper notice of an upcoming Orphan's Fair in 1874 was a reminder of one more charitable event among the many, but the newspaper suggested that there were potential political opportunities to be taken advantage of. The fair would be "a fine place at which to electioneer. Although the women have no votes, the majority of them have a 'voice' and often a voice which the head of the house finds wonderfully controlling."[57]

Not all found women's indirect political power satisfactory. Efforts to obtain the vote for women date back to the decades before the Civil War, when they were submerged in another reform campaign, that of the abolition of slavery. With the end of the Civil War the suffrage movement re-emerged. After 1867 eleven states considered the issue of women's votes, but there were only two successes during the nineteenth century.[58]

In 1869 Wyoming Territory became first among states and territories to authorize woman suffrage, garnering recognition as a leader in women's voting rights. There was support for similar legislation in Nevada.[59] During the year in which Wyoming enfranchised women, Representative C. J. Hillyer of Virginia City introduced a constitutional amendment in the Nevada Legislature to remove the word "male" from the voting article. Hillyer argued for the rights of women in the male bastion of the legislature, asserting that women were as intelligent as men and were subject to the same laws and taxes.[60] Hillyer's bill eventually passed in the Senate and the Assembly, but because the change would require a constitutional amendment, procedure required that it be considered again in the next legislative session.

Suffrage was controversial, with passionate adherents arguing both sides of the issue. Attaining the vote would involve women more directly in politics, including them in issues not traditionally within the domestic sphere. Whether supporting or opposing woman suffrage, however, women entered spiritedly and publicly into discussion of the issue.

Anna Fitch was the wife of Nevada's Congressman Thomas Fitch, and a poet and author in her own right. A former Virginia City resident, she was living in Washington, D. C. at the time that Hillyer attempted to gain the vote for women. Anna Fitch was vehemently opposed to enfranchisement for women. Representative Hillyer's vision of politically involved women did not conform to her understanding of a woman's proper role, a role that she publicly explained in a letter to Hillyer that was printed in the *Territorial Enterprise*. Anna Fitch first warned Hillyer that women were unprepared for a role in public life and would need to be educated to "fitness for a place in the councils of the people." With the Civil War just ended, Fitch questioned whether the nation was in the "exact condition to wet-nurse [women] at this moment?" If such political education were possible, Anna Fitch believed that it would be useless because, a "woman's nature is wholly emotional; she does nothing worth the doing except through the channel of the sympathies . . . She idealizes, and sublimates, and, in a word, deifies somebody; that is, if she is a whole woman."[61]

In Fitch's opinion, giving women the vote would not grant them political power; it would simply double male votes. Although she had heard women declare that they would vote against men if they thought they were right, Fitch denied the possibility: "In the end you would generally come to see that his way was right . . . Ah! a woman may hold opinions of her own, but place a bloodless ballot-box on one side and a warm hand and a loving heart on the other, and he is but a poor judge of woman nature who cannot predict the result."[62]

Anna Fitch articulated a perception of women that was common at the time. Many physicians believed that women had more sensitive nervous systems than men did, making them more prone to physical distress and hysteria. The physical differences between men and women were frequently emphasized to explain the need for women to remain within the home and to warn those who advocated education and careers for women. The human body was thought to be a closed energy system, with resources apportioned according to biological determinants based on gender. Thus a woman who developed her brain would find her reproductive organs shriveling, with resulting sterility and facial hair.[63] Clearly, if this were true, training women to function in the masculine world of politics could have disastrous consequences for women and society.

Anna Fitch believed she understood the true nature of women, and she cautioned Representative Hillyer and his supporters that, however well intentioned their efforts, the repercussions could be tragic. The proper place for a woman was the private sphere, in the home and caring for her family, as God intended; her public activities must, by nature's design, be limited to those domestic tasks for which she was suited. As Fitch wrote in her letter: "The true woman's sphere is . . . a charmed circle, radiating light and dispensing happiness, not by actions alone but by the simple fact of its existence . . . there she unburdens man of his wearisome cares and girds him anew for the battle of life. . . . Oh! who would thrust woman from her 'sphere' of which she is the centre and the sun."[64]

Laura de Force Gordon represented another side of the suffrage argument and had a different vision of the proper woman's role. She would have been more than happy to "thrust woman from her 'sphere,'" and she worked toward that end. Gordon had gained fame as a lecturer and journalist first, then later as a champion of spiritualism and women's rights. She came "overland" to Nevada in November 1867 with her husband, Charles H. Gordon, who practiced medicine in Virginia City for two years. Her first public speaking engagements were on spiritualism, and she traveled to other mining communities to present her views. She was well received and, according to the Reese River *Reveille*, her lectures were "well attended and enjoyed; the lady has an agreeable presence and speaks with fluency and grace."[65]

Laura Gordon rather quickly expanded her subject and by 1868 was residing in Virginia City but traveling to lecture in California and Nevada on the topic of suffrage. Her talk in Austin, Nevada, which had been presented before the California Assembly, attracted a "slender" audience. The newspaper account of the lecture noted that in Gordon's "judgement the ballot is the panacea for all the social evils under which woman labors." While she admitted that politics was a corrupt and dirty business, according to the reporter, "it did not seem to occur to her that woman could not enter the 'filthy pool' of politics without soiling her wings."[66]

Gordon spent considerable time in early 1870 helping the newly organized California State Suffrage Association present petitions for women's suffrage to the California State Senate. Nevada was indirectly represented at the California convention by Mary McNair Mathews, who was a guest of a delegate.[67]

Later in 1870 Laura de Force Gordon was back in Nevada to attend the first statewide suffrage convention in Nevada, called by State Senator M. S. Bonnifield and other northern Nevada men, on July 4, 1870, at Battle Mountain. The convention established a state organization with Gordon as president, and the Elko *Independent* was laudatory: "[It was] the most powerful address we have ever heard fall from the lips of a woman—and for convincing logic, spirit, animation, piquancy in recounting an anecdote or painting a characteristic, and for vigor of conception, we have never heard the effort excelled by man."[68]

Laura Gordon embarked on a lecture tour of Nevada, speaking to audiences in Galena, Battle Mountain, Carson City, and Elko. She arrived in Virginia City in late July to lecture at Piper's Opera House. She had been a close friend of Alfred Doten's sister Lizzie, and became friends with journalist Doten and his family during her sojourn on the Comstock. Like so many other people and events, Laura Gordon became a part of the famous journal, Doten's detailed and often titillating record of his life and times in Nevada. Of her suffrage lecture at the opera house, Doten described a "slim" attendance with only a few people willing to pay the one dollar admission; the issue of women voting was obviously not as popular as was communicating with spirits from the other side, a subject she had used on occasion to attract larger audiences.[69] The *Territorial Enterprise* reported: "It is evident, for some reason, that there is a lack of interest. . . . Mrs. Gordon appears to be thoroughly in earnest, and while no one can give any good reason why things should not be settled as she desires, the multitude do not appear to care much whether they do or not—apathy appears to be what she most finds to combat."[70]

An editorial in a Reno newspaper expressed disappointment in the oratorical skills of Laura Gordon, finding that she had "none of the qualities that are deemed requisite to an orator." Gordon had claimed intellectual equality for women, and the newspaper conceded, "We are disposed, not only to grant the claim, but, further, to allow the ladies moral superiority . . . as one of the natural consequences of their present condition of exemption from the contaminating influences of public life and political corruption."[71]

Undeterred, Laura de Force Gordon continued to lecture on women's enfranchisement in August and spoke at the Methodist Church in Virginia City the following month. While staying on the Comstock she organized a woman's suffrage society in a meeting at the Gould and Curry Mine offices. Alfred Doten recorded that he and a friend, Dr. Frederick Hiller and wife, were present with a dozen other men and women. That appears, however, to have been the only meeting of the group.[72]

During the 1871 Legislature, Gordon stayed in Carson City for several weeks to support Hillyer's suffrage amendment, now back for its second consideration. She spoke in the Assembly Chamber on February 1, 1871, when it was reported that: "The floor, the lobby, the gallery and even the recesses of the windows were crowded almost to suffocation with the beauty, wit, and the intelligence of Carson . . . [Gordon said] women were the vitalizing power of the country. . . . Wanted the question of female suffrage in this State submitted to the people, and appealed to the Legislature to sustain the proposed amendment striking the word 'male' from the Constitution of Nevada."[73]

The resolution failed, nineteen to twenty-five, and a motion to reconsider the following day failed by one vote. The question was voted upon without a call of the house, when several members friendly to the amendment were absent.[74] Gordon's final effort to garner support in Nevada for woman suffrage occurred later that year, when she brought Susan B. Anthony to Carson City on December 21 and to Virginia City on December 22, 1871. Alfred Doten remembered Anthony's talk on "The Power of the Ballot" as: "[the] most able lecture ever given in this section. It was totally unanswerable, even by the most obstinately bigoted speaker, writer, or satirist especially if too illiberal to attend and learn a little good, sound practical sense, on the matter of who have really the best right to the ballot, and would be most benefitted by it."[75] Able though Anthony might have been, her efforts had little impact. It would be more than four decades before constitutional change granted women political rights through the ballot box.

In her analysis of frontier women, Julie Roy Jeffries stresses the point that "freedom" and the liberating environment for women in the West has been overemphasized. The two territories that enfranchised women early, Wyoming and Utah, were acting conservatively, allowing the female vote because it fit their political agenda, not because they considered women equal to men. Woman suffrage in those circumstances fit within the domestic ideal. Anna Fitch might have disapproved of the vote in Wyoming and Utah, but she would have recognized and concurred with those states' justification for women voting. Jeffries also concluded that women were generally uninterested in joining the political process. This was certainly the case in Virginia City as is illustrated by the lukewarm interest in Laura Gordon's suffrage talks.[76]

The public activities of women on the Comstock bolster Jeffries' argument that the western experience was not necessarily a liberating one for women. The region's women's public sphere was similar to that of middle-

and upper-class women in other parts of the country. Those who traveled to Virginia City found themselves comfortable with the elegant rooms of the hotels and familiar with the accouterments of fine dining. They recognized the actors and plays of an evening's entertainment. Travelers to the Comstock, in short, found a niche that was comparable to the one they had left behind in the east or further to the west in San Francisco. Like other visitors who passed through, these women brought their baggage with them in the form of social and cultural norms and expectations with which they were familiar and comfortable. Whether working with charitable concerns, attempting to reform the undesirable behavior of those who haunted saloons or tippled secretly, or working for or against woman suffrage, the public sphere was proper and acceptable for the middle- and upper-class women who inhabited it. The women who lived in the mining communities of the Comstock both created and encountered "the advantages of ladies' society."

PART FOUR

❧ Ethnicities ❧

The international character of Virginia City impressed nearly all who came there to visit or stay. Eliot Lord discussed the range of ethnicities among miners in his 1883 publication, *Comstock Mines and Miners*. Other authors have followed in his footsteps, but a thorough treatment of diversity among women in the district has yet to appear. Although the following chapters begin the process of dealing with this important topic, they cannot claim to be comprehensive. This is largely because of their brevity and because sources remain elusive. Spanish-speaking women were some of the first to establish families in the mining district, but they never numbered more than a few dozen, and their relative importance declined after the early 1860s. Information about them is difficult to come by. There were even fewer African American women in most of Comstock history, and like their counterparts from Mexico and South America, little survives in the written record to document their lives. These and other groups need to be investigated in order to attain a complete understanding of ethnicity and immigration, but teasing insight from scarce sources will remain the task of future researchers. Perhaps the archaeological model proposed by Donald

Hardesty in chapter 12 will help increase the presently scant amount of evidence.

The largest ethnic group among Virginia City women looked to Ireland as home. To understand the district's Irish population is to go a long way toward placing women and immigration in perspective. The extent to which the Irish experience held true for other ethnic groups, however, is unclear. In contrast to the numerous Irish, Northern Paiute women were rare in the mining district. Documentation of their lives is equally rare, and ultimately it may be impossible to learn much about these women—the first females in the area. There were also few Asian women on the Comstock, either in Chinatown or in the community as a whole. More remains, however, to document their existence. Like the American Indians, the Chinese were an important component in the Comstock ethnic kaleidoscope.

While they cannot claim to represent the full range of Comstock ethnicities, the following essays can, at least, amply demonstrate the fact that ethnicity shaped both collective and individual experience for women of the district.

❧ 10 ❧

Their Changing World

Chinese Women on the Comstock, 1860–1910

SUE FAWN CHUNG

Introduction

Unbeknownst to the larger community, the Chinese women of the Comstock changed their lifestyles and values greatly between 1860 and 1910. The small size of the Chinese female population on the Comstock allows for a more detailed study of these transformations than would be possible in larger urban areas. An understanding of the psyche and experiences of the Comstock Chinese women requires a basic understanding of the values, role, and position of women in traditional China that accompanied them across the Pacific Ocean. As in many other traditional societies, Chinese women were subordinate to men. Like the men, they began to adapt to their new surroundings and to adopt some American ideas and practices due to the absence of a continual stream of new immigrants who would have normally reinforced traditional values; to the lack of the traditional family support system; and to the necessity, brought about by economic and social factors, of interacting with the Euro-American community in a frontier setting. This study examines the reasons the women came to the area, who some of them were, the differences between married women and prostitutes, and the situations the women encountered.[1]

Early Chinese Female Immigrants, 1860–1880

Chinese women emigrated to the United States in one of four ways: with their husbands, as *muzai* (literally meaning "little sister" but usually referring to domestic or bond servants, or prostitutes), with relatives, or as prostitutes. American-born Chinese girls were so few in number in the late nineteenth century that they had little impact upon the Chinese community until the turn of the century.

Most Chinese immigrants left their wives at home. This usually happened for one of several reasons: so the wife could fulfill her Confucian filial duty of caring for aging parents; to insure the return of the prodigal son;[2] or to obey the Chinese laws banning the emigration of women.[3] The expense of transportation and accommodations, and the reluctance of Chinese men to bring women and children into a hostile environment, also inhibited the emigration of Chinese women.

A few men brought their wives. Some of these left the women in safer Chinatown communities, while they worked in towns with a questionable reputation, such as the lawless Virginia City in the 1860s.[4] According to the 1860 census manuscript for Carson County, Utah Territory, there were no Chinese women and only 23 Chinese men in the area at that time. Of the men, 16 were laundrymen, 2 were cooks, 3 were laborers, 1 was a barkeeper, and 1 was of unknown occupation. All but the last were much needed service workers for a male-dominated frontier mining community.[5] By 1870 there were 103 Chinese females (among a total of 710 Chinese in Gold Hill and Virginia City), with an average age of 23.4 years: 2 "keeping house," 1 child, 1 mother living with her son, 1 (Mary Tresa, 40) of unknown occupation living with her brother, 4 of unknown status, and 94 prostitutes or "harlots" (90 percent of the female population). A special 1875 state census lists 104 Chinese women, 3 married and 101 of "unknown marital status," in Storey County.[6]

In the early 1870s the three prominent married women in the Chinese community on the Comstock were Mrs. Sing Tong Loo (noted as Loosingtong), age thirty-three, and Mrs. Hop Lock, age twenty-eight, the only women listed as "keeping house" in the census. They were the wives of two of the four Chinese physicians. The third was Mrs. Lee King, who was not listed in the 1870 census. Like the other two Chinese physicians, Drs. Loo and Lock had a declared worth of $1,000 and therefore were among the wealthiest in the Chinese community.[7]

TABLE 10.1   Chinese Population in Nevada, 1860–1900

Dates	Nevada	Storey County	Women in Storey County	Men in Storey County	Total U.S.
1860	23	0	0	23	34,933
1870	3,152	749	103*	647*	63,199
1875	—	1,362	104	1,258	—
1880	5,416	642	42*	577*	105,465
1890**	2,833	245	—	—	107,488
1900	1,352	76	3*	73	89,863

— means no data available.
*Based on census manuscript data by race and birthplace; figures differ from those in the U.S. census summary, which were usually lower, and totals do not agree since sources differ.
**Based on the U.S. census report; the eleventh U.S. Manuscript Census was lost in a fire.
Sources: Eighth, ninth, tenth, and twelfth U.S. manuscript censuses of 1860, 1870, 1880, and 1900, respectively; the 1875 Nevada census of Storey County, and the eleventh U.S. census report of 1890.

Due to the shortage of medical care available, Chinese physicians who spoke some English probably did well financially and usually catered to both the Chinese and the Euro-American population.[8] In communities throughout Nevada, physicians were most likely to bring their wives with them. Doctors were not regarded as part of the scholarly class in China, but they fell into the third of the four Confucian social classes, that is, artisans or skilled workers who contributed to the smooth functioning of society. Their social status and political power propelled them to the top echelon in the United States in accordance with American values, and their wives benefited from this new status. Apparently various Chinese organizations, or *tangs,* recognized the need to provide medical care for their members and sponsored Chinese physicians and, occasionally, their wives, in frontier communities like the Comstock. Dr. Loo and Dr. Lock probably represented rival *tangs,* which hired them to care for their respective members.[9] Two other Chinese physicians also practiced in Virginia City in 1870: Dr. Wing Sing, age thirty, and Dr. Ah Lang, age fifty. No other information is available on either physician. In the mid-1860s there was another Chinese physician on the Comstock. His wife and son were photographed,[10] but the family moved away by 1870.

Dr. Hop Lock, a well-known Chinese physician in Virginia City whose newspaper advertisements included testimonials of satisfied Euro-American patients, settled in Virginia City as early as 1864. As he became more prosperous, he moved his office and by 1871 was working at 130 C Street between the offices of two other Chinese physicians. Although Mrs. Lock is never mentioned in the local newspapers, her husband gained fame in February 1871. Dr. Lock was arrested for stabbing Lee King, who eventually died of his wounds, at the latter's home. According to the local newspapers, Lee King, who was well respected by the Euro-American community and liked by the local police, lived with his wife. The local press identify Lee King as the operator of a Chinatown gambling and lottery shop, but neither he nor his wife is listed in the 1870 census for Virginia City. The press also identifies another woman living with the Kings. This woman was probably a concubine or a *muzai,* because in China, "even a man of quite modest financial means might take one or two concubines who would help his wife with the housework."[11]

The complicated criminal case involved the robbery that took place several days earlier on February 7, 1871, of Mrs. King's jewelry, estimated at a value of over $300; silks and fine clothing, estimated at over $200; $1,300 in gold and silver coins; and $200 worth of opium. The value of the missing goods indicated that the Kings were people of some wealth. Dr. Lock and an accomplice, Chung Chow, stabbed Mr. King, who subsequently died of his wounds. According to the newspaper account, King's body was immediately shipped to San Francisco and then to China while Mrs. King, as well as Mr. King's brothers, remained in Virginia City. By examining arrival and departure notices elsewhere in Nevada, one can see that Chinese women did not travel alone at that time, so it is probable that Mrs. King could not leave Virginia City until one of her brothers-in-law was able to accompany her. In April 1871 the jury brought in a verdict of "not guilty" against Dr. Lock. The local newspaper indicated that this was the result of the fact that Dr. Lock had powerful friends in the Euro-American community and suggested that the blame for the fatal wound could be ascribed to the missing Chung Chow. Dr. Lock resumed his practice for a short time. Meanwhile, in 1873, Dr. Hop Lock's brother, also a physician using the same Americanized name as his brother's, moved into Virginia City to try to establish a practice. Apparently he was not as successful as his brother was, because the 1878 *Directory of Chinese Business Houses* published by Wells Fargo and Company lists Gin Hin as the only Chinatown physician.[12] By 1878 Mrs. Hop Lock and the two brothers had left Virginia City.

The other married Chinese physician listed in the 1870 census is Dr. Sing Tong Loo (or Loosingtong), age forty-eight. The household also in-

Wives of doctors and merchants often misidentified as prostitutes. (Courtesy of Special Collections, Getchell Library, University of Nevada, Reno)

cluded his wife, age thirty, and Lee Susa, age twenty-two and a "harlot," according to the census. As in the King family, Lee Susa was probably a concubine, or *muzai,* not a woman who sold sexual favors to other men. In a community with such a shortage of women, the doctor's wife probably kept a close watch upon the young woman. Their stay in the community did not extend to the 1880 census.

For the Chinese woman who had achieved some measure of wealth and independence in Nevada, the experience of returning to China could be unsettling. Madam Foo of Winnemucca had lived in Virginia City but always dreamed of returning to China. In 1873 she went home, only to discover that too little had changed in her village and that she was unable to resume a traditional Chinese female role.[13] Two years later she returned to

Winnemucca and transformed the Chinese community to meet her needs by organizing the more prosperous Chinese to build a new Chinatown with larger homes and stores for themselves on Baud Street. Her return to China made her realize that she was "too Americanized."

There may have been more than two or three married Chinese women on the Comstock. Women accepted the fate of having only one mate although their husbands could have relationships with more women, including several wives, concubines, and other females. A closer examination of the census data suggests that there were secondary wives or concubines whom the Euro-American census takers viewed as prostitutes. Of the 103 Chinese females among the 710 Chinese in Gold Hill and Virginia City, approximately 26 couples,[14] or one-fourth of the female population, may have been married. This calculation is based on the fact that a male and female were living in the same household with the same last name or some possible related name and had indicated "married" to the census taker. The listing of "harlot" or "prostitute" as an occupation was disregarded if there was the possibility that the woman was a secondary wife or concubine from the Chinese perspective.[15] Concubines were not responsible for the care of the ancestors, were not endowed with property, provided for the continuation of the male line (especially important when the wives gave birth to only girls or older male children died), and provided sexual gratification for the "husband" and for no other men. Since the marital status of most men is marked "unknown" in the census, it is possible that even more than one-fourth of the Chinese women on the Comstock at the time were married. A single woman living in a household with more than one male was not necessarily a prostitute, since kinsmen tended to live together under the care of a husband and wife team. The presence of these families support the idea, which scholars have recently proposed, that the Chinese in Virginia City were not simply sojourners but had made an effort, that was thwarted by anti-Chinese legislation, to establish an immigrant community.[16]

Seven wealthy Comstock Chinese married men (reported worth in parentheses) appear in the 1870 census. The married women living with them are identified as prostitutes. The couples are Chung Ching, thirty-four, a drugstore owner ($3,000) with M. Ching, twenty-four; Chung Loo, twenty-three, a merchant ($2,000) with Mrs. Loo, eighteen; Ah Mong, twenty-eight, a laundryman ($500) with Mrs. Mong, twenty-three; Ah Tuck/Fuck, thirty-nine, a laborer (probably a labor contractor, $1,000) with M. Tuck, twenty-three; Hop Gee, twenty-two, a shoemaker ($500) with Mary

Gee, nineteen; Ah Fong, thirty-four, a laundryman ($1,000) with Gee Fong, twenty; and Young Chung, forty, a gambler ($2,000) with Y. Chung, thirty. Because these couples lived in residences by themselves, it would not be surprising if these women were actually concubines or secondary wives who were mistaken by the census taker for prostitutes. Four other couples who reported no wealth but indicated that they were married and living as couples or with kinsmen could be added to this list: Fong Ching, thirty-three, gambler, with Su Ching, thirty two; Ah Sung, thirty-five, laborer, with M. Sung; Young Chung, forty, gambler, with Y. Chung, thirty; and Ah Tong, thirty, gambler, with M. Tong, twenty. All the women on the list are designated as prostitutes or harlots.

Several other Chinese whose marital status was "unknown" could be added to this list, but some of these would be more problematical: Si Gin, forty, cook, with M. Gin, twenty; Ah Kee, twenty-eight, gambler, with M. Kee, fifteen; Quang Ping, thirty, merchant, with Ah Ping, nineteen; Yee Yuf, thirty-three, gambler, with Cooly Mee, nineteen; Ki He, twenty-eight, laborer, with Mary Choy, twenty; Sam Wing, twenty-eight, gambler, with Ti Wan, twenty; Ah Boe, thirty, gambler, with Susan Rosa, twenty-four; Quang Sing, thirty-three, gambler with Hung Uen, twenty-two; Ah Chou, twenty-eight, gambler, with Ming Hee, twenty; and Cum Tang, forty-one, wood packer, with Ping Tang, twenty-eight. Some of these last names differ from those of the males with whom they resided, but this does not necessarily indicate that they were not married. In parts of South China, women often kept their maiden names and indicated their married status by the designation *shi* (*she* in Cantonese).

By subtracting the twenty-three women who were probably married and living with their spouses from the total number of women designated as prostitutes, the number of Chinese women involved in sexual commerce drops dramatically to seventy-one (70 percent of the former total). Assuming that there were married women who were missed by the census taker, the ratio of prostitutes to married women might be even lower. This would be significant, because it would indicate that there was an attempt to establish a "stable" rather than a transient community at an early date.

Another key to studying married Chinese women on the Comstock is to examine the number of children under the age of thirteen noted in the 1870 census. Altogether there are ten children listed, nine boys and one girl. One lived with a nuclear family, one with a family of possible relatives, two with single mothers, one with a single father, and three with related males.

The living situation of the remaining two is unknown. Mary Ching lived with her eleven-year-old stepson, Ching Ching, in a Gold Hill boarding-house. Ten-year-old Charlie worked as a waiter with Sam Sip, thirty-three, a cook. Charles Ching, twelve, lived with Ah Sam, forty, a laborer. Ah Sing, thirty-one, a peddler, lived with Ah Chue, a six-year-old, California-born male. Wing Sing, thirty-four, a laundryman, lived with M. Sing, twenty-five, designated as a prostitute (but more likely a concubine), and their daughter, S. Sing, five, who was born in Nevada. Ah Moi, twenty-nine, a prostitute, lived with her eleven-year-old son, Hop Moi, and a group of other prostitutes. Ah Tong, thirty, a gambler, was married to M. Tong, twenty, designated as a prostitute (but probably a concubine). Four-year-old Wong Yon, who was born in California and was probably their son, lived with them. Ah Moey, thirty-three, a laborer, lived with his son, Lee Moey, age three, who was born in Nevada. No wife is listed, but she might have been away from home on the day the census was taken. Hank Qua, five, male, born in Nevada, lived with wood packer Ti Loy, thirty-eight, and cook, Ah Sam, thirty, who were probably his relatives. Ah Choy, four, male, born in China, appears with no obvious parents. Of the ten children, three were born in Nevada, three in California, and four in China. The booms and busts of the local economy usually determined how long these children and their parents or guardians stayed in Nevada.

Unlike other parts of Nevada where there were female Chinese seam-stresses, miners, hairdressers, and other types of workers, the employed Chinese women of the Comstock were either laundry workers or prosti-tutes. Since women were an integral part of the textile labor force in China, it is not surprising that they also entered into the laundry business in fron-tier America.[17] The Reno *Crescent* (February 13, 1869) notes that the wife of cook Ah Tom took in washing. Chinese wives elsewhere also did this. The 1870 census manuscript indicates that three young women, ages twenty-five, sixteen, and eighteen, living together in Gold Hill, operated a laundry by themselves. Undoubtedly they had followed a relative to Nevada. One woman, Meary Sung, age eighteen, worked in a laundry in Virginia City with men who were probably relatives in a predominantly Euro-American section of town. None of these women were as prosperous as were Ah Fong and Sam Kee (1844), the wealthiest Virginia City Chinese laundrymen, or Ah Pow, Ah Mong, and Ah Bing. Chee Luck Chung immigrated to the United States in 1874 at age ten, successfully opened a laundry in Virginia City shortly thereafter, and saved enough money to own a laundry and house by 1910. Laundry work was labor intensive but became a major occupation for both Chinese men and women.

Finally, most of the early female Chinese immigrants came to the Comstock as prostitutes. There is no question that prostitution flourished in the late nineteenth-century American West. The gender imbalance was partially ameliorated by the importation of Chinese women for prostitution.[18] As in traditional China, Chinese prostitutes were obtained for brothels by being purchased from relatives, by kidnapping, by business contract between the company and the women, and by "free" choice (women who had no means of survival except to live by prostitution). These women were often under the control of a Chinese organization (*tang*) or pimp.[19] Prostitutes regarded their employment position as temporary, because in Chinese society all women were expected to marry.[20] In reality, many women found that they were tied to the profession as other debts were incurred or situations arose trapping them in prostitution.

There were several classes of prostitutes in the United States. The elite were skilled entertainers rather than providers of sexual services and in the western world were called "sing song girls." The "long three," who were named after a domino with two groups of three dots, each charged three *yuan* (Chinese dollar) for drinking and three more for spending the night and provided entertainment and sexual favors for a select group. The "one-two," who were also named after a domino, charged one *yuan* for fruit and snacks and two for drinking and were less skillful entertainers. At the bottom were the "salt-pork shops," which provided service on demand; "pheasants," or streetwalkers, who might or might not be connected with a brothel; and brothel employees, who could be subdivided into two categories: those connected with opium smoking and those who catered to members of the lower class, such as common laborers.[21]

In the American West, this hierarchical system translated into those who entertained, those who served the elite of the Chinese community exclusively, those who were available on demand for Chinese customers only, and the lowest class, which serviced both Chinese laborers and non-Chinese clients. Prominent Virginia City residents, including Dan De Quille and Alfred Doten, both influential newspapermen who often "cruised" Chinatown, knew only the lowest class of prostitutes. According to Doten: "I went with Sam Glessner down to Chinatown—drank at Tom Poo's—went to Mary's house—we were in her room with her—she gave us each a cake left over from the holiday of yesterday—filled with nuts & sweetmeats—we laid on the bed with her & smoked opium with her—a little boy some 2 years old sleeping there, belonging to one of her women—long and interesting chat with her."[22] This passage is of particular interest because a bordello owner whose name appears only as "Mary" had adopted the practice of including refreshments that were characteristic of the "one-two" higher-class prostitutes, but she was involved with the opium trade that usually characterized the bottom of the categories of Chinese prostitutes.[23] She operated her bordello next to a popular Chinese saloon in Virginia City that served all racial groups and had close relations with Tom Poo, a relative or friend, who often served as the interpreter in transactions between the Chinese and the larger community in Virginia City.[24] China Mary, as she was known, adapted to her new situation by combining traditional and innovative practices and perhaps, like Ah Toy (Atoy) of San Francisco a decade earlier,[25] strove to attain wealth and status through her different approach.

Prostitutes had a very hard life but probably were able to accumulate money. Based on Lucie Cheng Hirata's estimates, low-status Chinese prostitutes in North America generally charged thirty-eight cents per customer, served an average of seven men a day, worked an average of 320 days per year, grossed $850 a year, and paid out an average annual maintenance fee of $96, thus netting $754 annually.[26] Using these calculations it appears that a typical prostitute could earn more than twice the daily salary of the average Chinese railroad worker or miner.[27] Chinese prostitutes maintained their traditional dress, but those of the upper class dressed in silks and jewels while the lower-class ones wore black cotton and checked aprons with red or blue kerchiefs over their heads.[28] As in China, prostitutes could be released from service for "a fair market price," and there were several cases of men in Nevada who tried to buy or successfully purchased a prostitute for a wife.

Life was especially hard for the low-class Chinese prostitutes. They lived in segregated neighborhoods and had few, if any, relations with other non-Chinese prostitutes or Chinese women.[29] Of the nineteen prostitutes on the Comstock who committed suicide between 1863 and 1880, six (32 percent) were Chinese.[30] Thus, for example, Ah Gone, a twenty-three-year-old prostitute who lived in "a little Shanty opposite the Golden Eagle Hotel" in Lower Gold Hill, was found dead of an apparent overdose of opium in 1871.[31] In 1874 Guan Yook, a Eurasian prostitute, age twenty, and worth $750 to her owner, killed herself by swallowing morphine.[32] Guan

Yook, the offspring of a mixed marriage in a racist society, had few career and marriage options. Death was the only viable escape from this harsh lifestyle.

The 1870 census indicates that several prostitutes lived with their husbands (for example, laundryman Wing Sing, Mrs. Sing, and their five-year-old daughter) or a possible male relative (for example, merchant Bock Sing and several prostitutes with the same surname) in Virginia City. The shortage of women allowed some men to prosper from the labors of female spouses and relatives.

Most prostitutes probably lived in circumstances similar to those of their counterparts in China. In these cases, the women were viewed "more like things than people."[33] As their numbers declined due to the Page Law and the Chinese Exclusion Acts, their value increased. By the 1920s a teen-age prostitute could be worth as much as $6,000 to $10,000 in gold in San Francisco.[34]

The image of Chinese Nevada women as prostitutes was frequently publicized for both political and racial reasons. Prostitutes, by nature of their profession, probably would not have been missed by the census taker. U.S. census figures in 1870 supported the movement for the passage of the 1875 Page Law by listing almost all Chinese women on the Comstock as prostitutes. During her stay in Virginia City in the 1870s, the strident anti-Chinese writer Mary McNair Mathews often complained that "no respectable Chinese women ever came to America."[35] Once the Page Law passed, the stereotype of all women as Chinese prostitutes continued despite the fact that the 1880 census takers, who did not have the same political agenda as those a decade earlier, did not find many Chinese prostitutes.[36] On July 21, 1882, the Carson City *Morning Appeal* reinforced this stereotype with the following statement: "There are not a dozen respectable China-women on the coast, and the few that are here are never seen in the streets. In Carson there is not one of the latter class. The Chinese females here are all of the very lowest social order . . . sent . . . for immoral purposes." The stereotype that all Chinese women were prostitutes would not change until more second-generation Chinese American women reached adulthood.

Because of the shortage of women on the Comstock, there were several notable incidents of attempted or successful kidnappings of Chinese women in Virginia City. One of the earliest reported cases involved Ah Hong (also known as George Hong), who married Cow Kum in Judge Harris' courtroom in Downieville, California, on May 2, 1864.[37] Shortly thereafter the couple went to Marysville, California, where Cow Kum was kidnapped by Ah Wong and a Euro-American accomplice. Ah Hong settled in Virginia City and later learned from a close friend that his wife was kept in a brothel run by Ah Wong in Virginia City. Using the American judicial system, Ah Hong reclaimed his wife, and the court ordered Ah Wong to put up a bond of $1,000 for six months to ensure that he would not trouble Ah Hong and his wife. Not all kidnappers were treated with such lenience. Ah You and Ah Bau of Virginia City were arrested on September 18, 1875, for the attempted kidnap of Ah Toi (also known as Lan Rey), a Chinese woman from Reno.[38] Ah You and Ah Bau were tried and convicted and were then pardoned on July 10, 1876.

The kidnapping of Chinese prostitutes fascinated the Comstock reporters. Under the colorful headline, "A Mongolian Helen: The Tartar Trojans of Virginia City Threatened with Invasion by Pig-Tail Greeks of Carson," the July 13, 1876, edition of the Virginia City *Evening Chronicle* told the story of a "Celestial beauty from Carson [City]" who was kidnapped by Hung Wah of Virginia City, of the retaliation by abductor Ah Sin, and of the vow by Hung Wah to recapture his "loved one." On June 8, 1878, the same newspaper reported the abduction of a woman, "not yet 18," who was brought from China in 1877 for the purpose of prostitution and who subsequently fell in love with Ah Kim. Ah Kim tried to purchase her freedom from an agent of the Sam Sing Company, but he could not raise the $300 required. Thereupon the agent drugged the girl, put her in a wooden box, and tried to ship her on the Virginia and Truckee Railroad to Auburn, California. However, an alert station dispatcher discovered the boxed woman, freed her, and turned her over to Ah Kim, who had gone to the station in search of his loved one. The kidnapping of Chinese females as young as ten years old[39] by rival companies or individuals remained a danger up to the early twentieth century.

Anti-Chinese sentiment augmented the dangers for Chinese living in the American West. As early as 1863 disgruntled Euro-Americans, often unhappy over economic competition or expressing ideas of racial superiority, attacked Chinatown.[40] If anti-Chinese sentiment became too violent or if the town's economy "busted," the Chinese moved to a new location. It was more difficult for a family to move than it was for an individual to do so. Solitary Chinese men, whether married or unmarried, far outnumbered those with women in the American West.

Although Chinese women were in many ways the preservers of culture, changes occurred as they adapted to the Comstock community. This was evident in the education of children and in wedding ceremonies. In the American West, the strong bond that existed between mother and son in traditional China[41] was extended to all children, because in this highly mobile society, the one enduring factor was the love for and loyalty towards each other a mother and children had. In oral interviews with Chinese-American children raised in Nevada, mothers emerged as the most important, selfless, and hardworking figures in their memories.[42] Unfortunately, because of the fragile health, and even death, of the women, especially during the difficult period of infant rearing, many young Chinese children were left without their mothers. Fathers, uncles, or other kinsmen had to assume child-rearing duties.

Chinese parents also faced the difficulty of educating their children. Nevada's 1865 school law stated that Chinese children would not be admitted into the public schools, and this was reiterated in an 1867 law.[43] In 1872 the Nevada Supreme Court declared the laws unconstitutional, so children of all races had the same legal right to attend public schools.[44] However, many Chinese parents were reluctant to send their children to school because of racial discrimination and because the children had to help support the family by working. If they aspired to send the child back to China to find a bride or employment, or had sufficient finds to protect the child from teasing and bullying at the public school, they hired an American tutor or Chinese teacher. If this was not a viable option, the parents sent the children to public school, a practice that became more prevalent throughout Nevada around the turn of the century and that was a vehicle for the greater acculturation of the more traditional or isolated mothers.[45]

One of the gravest concerns for first- and most of the second-generation parents centered around the process of acculturation and assimilation. A balance between the preservation of traditional culture and practices and the evolutionary, often inevitable, process of acculturation and assimilation was difficult to achieve. Parents and children often found widely different solutions. A story in the Carson City *Morning Appeal* illustrated how the issues of acculturation were becoming prevalent. A Chinese-American father, born in Fiddletown, California, in 1855 and his wife, born in Yreka, California, in 1861, had a son born in Carson City in 1879. According to the reporter, the father, "who speaks very good English, stated that up to his fifteenth year his principal associates had been white boys from whom he acquired a taste for the life of the superior race," but he knew that he really would never be on equal footing with them and yet found the habits of his own people "repugnant to him."[46] The father hoped for a better future for his three-year-old son but was uncertain whether he would find one in Nevada. Other families, especially those with mothers raised in a more traditional Chinese setting, were concerned about the preservation of their cultural heritage in an American environment.

Marriages, marriage licenses, and newspaper stories reveal another aspect of life for the Chinese women on the Comstock. Between 1874 and 1882 twelve Chinese men took out marriage licenses in Virginia City. Only two of the twenty-four individual names appear in the 1870 or 1880 census.[47] Details of six of the marriages published in newspaper accounts suggest the changing values of the Comstock's Chinese community. The act of taking out a marriage license indicated that the Chinese men wanted the marriage to be recognized by the American community. On May 20, 1875, Ah Too took out a license to marry Wan Ho, age twenty-five, both of whom resided in Virginia City. They were married ten minutes after obtaining their license by Judge Knox.[48] The *Territorial Enterprise* discovered the interesting romance and published a long article about it. Ah Too was actually Chi Chow, who had fallen in love with Ky Slung in China but was forced to seek wealth in the American West before he could win her hand. Meanwhile, Ky Slung's family sold her to Chung Wan, who eventually brought her to Virginia City. Chung Wan had to flee from Pioche, Nevada because of a debt owed to Hi, "the pork man." By accident, Ky Slung saw Chi Chow, and this renewed their determination to get married. In order that they not be separated, they decided to change their names and marry in accordance with American law. This type of romance, beginning in China and finalized in the United States in the face of great odds, was unusual but not unheard of, because Chinese immigrants generally moved to places where their relatives, fellow villagers, or friends had already settled.

Another marriage demonstrated the possible dangers faced by Chinese living in the American West. On July 12, 1876, Shu Quong, also known as Ah Quong, took out a license to marry Toy Guem (Toy Cum), age eighteen. Both were from Virginia City. A minister performed the ceremony.[49] A year later, while traveling to San Francisco because he had heard that members of the rival *tang*, the Sam Sing Company,[50] intended to kidnap his new bride, Shu Quong, a member of the Hop Sing Company, and his wife stopped in Truckee. There Ah Sam, a Sam Sing member in Truckee, asked the court to issue a warrant for Mrs. Quong's arrest for grand larceny.

Shortly thereafter Mrs. Quong was arrested. While her husband sought help from some Euro-American friends, Ah Sam engineered her release and took her away. Mr. J. A. Stephens and Mr. E. Wood, representing Mr. Quong and the Hop Sing Company, asked the court to issue a warrant for Ah Sam's arrest for false imprisonment, but neither Ah Sam nor Mrs. Quong could be found. The incident added fuel to growing hostility between the two Chinese associations or *tangs* but did not result in a tong war.[51]

By the late 1870s Chinese weddings had become even more American-ized. In 1878 one couple decided on a ceremony combining what the re-porter considered to be "Chinese and American traditions."[52] In January 1878 Ah Wan took out a license to marry Nan Ying, age eighteen. Both lived in Virginia City. The ceremony took place at the home of Mr. Lan-nan, with Father McGrath and a Chinese interpreter presiding. If the wedding had been a true Chinese one, both sets of parents, not to mention grandparents and other kinsmen, should have been present. Instead, Euro-American men and women joined the Chinese guests in the celebration, during which the joyful Ah Wan revealed that he had paid $400 for his bride. In China, grooms generally sent a gift to the parents of the bride in appreciation of the expenses they had incurred while raising her, but at the wedding, Ah Wan learned that if he had married in the "100 percent American fashion," he would not have had to pay any amount of money. The marriage ceremony is important because it indicates that a major step had been taken in the acculturation/integration process of the Chinese in the Comstock.[53] The traditions of China were no longer important, nor were they followed.

Another wedding in August 1878 exemplified the practical and expe-dient approach Chinese on the Comstock were beginning to take to mar-riage. Chung Wing Cheung, also known as Ah Wing Chung, of Virginia City took out a license to marry Vek Ye, age twenty-two, also of Virginia City. According to the August 13, 1878, edition of the local *Evening Chron-icle*, Vek Ye was afraid of being kidnapped and turned to the police, who placed her in jail for her own protection. Several Chinese men tried to claim her, but she instructed the jailer to admit only Ah Wing Chung, who read and wrote English "very fairly." When he appeared, he asked for a justice to marry them. Justice Moses, accompanied by Chief of Police McCourt and Postmaster Adkison, arrived and instructed the postmaster to act as "bridesmaid," the chief as "father of the bride," and the jailer as "best man" in the jailhouse marriage ceremony. It was suggested that the justice kiss the bride in accordance with "American custom," and the groom indicated that it was agreeable to him, but the judge declined to do it. The thought of kissing a person of color was too repugnant to him. "Amid the mirth of the spectators" the group adjourned to the jail's reception room for "a huge bottle of Jack Bradley's good whisky, which was drained to the dregs." This kind of wedding ceremony would have been totally unacceptable in China, but in the American West, adjustments were necessary.

The June 23, 1881, edition of the Virginia City *Evening Chronicle* noted the marriage of another Chinese couple, Sam Soon of Virginia City and Towy Came, age twenty-five, of Carson City, at the Virginia City Methodist Church. Traditional Chinese weddings were usually held in the groom's home, but Chinese Nevadans married first in the American court, then in an American home, and finally in an American church. Traditional Chinese regard the wedding ceremony as one of the most important and festive events in their lives, and community acknowledgment of the union is extremely important. With this Methodist wedding, Chinese Nevadans had moved further in the direction of assimilation by adopting American religious standards. By accepting these changes, the Chinese women of the Comstock moved further away from traditional Chinese customs.

Chinese Women of the Comstock, 1880–1900

By 1880 the character of the Chinese population on the Comstock had changed. The 1880 census manuscript offers the following profile of the Chinese population: a total of 619 people with an average age of 30.8; 577 males, average age 31.1; and 42 females, average age 26.7. This shows that both Chinese men and Chinese women in the area were slightly older and fewer in number than they had been in the previous decade, when the boom was on. Degree of wealth is not indicated, as it had been earlier. The census enumerators appear to have more accurately listed the marital status of Chinese residents than they had in 1870. The types of occupations were more varied than they had been previously, suggesting that the community was more developed than it had been before. In addition to the cooks, laundrymen, laborers, servants, physicians, gamblers, wood cutters and packers, merchants, peddlers, clerks, barmen, saloon workers, waiters and dishwashers, restaurant owners, drugstore owner, and prostitutes of 1870,

A man caning a chair in Virginia City's Chinatown in the 1870s.
(Courtesy of Nevada Historical Society)

the 1880 census includes a tea merchant, joss house keepers, teachers, rag pickers, barbers, chair repairmen, vegetable store owners, a phrenologist (druggist), butchers, opium den owners, and a banker. Some of the 1870 occupations, such as those of porter, cigar maker, shoemaker, carpenter, and tailor, do not appear in 1880.

By 1880 more family units had developed among the Chinese community, a trend also seen in California.[54] According to the census, eighteen Chinese women were "keeping house." Two married women whose occupations are not listed and who were living with two married men in separate households brought the total number of married couples to twenty, and this in an overall Chinese population that had declined. The two most prominent women were the wives of two of the three doctors, Mrs. Choney Wing, twenty-six, wife of Dr. Song Wing, forty-eight, living at 33 Taylor Street, and Pooty Chin, forty, wife of Dr. Song Haong, forty-five, living around the corner at 22 South C Street. Unlike the two wives of physicians listed in the 1870 census, these women lived near each other, and

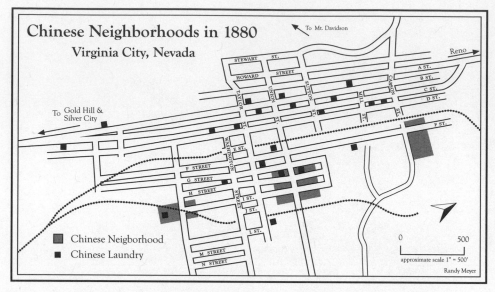

Chinese neighborhoods and laundries in 1880, Virginia City. (Base map from *Report upon United States G graphical Surveys West of the 100th Meridian,* by George M. Wheeler, 1879.)

it is probable that this was the more elegant section of Chinatown as far as the Chinese were concerned. Along with merchant Kee Chan, forty-one, and his wife Hese Chow (or Chan), twenty-five, these three couples probably formed part of the elite Chinese Comstock society.

About 56 percent of the Chinese men on the Comstock were or had been married, but their wives either resided elsewhere in the United States or, as is more likely, had remained in China to care for the husband's parents and for their own children. Wives in China often found solace in folksongs such as this one:

> *I am still young, with a husband, yet a widow.*
> *The pillow is cold, so frightening.*
> *Thoughts swirl inside my mind, chaotic like hemp fibers;*
> *Separated by thousands of miles, how can I reach him?*
> *Thinking of him tenderly—I toss and turn, to no avail.*
> *He is far away, at the edge of the sky by the clouds;*
> *I long for his return, especially since it's midnight now.*[55]

But these women were willing to endure the separation because of the dream of wealth and luxuries:

Right after we were wed, Husband, you set out on a journey.
How was I to tell you how I felt?
Wandering around a foreign country, when will you ever come home?
I beg of you, after you depart, to come back soon,
Our separation will be only a flash of time;
I only wish that you would have the good fortune,
In three years you would be home again.
Also, I beg of you that your heart won't change,
That you keep your heart and mind on taking care of your family;
Each month or half a month send a letter home,
In two or three years my wish is to welcome you.[56]

As in other western Chinatowns, prostitutes on the Comstock were clearly in the minority by 1880. Census takers listed only thirteen women as prostitutes (three of whom were married and therefore might have been secondary wives or concubines) as compared to ninety-four (or seventy-one, if women who may have been married are subtracted) in 1870. Two women, ages forty and thirty-five, who lived together in Gold Hill and whose occupations are not listed, may have been prostitutes, but the census taker was more cautious now and did not designate them as such. Interestingly enough, one prostitute, You Ah, is included in both the 1870 census at age twenty and in the 1880 census at age thirty, and therefore differed from the majority of prostitutes who followed the boomtowns as transients. At the same time, the departure of the prostitutes was an indication of the declining fortunes of the Comstock area.

During this period Chinese women branched out into other occupations. Four women were involved with lodging, and one, Lee Sou, thirty-five, married to cook, Ah Gook, thirty-five, operated an opium den on Union Street. These five women are noteworthy because, like the three women in the laundry business a decade earlier, they broke away from Chinese tradition to participate independently in the work force. Ah Chew, twenty-five, helped her husband, Ah Wah, thirty-two, in the laundry business in Gold Hill, thus remaining within the realm of Chinese tradition as a woman assisting her husband in his business. Undoubtedly there were more Chinese women in the work force, especially those who assisted their spouses—a practice common in China, but these women do not show up in any extant statistics.

In spite of the small, but growing, number of Chinese children, Virginia

City's Chinese community supported two teachers, Gung Go, forty-two, and Wy Ah, forty-eight, both of whom were married but were living apart from their wives.[57] If, indeed, teaching was their major occupation, then there must have been other children in Chinatown besides the three boys mentioned above. The increased number of families and children indicates that attempts were being made to establish a family lifestyle in Nevada, which, in turn, contributed to the higher birthrate of native-born Chinese Americans.

The re-creation of the family life that is the foundation of traditional Chinese society was possible in other ways, as well. About one-third to one-half of the 577 males in Gold Hill and Virginia City were married and, according to popular opinion, were more stable than were single men. The majority of these men left their wives in China, but some may have left them in "safer" communities in the United States or with relatives. These married "single men" had to save funds to support their families who lived elsewhere. Many of these men and women lived with or among relatives, or, if they were single and without relatives, they tended to live together in clusters. One example of this was the household of Toy Quy Ah, twenty-eight, which included her thirty-nine-year-old husband, Seng Gung, who was a laborer, as well as the unmarried joss house keeper, Ah Mung, sixty.[58] Employment also led to the creation of a "family" situation. Kee Chan, forty-one, who was a "China store" owner, and his wife Hese Chan, twenty-five, housed their clerk and two other men who worked in an employment office. Thus the family—nuclear, extended, or artificial—could be maintained.

The settlement pattern of the Chinese in Storey County indicates that by 1875 a growing number of Chinese resided outside Chinatown. By 1880 less than half of the Chinese population lived in Chinatown.

The economic prosperity of the Comstock was waning, and the Chinese began to move elsewhere. On December 19, 1877, the *Territorial Enterprise* reported that the Chinese population in Virginia City's Chinatown had dropped by half, with residents moving to Bodie, Belmont, Tuscarora, and other mining towns. Obviously this was an exaggeration, but the migration had begun. Many who remained on the Comstock had adapted to American society to the extent that they felt comfortable living outside Chinatown.

The shortage of women became more acute. The 1875 Page Law and subsequent anti-Chinese immigration laws prevented many Chinese fe-

TABLE 10.2   Storey County Settlement Pattern of Chinese[59]

Residence	1870	1875	1880
Chinese living with Euro-Americans*	30 (4.0%)	98 (7.2%)	107 (16.8%)
Chinese living next to Euro-Americans**	63 (8.5%)	363 (26.7%)	226 (35.5%)
Chinese in Chinatown	651 (87.5%)	901 (66.1%)	304 (47.7%)

*Usually servants or cooks.
**Usually laundrymen or cooks.

males from entering the United States.[60] This forced Chinese men to turn elsewhere to find spouses. An 1861 Nevada miscegenation law, amended and expanded in 1911 and 1919, prohibited interracial marriages and Asian cohabitation with Euro-Americans and made the violation of the law a misdemeanor punishable by one to two years imprisonment in the case of marriages and by a fine and a lesser period of imprisonment in the case of cohabitation.[61] Occasionally the laws were disregarded or the couples married elsewhere and tried to settle in Nevada. In 1877 a young Chinese man from a prominent Carson City family fell in love with and married a working class Euro-American woman in Carson City despite state laws against such a union. Her repeated physical mistreatment of him ultimately led to his death in 1880.[62] On the other hand, in 1889, when H. Warren of Winnemucca, Nevada, complained to the sheriff that a Euro-American woman and her daughters, ages sixteen and eight, lived with a Chinese man, a warrant was issued for miscegenation and unlawful cohabitation.[63] Mrs. Clara Yon Pow, who married her husband four years earlier in Idaho, did not deny living with her Chinese husband, and since the women were dressed "respectably" and were "by no means unprepossessing in appearance," nothing came of the charges. These cases indicate that there were occasions when the state's miscegenation laws were ignored.

By state law, Chinese men were allowed to marry Native Americans, Hispanics, and African Americans. The overwhelming number of Chinese men in Nevada, 2,817 in 1870 and 5,102 in 1880, compared to the small number of women, 306 in 1870 and 314 in 1880, forced many Chinese men to seek wives of another race.[64] In Nevada marriages between Chinese and Native Americans were not uncommon. One example of this was the marriage of Sam Leon (Wong Leong, Ah Sam, Sam Wong) and Daisy Benton, a Pauite from Schurz, Nevada, in 1910.[65] However, these cases were exceptions to the general rule, and most Chinese Nevada men suffered in an unnatural society of bachelors created by American laws. In the United States this situation lasted until the repeal of the Chinese Exlusion Act in 1943, which opened the door to some Chinese women.[66] A balance of the sexes would not be achieved in Nevada until 1980.

By 1882 the Chinese readily turned to American ministers to perform marriage rites. The last of the twelve marriage licenses taken out between 1874 and 1882 by Chinese men in Virginia City was taken out by Ah Yung, (b. 1858) identified in the 1880 census as a laundryman at 21 South B Street, to marry You Gang, or You Nan (b. 1861),[67] a prostitute in a bordello in Chinatown operated by Goon Gin (b. 1836) and his wife You Guin (b. 1856). You Nan probably was related to You Guin and unwillingly entered prostitution because of family financial difficulties. The March 29, 1882, and the March 31, 1882, editions of the Virginia City *Evening Chronicle* reported the departure of You Nan from Ah Low's (probably another name for Goon Gin) bordello and her quick marriage to Ah Yung by a local minister. After the ceremony, in keeping with American customs, You Nan wanted her picture taken. Accompanied by Camp Sing, a cousin of Ah Yung, she went to Beal's photography studio. Upon leaving the studio, Ah Low tried to kill You Nan but wounded Camp Sing instead. Ah Low was imprisoned for attempted murder, and the newspaper announced the possibility of another "Chinese war" because Camp Sing belonged to the Hop Sing Company,[68] You Nan to the Hop Wo Company (*Hehe huiguan*, a *Siyi huiguan*), and Ah Low to the Yung Wa Company (*Yanghe huiguan*, a non-*Siyi* organization).[69] From this information it is possible to infer that You Nan and Ah Low spoke different dialects of Chinese and followed different local customs. This probably made You Nan's life in the bordello more arduous. The "war" did not materialize, but the marriage indicated that for most of these late nineteenth-century Chinese prostitutes, the ultimate goal was still to find a husband and establish a family as they would have done in China. Since only an affluent man could purchase a Chinese prostitute, once she was married, she left the trade and never had to return.[70] The American religious ceremony insured these former prostitutes the protection of the American judicial system against abductions.

The small number of Chinese women on the Comstock forced many single men to postpone marriage or to go through life without a wife. By 1900 the total Chinese population in Storey County had dropped to seventy-six (all but one of whom lived in the Comstock area), of whom only three were women (two prostitutes and one married woman).[71] Of the seventy-three Chinese men listed in the 1900 census, thirty-five were married, and two were widowers (a total of 52 percent). Two men had been married for forty years but lived separately from their wives.[72] Lee Hi (b. 1840) had lived in the United States for forty-six years, arriving in 1853, and had married in 1860 when he spent the year in China. At the age of sixty, he owned a store and had two employees. Sam Kee (b. 1838) had married in China at age twenty-two and had then sought his fortune in the United States beginning in 1863. Unlike Lee Hi, he could not read, write, or speak English, so his job opportunities were limited, and he became a cook. Although Lee Hi was probably able to visit his wife by virtue of his merchant status, after the 1882 Exclusion Act Sam Kee probably would not have risked traveling to China, because by law he would not have been able to return to the United States. Perhaps Sam Kee's wife sang this folksong:

Just as we were married, you set out for the journey.
How was I to tell you my own feelings?
Wandering around a foreign land, when will you ever come home?
You've wasted many years of my precious youth.
The spring feeling of my heart has since died—
Because of poverty, I don't have a choice;
But, let it be known to all my sisters:
Don't ever marry a young man going overseas.[73]

By the late 1880s the richest Chinese family in town was the Yee family, who owned the general merchandising firm of Quong Hi Loy and Company at 100 Union Street, with an investment capital of $6,000.[74] Seven members of the Yee clan had shares in the store. In 1895 five were in China, one was in San Francisco, and only Dick Yee lived in Virginia City. By 1900 six resided in Virginia City and only Ying Yee, who had lived in Virginia City from 1888 to 1896, lived in San Francisco (1896 to 1900). The Yee family owned their own home in Virginia City, where two of the brothers lived. Dick Yee (b. 1865) had immigrated in 1880, then went back to China to marry in 1886, where his wife resided. His older brother (b. 1864) had followed him to the United States in 1885, leaving behind a wife he had married three years earlier. The separated families were typical of many of the Chinese immigrants, but, like other merchants, they could and did visit their families in China.

Ah Book (b. 1834), who called himself a "capitalist," had lived in the

United States since 1853 and had never married. He was typical of his generation. Having ventured to the American West, he was unable to find a bride and thus lived the lonely bachelor life. For men like Ah Book, Chinese community organizations provided one form of social interaction and filled their free time. Nevertheless, it was a solitary existence.

By the turn of the century the Comstock had "busted." The total population of Virginia City dropped from 8,511 in 1890 to 2,695 in 1900, and the Storey County Chinese population experienced a similar decline, falling from 245 to 70. In 1900 the Comstock Chinese continued to be engaged in some of the traditional types of employment, working as cooks, laundrymen, vegetable peddlers, laborers, lumbermen, faro dealers, placer miners, and farmers. The population was too small to support a Chinese physician, but since there were several in nearby communities, such as Carson City and Reno, the people were not without traditional medical care. By 1900 the oldest Chinese female resident (b. 1840) of Virginia City, a "courtesan" who had immigrated in 1861, lived in a single residence in a neighborhood that included descendants of Irishmen, Scots, and Englishmen.[75] She had learned to read and write English during her almost forty years in the United States and had probably acculturated to some degree, as is indicated by the location of her residence. Representing a much younger generation, the only other "courtesan" was Non Suey (b. 1870), who immigrated in 1887 and lived in a rented house in Chinatown by herself. She could not read or write English but learned enough spoken English to survive. The only other Chinese woman was Heng Long (b. 1860), who arrived in the United States in 1881 and two years later married cook Es Suey Wah (b. 1841) who had immigrated in 1860. Although her husband read, wrote, and spoke English, the couple lived in Chinatown, because Heng Long did not know English and therefore needed the familiarity and protection of Chinatown. The profile of these three women represents the development of the Chinese women of the Comstock in the late nineteenth century.

As the Chinese population declined, traditional activities diminished in importance for Chinese women. In addition to the merchandising firm of Quong Hi Loy, Chinese women of the Comstock could shop at On Lung and Company at 2 H Street. Chung Kee (b. 1839) was the manager of the store, in which four partners had invested $4,000. Two of these partners were in China in 1902, and one lived in Hawthorne, Nevada. Chung Kee, also known as Jung Gon Wah, had immigrated in 1853, had married in

1860, and lived with his clerk Ching Fong (b. 1861), who had immigrated in 1881 and had been married since 1880, thus creating the artificial "employment family" often found in Chinatowns. In 1920 Chung Kee still operated On Lung and Company, but now with the help of his cousin, Sam Huie (b. 1861), who had immigrated in 1886. The business served a limited Chinese population in Storey County. Chinese stores and restaurants became the "social centers" for the small Asian community.

Of the forty-four Chinese in Storey County in 1910, the census manuscript listed only one woman, Foy Choy (b. 1852), a widow who lived by herself, spoke English, and rented a house in Virginia City. Foy Choy had immigrated from China in 1870, and although there is no other information about her, it is probable that she had lived in Virginia City before and in her twilight years had decided to settle in somewhat familiar surroundings. The other possibility is that she had relatives in Virginia City. The decline in the number of Chinese women in the area parallels the decline of the population in the once-thriving Comstock.

Conclusion

The Chinese women of the Comstock lived between two worlds: the traditional and the new. Because they were so few in number and because they lacked traditional support systems—especially the support of the extended family—they had to make adjustments to their new environment much faster than did women in large urban Chinatowns such as San Francisco's. Because it was customary in China for women to remain in the home, most remained fairly isolated. Eventually they found that they had to interact with Euro-Americans because of the smallness of their community and the neighborhood in which they lived. This was especially true as families moved outside of Chinatown, and they made contact with others as they accomplished the normal tasks associated with living.

In the 1860s and 1870s the Euro-American population regarded most of the Chinese women as prostitutes. This is apparent from the census data. However, an understanding of the structure of Chinese society makes it clear that some of these women might have been secondary wives or concubines.

By the 1880s the majority of the even smaller number of women were married, and some type of family life, whether nuclear, extended, or artificial (employment), was evident. Acculturation accelerated and more

Chinese men and women had American-style weddings. Some Chinese sought the protection of the American courts to prevent their women from being kidnaped or falling victim to foul play. The number of children increased, and their presence, especially if they attended public school, also led to a greater interaction between Chinese mothers and other mothers.

As in China, women worked in assisting their husbands and in traditional types of employment, such as sewing. They also entered nontraditional occupations such as operators of boardinghouses, laundries, and even an opium den.

Some traditional values, such as the subordination of women to men and the discrimination against second marriages for women, persisted. However, without a continual stream of new immigrants to reinforce traditional values and the need for seeking the safety of a Chinatown, Chinese women adopted more American values and customs. One example of this was Foy Choy, a married woman who lived in the United States for forty years and had settled in Virginia City in 1910. The census taker noted that she was fluent in English, which indicated her interaction with the Euro-American community.[76]

Thus Chinese women who lived on the Comstock were changed by the circumstances and conditions under which they lived, eventually becoming a part of the greater community. But as their roles and the customs they tried to preserve changed, the decline in the economy of the region forced the women and their families to seek their fortunes elsewhere. What remains is the example of the perseverance, courage, and strength that these Chinese women demonstrated as they lived their lives despite prejudice and hardships.

11

"And Some of Them Swear Like Pirates"

Acculturation of American Indian Women in Nineteenth-Century Virginia City

EUGENE M. HATTORI

Mid-nineteenth-century Euro-American settlement of western Nevada was destructive to the traditional economy of the Comstock's aboriginal inhabitants. Faced with displacement from their homeland and almost certain starvation, some Northern Paiute Indians modified their traditional lifestyles and created a distinctive, urban ethnic group within Virginia City's cosmopolitan population. Their existence, however, was more than a model of mere survival. Northern Paiute women maintained their cultural identity as Native Americans and contributed substantially to a modified traditional way of life that allowed them choices in both Native American and Euro-American economies. Most significantly, the role of Northern Paiute women in Virginia City can be viewed as an extension of their traditional, pre-contact role, and they were crucial to the survival of their culture on the Comstock.

The Northern Paiutes were hunters and gatherers who traditionally exploited seasonally available plants and animals in different environmental zones ranging from wetlands on valley floors to the pinyon-juniper woodlands on mountain ranges. The oftentimes sparse and scattered food sources seasonally limited group size to as few as one or two nuclear families. For Great Basin Native Americans the basic division of labor was according to sex: women gathered plant foods, and men hunted game.[1] The women ventured out from base camps on daily forays to collect foodstuffs. Women's contribution to aboriginal subsistence was considerable and included gathering nearly all of the vegetal foodstuffs and participating in mudhen, rabbit, and antelope drives; cooking; and the production of tools and containers used in food collecting, processing, cooking, and storage.[2]

Larger social groupings appeared when foods that occurred in large quantities and within sizable patches were collected or hunted. The fall pinyon pine nut harvest was of particular political, social, and economic importance to the Northern Paiute in west central Nevada, because a large number of families could assemble to collect this vital winter staple. Because of their dependence on pinyon, the Northern Paiutes and other Great Basin American Indian cultures respected the pinyon pine and gave thanks to the tree for allowing them to utilize its seeds for their survival. Men and older children knocked the pine cones and pine nuts from the trees and gathered them in baskets. Women then processed the pine nuts into a flour by roasting, shelling, winnowing, and grinding.[3] Communal hunts for pronghorn, jack rabbits, and American coots also brought a number of families together for social as well as economic purposes. Women and older children were integral participants in these communal hunts.[4] Times of plenty were eagerly anticipated, because associated activities included reunions with family and friends, thanksgiving ceremonies, courtships and marriages, political meetings, dances, and games.[5]

The Great Basin's natural environment is dynamic and subject to rapid and severe fluctuations. Native Americans were especially adept at dealing with climatically induced changes in the quantity and distribution of the flora and fauna that they depended on for sustenance. Their semi-sedentary lifestyle had been honed over millennia to maximize yield and minimize waste.

The changes to the Northern Paiutes' environment following Euro-American settlement in 1851 eclipsed all but the most severe environmental changes that have taken place over the past 7,000 years in terms of both severity and speed. A traditional Native American economy based on exploiting native plants and animals in different life zones became increasingly difficult to pursue. In general, Euro-American mining activities claimed the wooded hills, and agriculture claimed the fertile valley floors.

The loss of these important resource areas and associated resources was potentially disastrous to the indigenous population. For the Native Americans of the region, the decline was exemplified by the fragmentation of the nuclear family, alcoholism, disease, low birth rates, and premature death. Descriptions of solitary, destitute Native American men and women wandering frontier towns in tattered clothing and begging for handouts from passing strangers provide testimony to the darker side of Euro-American contact: "Near a little road-side grocery (in southeastern Oregon), supported by a post and flanked by an empty cask, stood a Noble Red Man. Indifferent to his tattered clothing which afforded no protection from the sharp wintery nights—with his long black locks flying in the wind—his whole soul was wrapped in a whiskey bottle."[6] The Comstock's Northern Paiute families, however, fared better. They adapted to the Euro-American invasion by eventually incorporating Virginia City as a residential base in their seasonal economic cycle.

Northern Paiute men, possibly affiliated with *Had-sa-poke*'s band, worked as laborers on Henry Comstock's placer claim near Gold Cañon in 1858.[7] No mention of Native American women or children is made in the early-day accounts of Gold Cañon prior to the discovery of the Comstock Lode in 1859. Contact between the miners and Native Americans was probably restricted to males. Because the focus of Euro-American settlement during the 1850s was scattered farms in the valleys at the foot of the Carson Range, in Washoe Indian territory, and in the placer workings in Gold Cañon, it was still possible for the Northern Paiutes to pursue an aboriginal way of life by making only minor modifications to minimize contact with Euro-Americans. Women, children, and most men probably continued their native lifestyles, while a few Indian men worked for Euro-American placer miners for wages or trade goods.

The discovery of the Comstock Lode in 1859, and the ensuing "Rush to Washoe" in the spring of 1860, established the Comstock Mining district—highlighted by Virginia City and its mines, mills, commercial district, and residential neighborhoods, including those inhabited by the Northern Paiutes. The environmental impact of the mines reached far beyond the boundaries of the mining district. The pinyon-juniper woodlands surrounding Virginia City and stretching to the south were widely and quickly decimated to provide cordwood and timber for the burgeoning mining industry and boomtowns.[8] The continued destruction of the pinyon groves near Virginia City unquestionably contributed to the short-lived Pyramid Lake Indian War of 1860.[9]

Farmers and ranchers rapidly usurped remaining tracts of land on valley floors in order to produce food for the burgeoning mining communities and also introduced domesticated animals that competed for many of the very same plants upon which the Native Americans depended. Prospectors, meanwhile, flocked to the remaining regions in search of the next big strike. In light of these far-reaching changes taking place within their traditional territory, a significant portion of the region's Northern Paiute population chose to modify their aboriginal existence and settle on the Comstock. The Pyramid Lake Indian War of 1860 demonstrated to the Euro-Americans the potential cohesiveness and military strength of the Northern Paiutes. For the Northern Paiutes, it confirmed the fact that a seemingly endless number of soldiers awaited a chance to exterminate Native Americans. Hostilities between the Comstock's Northern Paiutes and Euro-Americans, however, were short-lived and did not extend much past the early 1860s.[10] Reenactments of a Northern Paiute "battle" charge became a high point of Virginia City's Fourth of July parades in the 1870s and were eagerly awaited by Euro-Americans and Native Americans alike.

By the mid-1860s Northern Paiute families were frequently noted as an integral part of Virginia City's cosmopolitan atmosphere.[11] Their treatment by other townspeople was mixed. There is little doubt that they were not afforded the same rights as were the various Euro-American ethnic groups, but there are instances in which Euro-Americans mistreating Northern Paiutes were punished.[12] The overall acceptance of the local American Indians contrasts with that of the emigrant Chinese, who were generally held in much lower esteem by the Euro-Americans during the era of anti-Chinese sentiment in the western United States.

Virginia City attracted Northern Paiutes along with thousands of Euro-American prospectors, miners, merchants, laborers, and mining speculators. Unlike the majority of the population, however, the goal of the Northern Paiutes was not to acquire wealth and return to their homeland; they were home. The Northern Paiutes were drawn to the Comstock, because it offered them far more than just refuge during a period when their very existence was in jeopardy.

As with many of the ethnic groups in mining camps and towns, Northern Paiutes occupied discrete "neighborhoods" consisting of ten or more habitations. Some Native Americans crudely modeled their dwellings after Euro-American wood frame structures, while others constructed fabric-covered domes of traditional design but with substituted construction materials.[13] These dwellings housed nuclear families and possibly extended

Northern Paiute homes below mine dumps. (Courtesy of Fourth Ward School, Virginia City)

families. Contemporary photographs, lithographs, and newspaper accounts reveal at least three settlements located on the northern, eastern, and southern edges of Virginia City. Two of the settlements were located at the foot of mine waste-rock dumps and the other on the outskirts of town in Six-Mile Canyon. All of the locations provided some privacy from the rest of the city. At least one and perhaps two camps were located in Gold Cañon near Gold Hill. These separate settlements may represent groups of families affiliated with different Northern Paiute bands from the immediate vicinity, Winnemucca and Pyramid lakes, the Carson Desert, Mason and Smith valleys, or Walker Lake.[14]

Northern Paiute women in Virginia City undoubtedly played a significant role in their overall economy. The heavy reliance upon a woman's contribution to the family's larder was probably not much different from aboriginal lifestyles; the setting, methods, and foodstuffs, however, differed greatly. The tasks performed by the women included wage labor as well as a modified form of collecting food on a daily subsistence round.

Native American women pursued the task of collecting cast-off, "day-old" fruits and vegetables from vendors within the commercial district.[15]

Surplus food from some Euro-American households was even set aside and given to the first Northern Paiute who happened along.[16] This may explain an incident that occurred in 1905 when a visitor to a Virginia City residence fainted upon seeing a Northern Paiute woman's face pressed against a window during the Native American woman's search for food.[17] This daily, urban collecting round was an extension of the aboriginal way of life, in which women struck out daily in search of ripening seeds and roots. The women also maintained traditional economic pursuits in the spring and fall, when families left Virginia City for the wetlands and mountains of the Basin and Range province.[18]

Native American women also contributed to the local economy by working for wages, most commonly as laundresses, seamstresses, and house-keepers.[19] As such, Northern Paiute women were noted as: "very patient and industrious, and [they] would make excellent servants with a little training they could be prevailed upon to take up their residence in any dwelling. . . . The woman are very apt, and ladies find but little trouble in teaching them to do washing or any common housework."[20] Training Native Americans in the "domestic arts" was a common theme in nineteenth-century America, and it was one of the goals of the Stewart Indian School, officially opened in 1890.

In addition to being excellent employees, the women's participation in the Euro-American cash economy provided a means to purchase clothing, fabric, and Euro-American foodstuffs. As the *Territorial Enterprise* reported in 1872: "The Indians living in the suburbs of the city are becoming more and more fastidious in regard to their food. . . . Now they buy nearly all their bread at Fitzmyer's and other bakeries."[21] Northern Paiute males served as woodcutters, scouring the hills for trees and even stumps to sell as cordwood to Comstock residents for cooking and heating. Cordwood was too valuable a commodity to use for their own consumption, so the task of procuring fuel also fell upon the shoulders of the women. Northern Paiute women scavenged discarded packing crates, sticks, and even wood shavings to fuel their family's campfires and stoves.[22]

A notable Virginia City Northern Paiute was a nameless berdache or *tüvasa,* a male who adopted the dress and behavior, including working occasionally as a laundress, of a Northern Paiute woman.[23] The acceptance and even encouragement of berdaches by the Northern Paiutes and other hunting and gathering cultures may reflect the value of female labor to the economy and the potential shortage of female labor.[24]

At the end of the day, Native American women generally returned to their camps on the outskirts of town rather than taking up residence in their employer's home.[25] An exception to this is noted in the 1880 Gold Hill Indian census, which lists forty-year-old Mary as working as a servant and living at 200 Dayton Road.[26] It was rare for Northern Paiute women to remain in town after dark. The social separation of the cultures was rather distinctive, and intermarriage between Northern Paiute women and Euro-American men was rare.

Despite their economic interaction with the Euro-American-dominated city and their adoption of some Euro-American cultural traits, Northern Paiute acculturation was by no means complete. Northern Paiute women in Virginia City continued the aboriginal practice of face painting by applying red, white, and blue lines to their chin and cheeks and sometimes a brick-red pigment over their entire face.[27] The use of glass trade beads, particularly robin's egg blue "pony beads," and, possibly, abalone shell ornaments is also attributable to Native American culture.[28]

The apparel worn by the Comstock's Northern Paiute women was made of Euro-American fabrics, but the styling was distinctively Native American. These women wore loose-fitting calico print dresses with two or three flounces.[29] Accessories included a fancy shawl, a scarf worn atop the head, and, rarely, skirt hoops. Virginia City's Northern Paiute berdache adopted the standard calico dress, gingham scarf, and red face paint of the women.[30]

Adoption of English by Northern Paiute women appears to have exceeded the boundaries of basic communication and to have included some idioms. As noted by the outspoken Mary McNair Mathews, "Most of [the Northern Paiutes] talk good English, and some of them swear like pirates."[31]

A *Territorial Enterprise* reporter addressed the issue of Native American acculturation. Upon witnessing a Northern Paiute girl playing with a Euro-American doll, the reporter suggested that this was a unique occurrence: "Never before in any country have we seen an Indian child with such a toy. . . . She pressed it to her breast with both hands and in the fierceness of her affection for the ridiculous, inanimate thing was in a perfect tremor all over, and most savagely did her great black eyes glare upon any one that approached her."[32] Fragments of a toy cup and saucer, jacks, and doll parts from an 1880s–1920s Northern Paiute encampment at Virginia City (Smithsonian site designation 26st2009) indicate that adoption of Euro-American toys by children was not as unusual as the reporter thought.

The 1890 special Indian census states that the Nevada Indians living off the reservations near mining camps were, "as a class, decreasing in population, as one rarely sees squaws with young babies."[33] Indeed, the Great Basin Native American population declined from about 21,500 in 1873 to a historic low of about 12,000 in 1930.[34] For a brief time in the 1870s and 1880s, however, the Comstock apparently defied the downward population trend: "A PROLIFIC RACE.—No person can visit Virginia (City) without being surprised at the numerous proofs of the fertility of the native women of Nevada, the Piute squaws. Children literally swarm about their camps, and when a Piute matron takes her promenade through the streets she might aptly be compared to an old hen quail at the head of her brood. Men, women and children present a most hardy and robust appearance. Some tribes of aborigines may be dying out, but the Piutes are not."[35]

Even the age structure of a portion of the population is reflected in a description of one Northern Paiute settlement on the southern end of Virginia City:

First in the line upon the trail was an old squaw, with face as wrinkled as a nutmeg. On her back were three gunny sacks filled with hay, and as she stooped half way to the ground, she poled herself along with a broom handle, she looked like a witch bent on some errand of evil.

After the old hag came a stout, middle aged woman, with a load of wood on her back and, sitting astride her neck a boy at least six years of age. She was playfully biting at the boy's fingers and was as jolly as though her load was not that of a kitten. With visage sedate and solemn her lord stalked at her heels, bearing upon his broad back not less than four pounds of "neck" beef.

Next in the line trudged a sad faced, sightless old woman, led by a very small boy, dressed in cast-off clothes that made him look as thick as he was long.

Then came half a dozen young women of the camp who had apparently not yet been placed over a household and made beasts of burden. These wore bright calico dresses, a sort of calico gown above and gay cotton handkerchiefs on their heads. They appeared to be belles of the village and were as chatful and giggling as need be.

After these more old and middle aged squaws, packed with wood and hay, with now and then a child mounted on a load; bucks loaded with nothing but their dignity, juveniles packing bundles of sticks or

American Indian woman with load on back, walking in Virginia City. (Courtesy of
Nevada Historical Society)

prancing along with empty hands, and stopping occasionally to fling
a pebble into the air.[36]

The 1880 manuscript census for Gold Hill Native Americans provides
additional support for the existence of a viable population, ranging from
infants to sixty-year-old men and women.[37] Fortunately for anthropolo-

gists and historians, an enumerator failed to merely summarize counts of Native Americans, as was standard procedure for the census, but instead provided varying levels of detail for 114 of 127 individuals enumerated. All but one of the 114 enumerated Native Americans resided in camps; the following information pertains to the 113 camp residents (table 11.1). Surnames are provided for the three smaller camps with 15, 17, and 19 residents, and "Indian" is used as the surname for all enumerated in the larger camp with 62 residents. The fifteen-member camp was entirely comprised of Johnsons, Marsh was the surname for 88 percent of the seventeen-member camp, and Jones was the surname of 53 percent of the people in the nineteen-member camp.

Women's first names are all listed as Euro-American ones such as Jane (n = 9), Mary (n = 9), Susie (n = 8), and Sallie (n = 9). All women (n = 39) fifteen years of age or older are listed as married or widowed, and all girls under fifteen—actually those who are twelve years old or younger—are listed as single (n = 25). This conforms to the traditional Northern Paiute pattern of marriage occurring shortly after the onset of puberty.[38] Over 38 percent of the enumerated population was younger than fourteen years old; slightly over half of these youngsters (57 percent) were girls.

A population pyramid for the detailed Native American enumeration in 1880 reveals a noticeable dearth of males between the ages of ten and twenty-nine years old and shows no males between the ages of fifteen and nineteen. There is a decrease in numbers of females between ten and nineteen years of age that corresponds to births in the decade between 1861 and 1870. This period was marked by explosive growth and development on the Comstock. Overall, there were 1.38 females to every male, which is in contrast to the Euro-American population on the western frontier, in which males greatly outnumbered females.

The varied ages reflected in the census and newspapers indicate that the reproductive potential of the group was intact, a measure of successful biological adaptation to culture change. The presence of the elderly blind woman and other elders sixty years of age or older reflects an extended family residence pattern as well as a degree of support for the disabled and the elderly. Elsewhere, frequent references to infants in cradleboards and young children further attests to the fertility of the Northern Paiutes.[39]

Archaeological representations of these social units include milling stones used by women to process food, ammunition used by men in procuring

TABLE II.I   1880 Census Data for Native Americans in
Gold Hill, Nevada

Name	Sex	Years of Age	Marital Status
Household (Camp) No. 122			
Lou Jim	Male	7	Single
Joe Jim	Male	30	Married
Annie Marsh	Female	6	Single
Susie Marsh	Female	25	Married
Jennie Marsh	Female	30	Married
Sallie Marsh	Female	30	Married
Mary Marsh	Female	35	Married
Sallie Marsh	Female	40	Married
Janie Marsh	Female	50	Widowed
John Marsh	Male	5	Single
Tom Marsh	Male	20	Married
Joe (Captain) Marsh	Male	30	Married
Sam Marsh	Male	30	Married
Jim Marsh	Male	40	Widowed
Sam Marsh	Male	50	Widowed
Jim Marsh	Male	60	Married
Sam Marsh	Male	60	Widowed
Household (Camp) No. 157			
Sallie Alpin	Female	16	Married
Captain Jack	Male	8	Single
Sallie Jim	Female	5	Single
Sam Jim	Male	10	Single
John Jim	Male	20	Single
Bill Jim	Male	30	Married
Jane Jones	Female	6	Single
Jennie Jones	Female	8	Single
Sallie Jones	Female	8	Single
Kate Jones	Female	15	Single
Susie Jones	Female	35	Married
Maggie Jones	Female	40	Married
Sallie Jones	Female	50	Widowed
Susie Jones	Female	56	Widowed
Charlie Jones	Male	4	Single
Joe Jones	Male	20	Married
Jane Marsh	Female	50	Widowed
Mary Marsh	Female	60	Widowed
Susie Marsh	Female	60	Widowed

TABLE 11.1 *continued*

Name	Sex	Years of Age	Marital Status
Household (Camp) No. 443			
Baby Indian	Female	0.25	Single
Kate Indian	Female	2	Single
Hannah Indian	Female	3	Single
Jane Indian	Female	3	Single
Jane Indian	Female	4	Single
Julia Indian	Female	4	Single
Josie Indian	Female	5	Single
Jennie Indian	Female	6	Single
Kate Indian	Female	6	Single
Susie Indian	Female	8	Single
Mary Indian	Female	9	Single
Sadie Indian	Female	10	Single
Young Indian	Female	10	Single
Jenny Indian	Female	11	Single
Kate Indian	Female	12	Single
Susie Indian	Female	17	Married
Mary Indian	Female	18	Married
Mary Indian	Female	19	Married
Hannah Indian	Female	20	Married
Jane Indian	Female	20	Married
Mary Indian	Female	20	Married
Susie Indian	Female	20	Married
Susie Indian	Female	20	Married
Sallie Indian	Female	22	Married
Mary Indian	Female	23	Married
Jane Indian	Female	25	Married
Mary Indian	Female	25	Married
Susie Indian	Female	25	Married
Sal Indian	Female	27	Married
Jane Indian	Female	30	Married
Sallie Indian	Female	30	Married
Sallie Indian	Female	37	Married
Susie Indian	Female	40	Married
Sallie Indian	Female	58	Widowed
Sallie Indian	Female	60	Widowed
Will Indian	Male	0.58	Single
Billie Indian	Male	1	Single
Jim Indian	Male	2	Single
John Indian	Male	4	Single

TABLE II.I *continued*

Name	Sex	Years of Age	Marital Status
John Indian	Male	4	Single
Jim Indian	Male	6	Single
John Indian	Male	6	Single
George Indian	Male	7	Single
John Indian	Male	7	Single
Sam Indian	Male	8	Single
John Indian	Male	13	Single
Joe Indian	Male	27	Married
James Indian	Male	29	Married
Bill Indian	Male	30	Married
Cap Indian	Male	30	Married
Captain Indian	Male	30	Married
Captain Indian	Male	30	Married
John Indian	Male	32	Married
Sam Indian	Male	33	Married
Mean Indian	Male	35	Married
Bill Indian	Male	36	Married
Jack Indian	Male	37	Married
Joe Indian	Male	40	Married
Mud Indian	Male	40	Single
Sam Indian	Male	45	Married
Sam Indian	Male	50	Married
Sampson Indian	Male	65	Widowed

Household (Camp) No. 474

Name	Sex	Years of Age	Marital Status
Baby Johnson	Female	1	Single
Baby Johnson	Female	2	Single
Susie Johnson	Female	5	Single
Mary Johnson	Female	7	Single
Jane Johnson	Female	9	Single
Kate Johnson	Female	10	Single
Mary Johnson	Female	23	Married
Jane Johnson	Female	26	Married
Susie Johnson	Female	30	Married
Mollie Johnson	Female	35	Married
Jack Johnson	Male	8	Single
Sam Johnson	Male	11	Single
Jim Johnson	Male	37	Married
Billie Johnson	Male	38	Married
John Johnson	Male	40	Married

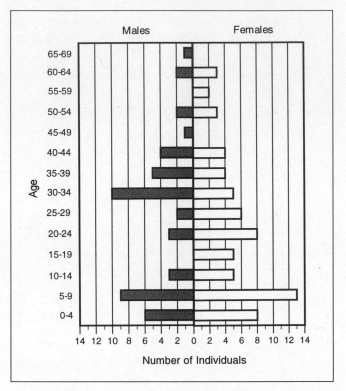

Fig. 7. Population numbers for Northern Paiutes in Gold Hill, Storey County, by sex, 1880. (Source: Tenth U.S. Manuscript Census, 1880)

game, and children's toys. Other artifacts attributed to women include Euro-American jewelry, trade beads, and ornate buttons.

Wage labor and the abundance of collected Euro-American food afforded the Paiutes increased leisure time to congregate on a daily basis in large groups within the city, where they watched the events in the city and gambled or socialized with other Northern Paiutes. These gatherings occurred in the city proper, outside the boundaries of the Native American settlements. Locales chosen for congregations of Northern Paiutes included vacant city lots, industrial storage yards, and the edges of city streets. Some of these meeting areas may actually have represented a neutral ground for members of different neighborhood groups, possibly from different Northern Paiute bands. Card games were one of the principal pastimes for Northern Paiute women.[40]

Larger Native American meetings described as "fandangos," or dances, included Northern Paiute and Washoe participants from neighboring towns, farms, and the Walker River Reservation. Participants advertised these events weeks in advance with bonfires on hillsides visible for tens of miles. From the early 1870s their frequency increased, and some of these gatherings were undoubtedly related to the 1870 Ghost Dance of Wodziwob.[41] One of the 1871 round dances is described in the following excerpt:

> The ring, or great wheel of dancers begins revolving with the first notes from the singers, and, like a huge laboring water wheel, creeps slowly around for hours on a stretch, always moving from right to left. The men are all at the head of the line, with the women coming after by themselves. . . .

Nineteenth-century American Indian woman with infant in cradleboard on B Street, Virginia City. (Courtesy of Fourth Ward School, Virginia City)

Nineteenth-century American Indians playing cards below D Street, Virginia City.
(Courtesy of Fourth Ward School, Virginia City)

Only the feet move—the shoulders and bodies circle around as a solid mass. So slight is the motion of the bodies of the dancers that we saw half a dozen women dancing with a single blanket stretched over and doing duty on the shoulders of the whole number, and in many instances three squaws were dancing under the protection of the same shawl . . .

No levity is observable among the dancers. Some of the Indians informed us that the songs were of a religious character, in short, as they expressed it—"All prayers."[42]

As the economy of Virginia City waned during the borrasca of the 1880s, the town began its downward economic spiral, never again to regain the prominence reached in the 1860s and 1870s. While most of the Euro-Americans readily abandoned the region for other mining camps or their homelands, the Northern Paiutes stayed home. They remained in their encampments well into the twentieth century, and in one instance, an Indian cleaning woman named Julia acquired an abandoned house in Virginia City.[43]

Northern Paiute women on the Comstock were frequently praised for their virtue and good demeanor.[44] Little, however, is written about specific Virginia City Native American female personalities. Although not a typical member of nineteenth-century Northern Paiute society, Sarah Winnemucca Hopkins was a seasonal Comstock resident before and after its discovery.[45] On her visit in September 1864, she accompanied her father, Chief Winnemucca, to Gold Hill, where she translated his speech into English. It was noted, with uncharacteristic non-bias, that she spoke "very plain" English and was quite intelligent. An elderly Northern Paiute woman, known as Eve, was also a well-known Virginia City figure and Catholic parishioner. Father Manogue baptized her in 1863, and her children and grandchildren were subsequently baptized Catholics.[46] One of her sons was Captain Bob, a well-known local figure and spokesman for the Virginia City Paiutes. Tragically, Eve succumbed to the effects of a dog bite despite medical care provided by the Daughters of Charity.[47]

In conclusion, the Northern Paiute women on the Comstock were crucial to the survival of their culture. Confronted with environmental degradation and loss of former territory, the Comstock's Northern Paiute women continued the practice of providing the majority of the family's sustenance, but, in this case, through urban foraging and wage labor. Native Americans became willing participants in the Euro-American economy, yet they maintained their social distance and preserved elements of their pre-contact culture.

Euro-Americans on the Comstock were, for the most part, sympathetic to the plight of the Northern Paiutes. They also benefited from a source of inexpensive part-time labor. Euro-American families employed Northern Paiute women as housekeepers, laundresses, and seamstresses. Cash obtained from wages and sale of native foodstuffs reentered the Comstock economy, and the Northern Paiute labor filled an economic niche in the community. Perhaps the greatest contribution of the Northern Paiute women on the Comstock, however, was the role they played in maintaining the nuclear family, thereby ensuring the survival of the culture.

❦ 12 ❦

Erin's Daughters on the Comstock

Building Community

RONALD M. JAMES

In 1846 a girl named Margaret was born in California to Irish immigrants in the wake of the Mexican-American War. Although she had the distinction of being one of the first children of northern European extraction to claim West Coast nativity, Margaret was to live to a certain degree in anonymity. By 1865, at nineteen, she was the wife of an Irish shoemaker named John Cronin and had a son in Grass Valley, a mining community in the Sierra's western foothills. Six years later the couple had another child. In 1872 they moved to Nevada's Virginia City, where John continued his trade.

The Cronins had three more children in Nevada. After the Great Fire of October 26, 1875, they lived in a South D Street lodging house owned and occupied entirely by people of Irish extraction. Considering where they lived, the Cronin children probably attended the Fourth Ward School, newly opened in 1877. They may have belonged to the Savage Gang that scavenged firewood from the local mine dump.[1] Margaret was pregnant during the Great Fire and possibly watched the flames as they decimated Saint Mary in the Mountains Catholic Church and the Irish neighborhoods surrounding it, feeling a deep sorrow for the losses of her friends. Perhaps she watched from her home as work progressed on the new church, glad to see the community reestablished and looking better than ever. It is easy to guess at the details that probably filled Margaret Cronin's life, because there is strong evidence that many Irish and Irish Americans were committed to their community and regarded its preservation as a high priority. While Irish women became numerically important on the Comstock, their true significance rested on this aspect of their character: together with their fellow immigrants, they worked to build homes, placing values on ethnicity and stability while pursing lives that further reveal the complexity and diversity of their group.

Like many American children of Irish extraction, Margaret Cronin picked a spouse from the Irish community, reinforcing her ethnic identity. Of greater importance, however, is the Cronins' tie to the mining community. They traveled from Grass Valley, California, to Virginia City. When they left the Comstock, they lived in Austin, Nevada, and then traveled to Eureka, Utah, yet another mining town, flowing with the stream of miners who followed bonanzas in search of work. This is typical of the Irish and Irish American women's experience on the Comstock and perhaps throughout the mining West.

Although immigrants of Irish ancestry arrived in a place dramatically different from the lush greens of Ireland, and although mining towns have always been notorious for their transience, Margaret Cronin and others like her sought to make the place home and struggled for permanence. After arriving in a town, she settled longer than did most of her non-Irish counterparts. When she left, she attempted to maintain her sense of community by moving with her neighbors, even though the shoe-making industry did not restrict her husband's employment to a mining town. Clearly, her husband had a say in these decisions, and his desire to find employment was also a factor. Nonetheless, there is compelling evidence that the women played a pivotal role in promoting an Irish ethnic framework within which to live and work. Margaret Cronin may have only a scant profile in the historical record, but her life serves as an example of how Irish American women of the West viewed themselves and their families and community.[2]

The presence of thousands of Irish and Irish Americans on the Comstock made the task of building an ethnic community easier than it might otherwise have been. There were over 16,000 people living in the Storey County part of the mining district during the 1880 U.S. census, just as the county perched on the edge of decline. Irish or Irish Americans represented one-third of this population.[3] Women of Irish ancestry were so numerous that they constituted nearly half of their ethnic group, and they also represented almost half of the women on the Comstock. The high proportion of women among Irish immigrants corresponds to a national trend unique to the Irish, a group in which the number of males and females were nearly equal in North America. This was unusual for most major immigrant groups of the time. Consequently, an understanding of the Irish women's Comstock experience can serve as a bedrock for research on the area's social history and gender issues and can contribute to Irish immigration studies. In fact, much of the overview in chapter 2 applies by default to Irish women, because they were so numerous. Taking such a discussion further is problematic because so much of the primary documentation deals mostly with men. Nonetheless, there are opportunities to develop a profile of this part of western history.

Historians have long noted that the Irish in the region often arrived from elsewhere after leaving Ireland.[4] The immigrants typically spent several years in the eastern states, Australia, or Canada before traveling to the western United States. This pattern meant that the region's Irish were usually acclimated to a non-Irish, urban setting. They were not, in short, "right off the boat."[5] In addition, where the Irish found rigid, prejudiced communities in the East, the West offered a more fluid society. The Irish who settled in western boomtowns frequently came to represent a large part of the population, and, without many of the barriers they found elsewhere, they could succeed economically and politically. These observations appear to apply to the Comstock without reservation. The example of the Comstock Irish, however, provides an opportunity to add to this understanding. A mining district noted for its complexity and diversity, it included thousands of people from throughout the world as well as a sizable Irish population. Comparing groups is consequently possible. By looking at the Irish women in their context, an image of their commitment to their group and community emerges.

The Irish pursued diverse occupations, followed any number of marriage patterns, and resided in dwellings scattered throughout the mining district. Still, the majority of Virginia City's Irish women lived in distinct neighborhoods, married fellow countrymen, sent their children to school, and supported Irish organizations and the Roman Catholic Church. Unwilling to disrupt their ethnic community, they tended to remain on the Comstock longer than did most who came to that fluid society. A comprehensive look at Irish women of the mining district reveals a group of people who, for the most part, worked towards building a place that nourished their sense of ethnicity and values. These women were not aimless exiles drifting from town to town. In spite of the fact that they had traveled far from home, those who came from Ireland to the Comstock left evidence of intending to stay. An overview of the Irish experience on the Comstock makes this clear.

In 1860 the census enumerator listed fourteen Irish women as living in the Comstock Mining District.[6] Recorded thirteen months after the initial gold and silver strike that inspired the famous Rush to Washoe, the census captures the community in an early stage of development. With a total of only 111 women sixteen years or older on the Comstock, it is not possible to arrive at many statistically valid conclusions about this group. Nonetheless, the record of these few daughters of Erin provides some insight into the earliest foundation of the Irish community in the area.[7]

Ten of the fourteen women were married, and the enumerator lists only the four without husbands as having occupations; these occupations were evenly split between washing and sewing.[8] What is striking about the Irish women in general in 1860 is that they made up a much smaller proportion of the population as a whole than would later be the case. By the time of the 1870 census, the Irish had risen to considerable importance within the mining district, representing roughly one quarter of the population. Furthermore, the number of Irish women had increased, so that they constituted one-third of that immigrant group.[9]

One of the clearest indications that the Irish of the Comstock eventually attempted to establish a community within the transient mining district is the fact that they chose dwellings close to one another in distinct neighborhoods. The tenth U.S. manuscript census of 1880 was the first to list precise addresses, and these correspond, in turn, to the 1890 Sanborn Perris Fire Insurance Map for Storey County.[10] As is illustrated in map 4, the Irish of Virginia City clustered in several distinct areas. The largest of these, which is also the place of the greatest homogeneity, occurred in the blocks extending east and south from the Catholic Church.

Since populations in the mining West could change rapidly, there is some question as to whether the neighborhood retained integrity over time. With limited reliability, it is possible to identify this vicinity in the ninth U.S. manuscript census of 1870, and it appears that the area was densely populated by the Irish at that time, as well.[11] In addition, Storey County

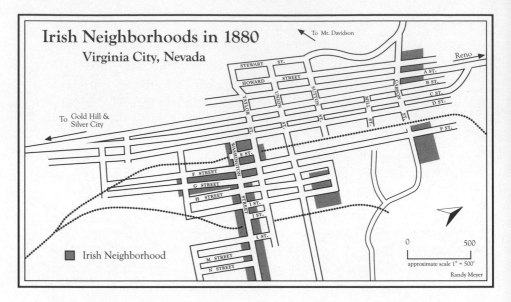

Virginia City Irish neighborhoods in 1880. (Base map from *Report upon United States Geographical Surveys West the 100th Meridian,* by George M. Wheeler, 1879.)

records from these blocks indicate that a high percentage of the places inhabited by the Irish were owner occupied.[12] This evidence points towards stability in the Irish neighborhoods of Virginia City. With that in mind, it is possible to ask a series of questions about the Irish community, and particularly about its women, during the heyday of Virginia City from the 1860s to 1880.

The daughters of Ireland provided the Comstock with some of its youngest workers. In a community that stressed free public education and that was surprisingly devoid of child labor, a handful of Irish and Irish American girls earned salaries. In 1870 Kate Welch worked as a servant and yet was only six years old. Mary O'Grady was only seven years older when she assisted in a millinery shop. Nine-year-old Bessie Kelly was a servant in 1880.[13] Although shocking by today's standards, such examples are more remarkable for their scarcity, considering the nature of many nineteenth-century industrial cities.

In contrast to these few cases, most of Ireland's children on the Comstock, like their youthful counterparts in other ethnic groups, appear in the census as "at school." In 1880, 95 percent of the Irish and Irish Americans between ten and fourteen years of age were pursuing an education. For the older fifteen- to nineteen-year age group, the number drops to below half, with 37 percent of the girls and 34 percent of the boys attending school. These percentages are still strikingly high, since at that age young people often began pursuing occupations or establishing families of their own. Even then, only 14 percent of the Irish women in this age group appear in the 1880 manuscript census with a paid occupation. Most of these young women worked as servants, but there were also seamstresses, a teacher, a cook, and a nurse. The others were either keeping house with a husband or remained at home with parents, presumably waiting for the right opportunity for marriage or employment.

As with most ethnic groups at the time, a high percentage of Irish and Irish American women on the Comstock were listed as "keeping house." Such a designation, of course, does not preclude the possibility that individuals were also making money through various means; but since the census cannot address such occurrences, it is useless in this respect. Women pursuing occupations other than those listed in the census are accessible only through a handful of anecdotes based on a few letters or newspapers. An assessment of the Irish and Irish American women (whether they were keeping house or not) as they appear in the various census manuscripts of the Comstock provides a means by which to examine Irish women's adaptation to the Comstock. Once again, the evidence reinforces the idea that these women worked to preserve their community and sense of identity as a group.

In 1880 most wage-earning women worked as domestic helpers, so it is not surprising to find that servitude was the principal occupation of the women of Ireland. Even in 1870, when there were fewer Irish women and fewer servants on the Comstock, almost three-quarters of the Comstock's fifty-three women working as domestic helpers were from Ireland. There is ample reason to expect the Irish to dominate the ranks of servants, but such an assumption is based on eastern stereotypes. Bridget, the Irish parlormaid, was a stock character of nineteenth-century theater and literature, and for good reason. Many women despised the idea of domestic servitude, but for the immigrant daughter of Ireland, the occupation represented an attractive opportunity for steady employment. Hasia Diner, in her classic study of Irish immigrant women, outlines several reasons for which servitude was desirable for Irish immigrants, but as always, her study employs an eastern perspective. In fact, only 131, or 8 percent of the adult Irish and Irish American women in Storey County, worked as servants in 1880.[14] In addition, 67 of these served Irish households, meaning that only 4 percent of the Irish and Irish American women actually left their ethnic community to work as domestics.

Two neighborhoods provide contrasting examples. On North B Street, outside the traditional Irish neighborhoods, four houses in a row employed Irish servants. Forty-one-year-old A. Coleman worked in the Fitzmier household; Celia Connely, at age seventeen, served the large Armor family, the head of which was a Prussian tobacconist; twenty-two-year-old Hanna Weltmore helped a clerk manage his household; and Bridget Costello, at age forty-five, was the servant for an American-born grocer and his family.[15] In contrast to this block of houses, two Irish domestics worked on the south end of the Irish neighborhood near the orphanage of the Daughters of Charity. The Harrigan family employed forty-nine-year-old Hellen Hogan. A laborer, Patrick Harrigan, like his wife, was born in Ireland. The household included their two children and two lodgers, one from Ireland and the other from Maine, born to Irish parents. Nearby, Maggie Forestal, aged twenty-seven, served the Quinns and two lodgers, all again of Irish origin. Indeed, the 1880 census demonstrates that the Irish neighborhoods of the Comstock almost without exception employed only fellow members of their ethnic group as servants.

The principal competitors of any women who chose domestic service as an occupation on the Comstock were Chinese men. Finding someone to work as a servant in a booming mining district with many employment options was difficult. Many people consequently hired men from the Asian community, even though the racist rhetoric of the time discouraged contact of any kind, let alone the intimate relationship required by in-house help. Experience with Chinese servants created a positive stereotype, however: even the fanatically anti-Asian Mary McNair Mathews had to concede that "They do things for us I would not like to ask a white person; besides, they never tell any family affairs like white girls do."[16] A reputation for hard work made Chinese men potentially effective competitors of Irish women seeking employment in domestic servitude. This was particularly problematic for the Irish, since Bridget the parlormaid had a reputation for not working hard, for temperamentally storming off to other sources of employment, and, as Mathews pointed out, for gossiping about the household she served.[17]

Viewing the situation objectively, one must wonder how Irish women could compete against Chinese labor. The answer rests somewhere between racism and the relative prestige of the help. While an Asian servant might yield more work, an Irish maid procured more status for the household. Some Comstock households employed one of each. Nonetheless, the Irish chose not to hire the Chinese, drawing almost exclusively from their own ethnic group for domestic service. From this fact one can read into the Irish motivation either ethnic solidarity or racism against the Chinese, and indeed both were probably factors.

Just as the Irish were nearly the only women employed as servants in their ethnic community, so, too, was laundry work a protected industry. Most of the laundresses on the Comstock were Irish or Irish American. This was one of the few occupations, along with the domestic service, in which the Chinese shared. In contrast to her experience in the area of domestic service, the Irish laundress had no more prestige than did her Chinese competitor. The Chinese were able to provide inexpensive, efficient laundry service on a large scale. Laundresses, by comparison, almost always worked alone, and it appears that they generally charged more than Asians did.[18]

Most of the Irish women who pursued laundry work were widowed with children, and almost all lived in areas dominated by their fellow immigrants. Irish-born Ellen McGinty, for example, was a forty-year-old widow in 1880, earning money with washing and ironing. She had been widowed only shortly before, her husband having left her with five small children, three of whom attended school. They lived on South F Street in the heart of the Irish neighborhood. For additional income, McGinty rented a room to Cornelius and Nora Cinners, also natives of Ireland. Ellen McGinty was typical of the laundresses on the Comstock. Piecing together a living, she undoubtedly worked hard to earn just enough money to survive.[19]

With similar examples furnished by the other laundresses living in Irish neighborhoods, it is easy to see a form of community welfare, an ethnic safety net providing for those widowed in a place where mining accidents were all too frequent. Together with the Irish tendency to hire servants from their community, it appears that employment practices and occupational choices were linked to the goal of supporting the success of the group. Indeed, the need was apparently great. The issue of supporting widows, for example, appears to have been exaggerated among Irish women. Thirteen percent of these immigrants were widowed in 1880, compared to 9 percent of the non-Irish community.[20] It appears that Irish widows' fellow immigrants did what they could to provide assistance.

With all the occupations available on the Comstock, it seems that Irish women rarely served as prostitutes. Anne M. Butler, in *Daughters of Joy, Sisters of Misery*, asserts that Irish women were drawn to the occupation elsewhere in the West. She credits this to a "social malaise" created by the "grinding poverty" experienced in Ireland and by the difficulties the immigrants faced in America.[21] While there is no reason to doubt Butler's observations about other states, Virginia City's situation appears unique. Almost no Irish women lived in the red-light district there or claimed to be prostitutes in either 1870 or 1880. Perhaps their number and sense of community served to discourage young Irish women from falling into the red-light district, an option pursued elsewhere in the West, where pressure by their ethnic group and by the Catholic Church may have been less easily exerted.

Similarly, patterns related to marriage serve to reinforce the idea of community maintenance. The majority of married Irish people found a spouse within their ethnic group. If marriage of a native-born Irish immigrant with an Irish-American were classified as exogamous, the rates for marrying outside the ethnic groups would have been 22 percent for women and 19 percent for men in 1880.[22] Of the men born in Ireland who married women other than those also from the homeland, almost three-quarters found an Irish American woman for a spouse. The corresponding statistic for Irish-born women is less than one-quarter. Indeed, Irish women were much more likely to find husbands among a broad cross section of the community. In fact, almost one quarter of the married women had spouses without Irish parents, in contrast to fewer than 10 percent of the men who followed this exogamous pattern (See table 12.1).

This difference between marriage choices of men and women was partly due to the fact that on the Comstock nearly half of the women were Irish or Irish American, while less than one-third of the men had a similar ethnicity. Random chance would have caused more women to be exogamous. On the other hand, there were sufficient available Irish men for every woman to find a spouse within her ethnic community. Women who married outside their group raise questions about the solidarity of the Irish neighborhoods. Nonetheless, nineteenth-century stereotypes may have encouraged exogamy rates for women: art, jokes, and literature typically characterized Irish men as ignorant, slovenly drunkards.[23] Many regarded Irish women, on the other hand, as industrious, fierce protectors of their families. In the mining West, where there was usually an imbalance between the

TABLE 12.1   Exogamy for the Irish on the Comstock,
1880

Nativity of Spouse	Wives $N=93$ (%)	Husbands $N=81$ (%)
Irish American	20.4	74.1
Euro-American*	18.3	11.1
English or English American	20.4	8.7
Scottish or Scottish American	11.8	2.5
German or German American	9.8	1.2
Canadian	5.3	1.2
Others	14.0	1.2

Four hundred and sixteen Irish women and the same number of Irish men
appear in the 1880 tenth U.S. manuscript census as married in Virginia
City; of these, 298 or 71.4 percent were married to each other. Twenty-
seven of the remaining women were not living with a spouse, leaving a
total of ninety-three for whom the nativity of a spouse is identified.
Twenty-nine of the remaining Irish men were not living with a spouse,
and three had died during the census year, leaving eighty-one for whom
the census identifies a spouse. If marriage to Irish Americans is regarded as
endogamous, the exogamous rate for Irish women of Virginia City is 18.9
percent; for Irish men, it is 4.7 percent.
*The term "Euro-American" in this context signifies people of European
ancestry who were, together with their parents, born in the United States.
Source: Tenth U.S. manuscript census of 1880 for Virginia City, Nevada.

sexes, women were able to pick spouses from a wide range of possibilities,
while men could not be as selective. As a result, some Irish women married
out of the group, possibly influenced by stereotypes regarding the nature of
Irish men. In spite of the higher rate of exogamy among Irish women, two
observations reinforced the idea of the community. The first is that the
overwhelming majority of women married within their group. The second
is that among those who selected husbands from outside their ethnicity, almost half remained in neighborhoods that were predominantly Irish.[24]

Records related to the children of Irish mothers reveal much about the
nature of their families and their sense of community. A comparison of the
ratio of adults to children among various ethnic groups on the Comstock
demonstrates that the Irish community had far more offspring than did their
British counterparts, for example. While the proportion of men to children
among the Cornish was one to one, the corresponding figure among the
Irish was almost one to three.[25] Children in 1880 yield a much more impressive statistic, however. By using the nativity of offspring recorded in

the manuscript census it is possible to demonstrate that Irish families remained on the Comstock longer than others did. Figure 8 shows, with the distribution of ages, by percentage, that the oldest child born in Nevada in English families tended to be much younger than did the oldest child born to Irish families. This simple observation suggests that the Comstock Irish sank deeper roots than did their English counterparts.

The history of Irish confederations on the Comstock also sheds light on the nature of the community. During the early 1860s there was an organization called the Sons and Daughters of the Emerald Isle.[26] This group arranged the annual St. Patrick's Day celebration on the Comstock, but it soon dissolved, giving way to associations with exclusively male membership. The Comstock eventually boasted four Fenian circles, as they called each unit of the 1860s revolutionary effort. In addition, there were several Irish military guards, a lodge of the Ancient Order of the Hibernians, and a chapter of the Irish Land League. Most of these were exclusively male, removing women from some of the principal means by which to celebrate ethnicity. Even the many saloons that catered to an Irish clientele tended to restrict access to most women, making the institutions yet another means for Irish men to gather. Nonetheless, women had at least two important ways to proclaim their Irish ancestry: these consisted of the Roman Catholic Church and the annual St. Patrick's Day celebration.

By 1861 Father Hugh Gallagher and his Virginia City congregation had built a Catholic Church dedicated to the Virgin Mary. It blew down, but the determined parish replaced the building immediately. Virginia City rebuilt its Catholic church twice more, finally in 1876 reconstructing the remarkable St. Mary in the Mountains, which still boasts an active parish and serves as a cornerstone of the National Landmark District. The precise relationship of women to the institution has been largely lost from the historical record, but clearly this was a place where the Irish, regardless of gender, found access freer than it was in the quasi-military Irish revolutionary organizations.[27]

Irish-born Father Patrick Manogue, who served as parish priest for much of the history of Virginia City during its first years, gave his church a particularly vivid ethnic stamp and also provided the Comstock with one of its best subjects for folklore.[28] In the patriarchal institution of the church, Manogue's role must never be underestimated. At the same time, however, the Daughters of Charity played a crucial part in founding the Catholic hospital, an orphanage, and a school. Over three quarters of the sisters were

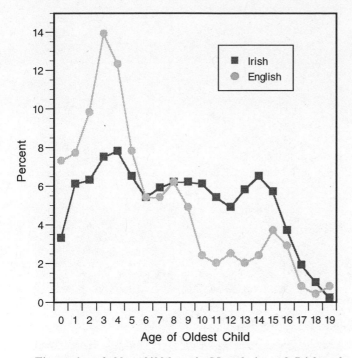

Fig. 8. Age of oldest child born in Nevada in each Irish and English family, Virginia City, Nevada, 1880. (Source: Tenth U.S. Manuscript Census, 1880)

Irish, so they, too, contributed to the ethnic character of the community's Catholic church.[29] Similarly, the vast majority of the congregation was of Irish origin or descent. In spite of these observations, it would be inappropriate to view the Catholic Church as an all-Ireland private club. Germans, Hispanics, French, Italians, and Portuguese, to name only a few, also lived on the Comstock and contributed to the congregation, giving the parish an international character. Consequently, it is not possible to equate church activities with the Irish community exclusively or even predominantly.

Some of the best-documented events involving women and the church were fund-raisers called ladies' fairs. Newspaper accounts tell of these events lasting over a week and raising thousands of dollars to pay church construction debts or to support general maintenance. Those who attended purchased refreshments and letters written as though to a loved one. The principal source of money consisted of raffled items donated for the occa-

St. Mary in the Mountains Church, ca. 1880. (Courtesy of Fourth Ward School, Virginia City)

sion. Women obtained the articles for distribution and arranged them on
tables, competing not only for the most appealing array of goods but also
for the most raffle tickets sold. The *Territorial Enterprise* published the results, including a description of the prizes and the amounts of money each
group of women collected. With only a few exceptions, the women involved
had Irish last names, and cross checks with the manuscript census confirm

their Irish origins.[30] It appears, therefore, that Virginia City's Catholic "ladies' fairs" were almost exclusively Irish affairs.

Theresa Rooney Fair, wife of James Fair, Irish immigrant and bonanza king, serves as a good example of a benefactress of the Catholic Church. She was born in New York to Irish parents and married in her early twenties in California. In Virginia City she donated the baptismal font and candlesticks for the Church. Her generosity, which contrasted with her husband's reputation for stinginess, was well known.[31] Theresa Fair represented an extreme example of many Irish women's involvement in the Catholic Church.

There was perhaps no better occasion than St. Patrick's Day for all Irish immigrants and descendants to gather and reinforce their ethnicity and sense of community. As early as 1864 the Comstock had a St. Patrick's Day parade. Newspaper accounts indicate that both men and women marched in the processions.[32] The Irish used some of the subsequent balls to raise funds for paying church construction debts, but most of these celebrations served as a means to garner support for the nationalist cause. In a way, the ultimate destination of funds mattered little. Perhaps the most important aspect of St. Patrick's Day was that it consolidated the Irish regardless of gender and reminded them of their roots. That is not to say that there were no other opportunities for the Irish community to express its solidarity. In July 1869 the Sweeney Guard held a ball "for the benefit of the widow and orphans of our lamented deceased Brother, Thomas M'coy, who lost his life by the greatest calamity at the Crown Point Mine, on the 7th of April."[33] These occasions certainly added to a sense of community. Still, participation in such events was not restricted to one ethnicity, nor did it directly serve one group. In contrast, St. Patrick's Day was and always will be unique on the Irish calendar.

Although the fight for freedom is a major theme in Irish and Irish American history, little is written about the role of women in the struggle.[34] Documentation is scarce, and yet fleshing out this important aspect of the Irish experience from the perspective of women's history warrants further research. The extent to which nationalism was exclusively or predominantly the experience of men as opposed to women remains largely unaddressed. Arriving at a better understanding of the relationship of women to the cause of nationalism can only serve to illuminate this complex subject and to provide a better framework in which to understand the subject of gender and revolution. Since the struggle against England clearly served as one of the most important cultural and psychological threads for the immigrant Irish population, assessing the role of women in this context is crucial.[35] Unfortunately, sources are wanting, and little can be said about this subject from the point of view of the Comstock.

Echoing this unfathomed aspect of the immigrant experience is the grave of Julia O'Connell, a woman of Ireland, and her baby, in the sagebrush-and-pinyon-strewn Virginia City cemetery. Her tombstone indicates that she was the wife of Thomas Devine, and was born in Cahirciveen, County Kerry, in 1838. She lived twenty-six years before dying ten days after childbirth in 1864. Mary, her first and only child, died eleven days after Julia did. Apparently the trauma of birth had been too much for both mother and child.[36] Three years later, Julia's hometown of Cahirciveen was to see one of the few armed insurrections in Ireland of the Irish Republican Brotherhood, more commonly known as the Fenians. The insurrection failed miserably, but the incident, in relation to Julia's life, calls to mind once again the question concerning women's role in the nationalist movement. Like other sources, the tombstone can only raise questions in this regard. Instead, the monument provides more direct indications of how this immigrant lived and died on the Comstock.

While Julia O'Connell found her fate in Virginia City, Margaret Cronin, mentioned at the beginning of this chapter, lived there for a decade and then moved on. The two women typify extreme alternatives for the Irish on the Comstock. In spite of obvious differences in their lives, both shared experiences intimately linked with building a community. After leaving the Comstock and Austin, Nevada, the Cronins reemerge in the historical record in 1891 in Eureka, Utah, yet another mining camp. At that time their eldest son, having married one Sarah Fergusson, gave them a granddaughter, named Annie after an aunt. Either alone or with her family, Annie Cronin moved to the turn-of-the-century mining boomtown of Goldfield, Nevada. There she met and married Matthew Murphy, an immigrant from Cork attracted by the industry's latest bonanza. Like Margaret Cronin two generations before, Annie traveled with the mining community and renewed her ethnicity by marrying an immigrant.

Persistence is at the heart of the Comstock Irish women's story. Much of the literature dealing with mining towns of the West emphasizes boom and bust economies and the temporary nature of the subsequent societies. It is all too easy to see lives there as being without permanence, continually in flux. In the case of the Comstock and of the Irish in particular, there was

Grave of Julia O'Connell, Catholic Cemetery, Virginia City.
(Photo by Kelly J. Dixon; courtesy of Comstock Historic District Commission)

considerable continuity and stability in the community they built. They were not the temporary exploiters of the resource found in western lore. Instead they settled on the Comstock, hoping to find respite from their travels. While it is common to view the Irish immigrants as forlorn exiles, the evidence from the Comstock suggests that building a new home was at least as important as was mourning over the loss of the old one.[37] Because Irish women were so important numerically to their group, they clearly played a unique role in this process. While Irish men joined revolutionary organizations and quasi-military units, women expended much of their community effort in raising funds for churches and organizing fairs and dances. Although men certainly had a role in domestic choices, households with Irish wives found neighbors of similar ethnicity and settled longer than others.

Hasia Diner sees economic prosperity as one of the major themes of Irish women's eastern immigration experience. While their Nevada counterparts do not contradict this observation, one must not underestimate the role of community, stability, and continuity in defining achievement from the point of view of these women. Community and the values that inspire one to strive for it are ambiguous and difficult to define. Nonetheless, Virginia City and its environs were far more of a community than the stereotype of the mining West might imply. In view of limited comparative research with regard to English-born women on the Comstock, for example, it appears that the Irish may have strived for continuity and a sense of neighborhood more than others did. With this perspective in mind, one can gauge the true success of the daughters of Ireland who came to settle on the Comstock.

PART FIVE

⇾ Image and Reality ⇽

The Comstock has always been a dynamic place where rapid change caused communities to take on radically different appearances and characteristics from one decade to the next. Evolving perspectives are both byproduct and agent of these transmutations: as each generation of Comstock residents faced new challenges, it reinvented its history, ultimately redefining local identity. The last two chapters of this book deal with this process in disparate ways.

During the mid-twentieth century Virginia City became home to a collection of veteran Comstockers and artists, writers, and aristocrats from all over the nation. The West had come into vogue. The mining district was fashionable and its fashionable visitors enjoyed it, but the new residents viewed it through a clouded lens, seeing what they wished and inventing the rest. When it came to the popular understanding of women in a western mining town, the effect was profound. Elements of their story conducive to the creation of myth came to the forefront. This new look at the Comstock exaggerated the importance of prostitution, for example. Revisions of this perspective have occurred ever since its inception in the 1940s, but in spite of attempts to make corrections, popular images of the past persist.

The last chapter of the book deals with a methodology possessing the potential to change once again the way we view Comstock women. Archaeology offers a means by which to bring people who might not appear in the written record more sharply into focus. Looking at the relationship between gender and material culture may result in yet another transformation of the popular view of the mining district's women. Consequently, the book's last chapter, while dealing with the past, may shape future perspectives, adding yet another layer to the ongoing history of the ways that authors and residents have viewed Comstock women.

⫷ 13 ⫸

Girls of the Golden West

ANDRIA DALEY TAYLOR

T he lives of women on the Comstock Lode have been preserved and recounted in numerous ways, not only in academic histories, but also in
literature, museums, and popular culture. One particularly colorful period for the region was the 1940s, when two modes of writing about
the Comstock emerged that would fundamentally change perceptions of
the region and its women. One could be called the Wild and Woolly
School. The other—perhaps best called the Real West school—was based
in history and memory, drawing on the experiences of those who actually
lived on the Comstock Lode. Of these two, the Wild and sometimes
Woolly School became the most important in defining an identity for the
mining district. Initiated by, among others, the notable folklorist Duncan
Emrich, this burlesque approach to the period of the big bonanza includes
such writers as Lucius Beebe and Charles Clegg, Roger Butterfield, Vardis
Fisher, Helen Holdridge, and Katherine Hillyer and Katherine Best (the
"Katies"), to name a few. Their writings are homogeneous and helped
manufacture a Comstock legacy that never actually existed.

Today, much of the work of these 1940s writers seems hopelessly flawed.
The literary restraints they placed upon historical Comstock women forced
their subjects into edited lives. Ironically, the women's actual lives were

larger than the fictional or historical boundaries that restricted them. Women such as Louise Bryant Mackay or Mary Ellen Pleasant were much more impressive in reality and certainly larger than the tired clichés used to encapsulate them in historical resin. Unfortunately, due in part to the efforts of these 1940s writers, the Comstock has become a place of congealed history. The greatest consequence of the 1940s writings was to alter the place so profoundly that the authors who had found themselves drawn to it were eventually repulsed by what it had become—in a large part through their own creations.

The Wild and Woolly School of historical and fictional writing about the Comstock Lode had its roots in the writings of Harvard historian, Bernard DeVoto, who became the Comstock's first twentieth-century iconographer. Beginning in 1929 DeVoto was first to plant the seeds of the Wild and Woolly School of history with a piece he wrote for *American Mercury* just after joining the Harvard faculty. By the time of the publication of his 1932 social history, *Mark Twain's America and Mark Twain at Work,* DeVoto had grown an entire garden. He certainly fell captive to mythologizing Comstock Cyprians in his articles and Twain books. He writes of Virginia City prostitutes, for example, as if he offered the truth: "They drove through the streets reclining in lacquered broughams, displaying to male eyes fashions as close to Paris as any then current in New York."[1] The image is far from accurate. As the Irish novelist Edna O'Brien once remarked, "myths become a law unto themselves; they begin imperceptibly, propagate like dandelion seeds and live forever."[2]

In many ways DeVoto's Wild and Woolly imagery is but a continuation of nineteenth-century dime novels and Buffalo Bill's Wild West shows. The one important difference is the open treatment of prostitution. It is with DeVoto that the Comstock myth of the demimondaine first emerges: prostitutes are romanticized, with exaggerated descriptions of their existence and acceptance in society. DeVoto does not couch prostitution in Victorian platitudes, nor does he paint the bleak reality of these women's lives for what it was. Somehow the harshness of their daily existence ran counter to his mythical abundant West. DeVoto's interest in western bawds spawned an entire school of imitators who, like DeVoto, preferred a reality hovering seductively between fact and fiction, like a shimmering desert mirage.[3]

DeVoto was a cultural mammoth, a teacher and writer, a national figure who seemed as comfortable writing fiction (his pen name was John August) as he was writing history. He was awarded the Pulitzer Prize in 1947 for his history, *Across the Wide Missouri,* and was accorded numerous other accolades over his lifetime. A transplanted Utahan to whom both western schools—the Real West and the Wild and Woolly—looked for validation, DeVoto joked with fiction writers that one could serve the Real West or the mythical West, but not both, at least not at the same time.

Why was DeVoto so drawn to cliché? Perhaps confined by those brick walls that enclose Harvard Yard, he sought to make the West more fabulous than it was, to let its history break forth in a radiant way, to find the men to match those mountains. DeVoto was essentially a westerner with all the commiserative cultural inferiority experienced by westerners ensconced in the eastern intellectual establishment.

The child of an apostate Catholic and an apostate Mormon, DeVoto's life evolved as a struggle between opposing poles, not only in his belief system, but in his life's work. He could not decide whether to be a historian or a novelist of the purple sage, so he did both. Even his obituary in the Boston *Globe* alluded to his struggle: DeVoto was "regarded as being a reactionary by the radicals and by the conservatives a wild left winger." He was a man who graduated from Harvard with a "seething resolution to write" and therein lies the key to DeVoto. He was a writer. To write well, or even divinely, was his ultimate mission.[4]

DeVoto's own erudite prose style, heavy with Gothic ornament and filled with profound nostalgia, set the standard for a generation of writers, including Lucius Beebe and A. J. (Joe) Liebling. Beebe was by far the best-known writer of the Wild and Woolly School. The scion of a Boston banking family, Beebe attended both Yale and Harvard, graduating from the latter in 1928. He joined the New York *Herald Tribune* as a cub reporter in 1929. Beebe became the chief practitioner of top-hat journalism in the chic, glittering world of 1930s New York City. He was the *Trib's* drama critic, writing two columns, one of which was nationally syndicated. In addition, Beebe developed a wide national following through his magazine articles about popular history, gourmet food, and trains. He had published three poetry books and a poetry quarterly before he tackled his first major book, a history of his native Boston.[5]

Beebe began making frequent trips west in the late 1930s with the Colorado heiress Evelyn Walsh McLean to attend the opera in Central City, Colorado. It was on a 1940 Warner Brothers' publicity junket for the movie *Virginia City,* when the studio sent trainloads of writers and stars thundering across the heartland from New York to Reno, that Beebe met his future Virginia City neighbors. Those aboard included Duncan Emrich, then an English instructor at Columbia University; Emrich was sent along by Warner Brothers to instruct the reporters on the ways of the West. Although the director had not shot a single scene in Virginia City, Warner Brothers was in the habit then of hosting elaborate film premieres in their actual historic settings. The studio put on a three-day extravaganza in Reno and Virginia City that remains unparalleled, with "an eighteen-car Southern Pacific special, including two baggage cars of horses, saddles, Prince Albert coats, trousers and Remington derringers . . . wafting a posse of stars and executives."[6]

Beebe recalled the fascinating disorder of the occasion, when film people, in their cups, met head-on with the reality of Virginia City. The film was "a real stinker and even Comstockers, not notably sensitive to historic record, took offense at it and expressed their opinion in a variety of stimulating ways," recounted Beebe.[7] Virginia City natives, furious at some assumed snubs, threw rocks at Randolph Scott. Artist Sheldon Pennoyer quipped, "I wouldn't blame them if they killed him. It's the world's worst film and I think everybody connected with it should be shot."[8] Beebe returned to Nevada in 1947, along with his partner, the photographer, Charles Clegg, to write about the Comstock's fabled Virginia and Truckee Railroad for a series of railroad books. The celebrated pair arrived aboard their own railroad car, *The Gold Coast.*

Yet another follower of the Wild and Woolly School was Duncan Emrich, a brilliantly schooled intellectual who had summered in Virginia City since 1937. He came to Nevada to obtain a divorce and fell captive to the Comstock's abundant, free-spirited ways. In addition, there were the Katies, Katherine Hillyer and Katherine Best, well-seasoned, slick magazine writers, most recently London-based war correspondents. They were freelancing for the Curtis and Hearst stable of magazines—*Saturday Evening Post, Gentlemen's Quarterly,* and *Ladies Home Journal.* Much the same can be said for another Virginia City import, Roger Butterfield. The writers were well acquainted from New York days during the 1930s. Katie Best, like Beebe, had written for *Stage,* Raoul Fleischmann's other New York magazine (Fleischmann provided the greatest share of financial backing for Harold Ross's *New Yorker*).

The writers coming to the Comstock during the 1940s were continuing the moveable feast begun in Paris during the 1920s and Manhattan of the

Virginia City's C Street in the 1940s. (Courtesy of Comstock Historic District Commission)

1930s. The Comstock writers' close associations with New York, London, and Parisian friends continued throughout these years. Beebe remained aligned with *The New Yorker, Town and Country, Life, Saturday Evening Post,* and *Gourmet* magazine brethren, many of whom visited Nevada on a regular basis, if only to establish the six weeks' residency necessary to obtain divorces; many came and stayed, and "a sense of place" developed in their writings about the state.

When twentieth-century writers and historians first looked to Virginia City's Comstock Lode, they found male roles conveniently and conventionally defined and larger questions of race, gender, class, and family unanswered. The magnificent wealth of the Comstock was long seen as a panacea. While the view of the West as a land of inexhaustible resources differs greatly from the despoiled reality of today, the writers who arrived during the 1940s saw the Comstock and the West as a place of wonder, legend, and abundance.[9]

Ironically, perhaps the finest fiction writer in their midst, Walter Van Tilburg Clark, never wrote about the Comstock. With such western classics as *The Ox-Bow Incident* (1940) and *The Track of the Cat* (1949), Clark's novels had settings other than Virginia City, although it is clear that he had a love for his place of residence.[10] In an article for *Nevada Highway and Parks,* Clark is quoted saying about the Comstock: "[In] such a place so much of what happens affects you: tragedy, illness, good luck, and death—the elements of life. When you're on these kinds of terms, you're never superficial. The old mining spirit, the enormous view, the buildings, these things are enduring. They all count."[11] Clark seemed the odd man out. He was distrustful of new Comstock writers, seeming at times almost fearful of their intensity. Clark spent his final years editing the nineteenth-century journals of Comstock newspaperman Alfred Doten. While this, too, was a tremendous contribution, Clark failed to produce the piece of fiction that might have given Virginia City its definitive literary treatment.

Adding to the irony of the situation, it appears that DeVoto, who did so much to inspire western authors, may have had a part in ending Clark's fiction-writing career. DeVoto once stung him so badly in a review published in the *New York Times Book Review* that Clark was never known to publish fiction again. DeVoto slammed Clark's short story "Hook," characterizing Clark's writing as "a demonstration of what happens when arty writing moved to the Western landscape—low grade Western mysticism."[12]

The task of celebrating the Comstock in writing was left largely to the newcomers. Beebe and Clegg became permanent residents of Virginia City in 1949, once they found a suitable house with an appropriate theatrical and writer's pedigree. The house they purchased was associated with the West Coast opera impresario, John Piper. Once on the Comstock they wholeheartedly joined a growing artistic and literary colony. Roger Butterfield lived up the street, as did Clark. Butterfield had given up an ideal job, an editorship at the *Saturday Evening Post,* to live on the Comstock in 1945. Here, he and his wife, Lynne, put together a publication, *The American Past,* that the New York *Times* hailed as a "leviathan of a picture book."[13] He worked hard to popularize the Comstock; some of his best writing about the mining district appeared in *American Heritage,* the quarterly he helped revamp in 1954. Butterfield was evenhanded about his approach, never one for hyperbole nor cliché. Unfortunately, he never seemed to flesh out a full-blossomed Comstock historical woman.

The Katies lived on D Street in the Spite House, a Victorian painted brick red. According to local folklore, the house's builder spited those living next door by constructing on the property line, cutting off his neigh-

Café Society gone West; left to right, folklorist Duncan Emrich, writer Walter Van Tilburg Clark (kneeling), historian Roger Butterfield (standing), poet Irene Bruce (sitting), writer Charles Clegg (background), folklorist Marian Emrich (standing), and writer Lucius Beebe at an authors' August 1949 book-signing party at the Delta Saloon, Virginia City. (Courtesy Don McBride Collection)

bor's southern sunlight. The Katies' two-story house stands across the street from the mighty "Mackay Mansion," whose offices and upstairs apartments once housed silver bonanza kings John Mackay and James Fair. The Mackay Mansion had been built in 1861 by George Hearst, whose sudden Comstock fortune later helped to fund the publishing empire of his son, William Randolph Hearst. In a remarkable feat of historical continuity, this was the same fortune that, in turn, nine decades later, supported countless freelance writers living on the Comstock, the Katies included.

Both the Katies were good writers. Their fanciful piece on Las Vegas "press agentry" was included in a book about literary Las Vegas. Probably because they wrote in the pre-feminist days, much of their work was relegated to women's magazines, and they, too, came to think of themselves as less than serious. They both had been London-based war correspondents during the Second World War, but women at the time covered other women, leaving men to male reporters. The Katies were at times rushed and sloppy in their work on the Comstock, but they nonetheless were wonderful writers who made a contribution.[14]

These were the cadre of writers who settled on the Comstock, affecting the local cultural climate and responding to the swirl of national events and trends. The bright lights of Manhattan's literary world had dimmed following the conclusion of the Second World War. Stage people and literati began taking up residence in small-town America. Alfred Lunt, for example, teased Beebe that Virginia City's population of 400 was a metropolis compared to Lunt's new abode in Genese Depot, Wisconsin, with a population of ten.[15] The exodus to small-town America affected many places; the Comstock, because of its beauty and genuineness, became a magnet to the post-war writers. The newcomers' impact upon sparely populated Nevada, and on the West in general, was profound. It was on the Comstock that many writers of the purple sage met up with old money—a Dupont here, a Vanderbilt there—as well as with the progeny of the Comstock's first millionaires, including representatives of the Millses, Mayres, and Mackays.

This migration of troubadours, followed by breathlessly beautiful debutantes, camp followers of the rich, changed Virginia City's sensibilities forever. Many of the émigrés had prepped at eastern schools and universities and expected the same culture they had left behind. They were attracted to the Comstock by its aura of bohemianism and freedom. Lucius Beebe defined the Comstock as "a pattern of unrelated but felicitous circumstances." With much braggadocio he once told the Welsh travel writer Jan Morris that Virginia City housed "the greatest density of peerless and unblemished nonconformists in the world," adding that he, of course, included himself in the counting and wasted no time joining up.[16]

Another group of people attracted to the Comstock were divorcées. This contingent, portrayed in Nevada-related fiction as pale blondes with soft centers, began appearing in the 1920s.[17] The town's economy, like that of neighboring Reno and Washoe County, was heavily underwritten by a brisk divorce trade, as would-be divorcées established the required six weeks' residency. Virginia City afforded the tantalizing possibility of rubbing elbows with someone rich or famous. It was not unusual to meet Saul Bellow or Joe Liebling, Salvador Dali or Robert Caples, at the Delta Saloon. It was Liebling who told the playwright Arthur Miller about a story he planned to write for *The New Yorker* about a debonair cowhand named Hugh Montbank who squired pretty blonde divorcées about town. During the thin times, Montbank earned his living as a mustanger, catching the free-roaming wild horses and selling them to dog food companies.[18] Eventually, Liebling's articles became the genesis for Miller's movie *The Misfits*, a culmination of divorcée genre fiction. Miller's marriage to the actress Marilyn Monroe was in the final stages of disintegration during the 1960 filming, and he was staying at the Mapes Hotel in Reno, preparing the daily rewrites for the *Misfits* as the cast and crew played in Virginia City.

While the 1940s Comstock writers were sophisticates, they were fascinated by lowlife figures—the drunks and the prostitutes. As was so often the case, Beebe, who never bothered with low life, was the exception; even his demimondaines wore gowns from the House of Worth. Like Liebling, Emrich worked hard to maintain a cachet of a certain anti-intellectualism and provincialism. Instead of writing for the quarterlies, where Emrich would have been forced into some restraint, he chose to reproduce the barroom vernacular in books published for popular eastern houses—Bonanza Books and Vanguard. Guiltily, Emrich warned readers in his acknowledgments that true scholars of folklore and history should consult the *Western Folklore Quarterly* and historical quarterlies.

As historians, the Comstock writers felt little compulsion to restrain their writing within a scholarly format, nor did they feel a need to adhere too closely to facts. Creating fiction within this school were people such as the novelist Vardis Fisher, who stopped midstream in writing his epic life's work to sink his line into the Comstock pond. The result was *The City of Illusion*, a novel about an indefatigable Comstock woman, Eilley Orrum Bowers.[19] Joe Henry Jackson, who edited *Sunset* magazine and became the West Coast's most influential book reviewer through the *San Francisco Chronicle*, said about Fisher's treatment of Bowers, "the student of the period will time and time again come across dates that do not match up, characters out of their time, incidents lifted bodily and used in other settings than their own. But what of it? Mr. Fisher is not writing the truth in the limited historical sense. He is, however, writing essential truth, which is far more to the point."[20]

The writers of the Wild and Woolly School felt it was more important to convey the essence of a situation rather than hard-boiled facts. Beebe was a great practitioner of this cause, and refused to limit himself in any sense, historical or personal. There is a marvelous story, retold by a Howell-North house editor, of working with Beebe on a railroad book. When the editor complained that a certain photo caption did not fit the historical photo supplied by Beebe, that it was the wrong train, the wrong station, even the wrong century, Beebe bellowed, "It's the story, the ambiance, the essential truth I am looking for, not the facts." Beebe then marched to the paper cutter and whacked off the offending part of the photo. "There," he said.[21]

Similarly, Duncan Emrich's departures from fact are legendary. Emrich was a golden boy, handsome and reckless, with a wacky sense of humor. Born to American missionary parents in Turkey in 1909, he was among the first of the eastern intelligentsia to arrive on the Comstock in the late 1930s. Like Emrich, a great many were attracted to Virginia City because it was naughty. This "naughtiness," left over from the exclusively male camp days, became a heady wine to prim easterners. Emrich became tipsy with the concept of the "whore with the golden heart," and made Julia Bulette the most important Comstock cultural icon of the 1940s.

The Comstock is celebrated in this literature not as a place of nineteenth-century technical or entrepreneurial leadership, or for its journalistic excellence, but as a town made famous because of Julia Bulette. As a result of the efforts of the Wild and Woolly writers, the great body of writing about the Comstock today is part of the soiled dove genre, quasi-historic and sensationalized.

While many historians of the 1940s found the soiled dove an enduring symbol of the West, the popular press of Julia Bulette's day, the *California Police Gazette,* was not sympathetic. Instead, it referred to the girls as "claptrap sufferers." In 1936, with the publication of Margaret Mitchell's *Gone with the Wind* and her creation of the unforgettable Belle Watling, a madam became an acceptable mainstream American fictional character.[22] Watling's western counterpart was similarly successful. According to Emrich: "Julie's legend has grown so that she has become the symbol of the western 'lady,' and her reputation has spread among the girls so that they look upon her almost as a saint of the profession, a more contemporary Magdalene."[23]

So, in addition to becoming socially acceptable in popular fiction and movies, there is another aspect to the *belle du jour* scenario. It became fashionably chic to slum; Emrich and his wife enjoyed cocktails at the D Street brothel with the town madame, Gertha, a black woman, who had set up shop a stone's throw from where Julia Bulette's cabin once stood. Appointed chief of the Archive of the American Folk Song Section of the Library of Congress in 1945, Emrich developed and became the first chief of the Folklore Section (now the American Folklife Center) there in 1946. Emrich took it upon himself to create, define, and write the folklore of the West, with particular emphasis on Virginia City. Rather than expend his efforts in documentation, Emrich outfitted himself with the latest recording equipment and taped his Comstock lore and legend, choosing, of all places, the continuously noisy Delta Saloon. The hilarious result was hundreds of hours of garbled tape punctuated by the sounds of the spinning roulette wheel, shrieking children, and always the merry tinkle of bar glasses.[24] Emrich aggressively interrupted his subjects, repeatedly asking about the whorehouses and his beloved icon, Julia Bulette. Ironically, old-timers remembered little about her; she appears to have been an inconsequential figure.

Essentially, Emrich failed to see the whole picture. Beebe and Clegg's revival of the legendary Virginia City newspaper, the *Territorial Enterprise*, provides an illustrative example of this failure. They asked prominent writers and historians to serve as contributing editors. It was their desire to have the newspaper reflect a rich western heritage and to create a scholarly quarterly with important historical writing and important book reviews on regional subjects by authorities in their fields. And they wished to accomplish this without losing the colorful nature of Virginia City and its cast of characters. In their selection of editors, Roger Butterfield suggested that Emrich might be honored. The result was disastrous. Emrich so exasperated Beebe and Clegg as they were putting together the newspaper during the spring of 1952 that Beebe, in letters to Butterfield, asked Emrich to resign. Beebe wrote about Emrich's ideas of "saloon carnivals, phony societies, hoaxes, gags, the reprinting of his own somewhat inappropriate works and promotion generally that would alienate everyone we are trying to solicit . . . he imagines we are reviving the *Enterprise* as some sort of juvenile gag in which he can ride to promotion and publicity on a burro." Beebe and Clegg did not need to worry. Disappointed by the lack of funding for his "folk industry" project, Emrich left the Library of Congress two months after the *Enterprise*'s launch. It was not only professional difficulties which plagued him, but those in his personal life as well. This was particularly true after 1953, when he was involved in a distressed marriage and when his "avid social" drinking appears to have found its way into his workplace. His enchantment with the Comstock passed on, too, although its manifestation, the canonization of Julia Bulette, has come today to define the mining district.[25]

All of Emrich's circle—Beebe, Roger Butterfield, and the "Katies" (Katherine Hillyer and Katherine Best)—fell captive to the myth-making. Even the Virginia & Truckee Railroad, looking for a publicity angle, tarted up baggage car No. 13 and renamed it the *Julia Bulette*. Schoolbook author Effie Mona Mack took a turn at the story in *The Life and Death of Julia C. Bulette*, under the nom de plume of Zeke Daniels. Her book was published by Lamppost productions, an imprint of the Bucket of Blood Saloon in Virginia City.[26]

Beebe referred to Bulette at various times as a "courtesan," "a madam," and the "Comstock's most compassionate strumpet." He could not allow her to slosh through history as a mere prostitute plying her trade from a small, plain, shotgun house on D Street; he felt compelled to embellish her memory with Parisian finery. His portrayal of her in his 1950 *Legends of the Comstock Lode* differs greatly from his more seasoned approach taken sixteen years later in his "Julia Bulette: The Comstock's First Cyprian." This second essay was published posthumously in the *Lucius Beebe Reader*, a collection edited by Charles Clegg and Duncan Emrich. In this work, Beebe concedes that Julia became an "aspect of regional folklore" and that only in death did she achieve folklore status. He could find no account of Julia's arrival in 1860s Comstock newspapers; had she been of sufficient consequence, Beebe reasoned, "the *Enterprise* would have certainly hailed her coming with a suitable nosegay of agate or long primer." Beebe ignored the fact that the daily roster of train, stage, and hotel arrivals and departures appearing in Comstock newspapers seldom included people of color or women of certain professions. Beebe, in his final tribute to Julia, concluded in a moralistic tone. He wrote that Julia lived the consequences of her own character. Yet even as he began to paint the bleak picture of her life, Beebe suddenly pirouetted in mid-sentence. With wild hyperbole, he wrote that Julia's "dinner table groaned under its load of Lucullan wonderments." Beebe said that "her fees were high, legend sets them at $1,000, but even if this figure is perhaps apocryphal, she still achieved overnight success both socially and economically." Julia, he concluded, emerged as Virginia City's "most controversial figure."[27]

Beebe and Clegg's grandiloquence on Julia's behalf and their sympathetic treatment of the soiled doves was calculated to garner national attention for their revitalization of the *Territorial Enterprise* newspaper. Beebe had spent a formative year between his expulsion from Yale and his acceptance to Harvard writing for the scandalous Boston *Telegram*. There, Beebe learned to compose a wicked Hearstian headline. He ultimately used his position at the newspaper to blackmail his way into Harvard, promising President Angell to quit the *Telegram* and stop writing about potential Harvard donors' Back Bay debutante daughters in compromising headlines.

Beebe and Clegg knew that their tolerant stance regarding prostitution did not set well with the Virginia City folk. The old-timers remained decidedly Victorian in their outlook. As John J. Count Mahoney grumbled in a letter, "I am of the opinion that one who eulogizes prostitutes for commercial, literary reasons does more harm than a pimp."[28] Beebe's publicity became a mixed blessing for Virginia City. Somehow the town became cemented in the national psyche as a place made famous because of its prostitutes. Like so many who came to the Comstock colony, Beebe and Clegg did not mean to evoke the flamboyant Virginia City that their writing created. When their mythologizing was manifested in the television series *Bonanza,* they were appalled. Trampled by tourists, the Comstock transformed itself to appeal to the new economic bonanza. Beebe, Clegg, and those of like mind were horrified as they saw the charming but delicate remnants of the nineteenth century yield to these new forces.[29]

An assortment of writers of both fiction and nonfiction also participated in the myth-making of the Comstock. Among the most recognized novelists was Vardis Fisher, who wrote of local heroine Eilley Orrum Bowers in his 1941 work, *City of Illusion*. Bowers captured popular imagination by winning and then losing millions. She finally died in poverty after spending her last decades marketing herself as a fortuneteller. Today, Fisher is known primarily for providing the inspiration for the film *Jeremiah Johnson,* which was loosely adapted from his book, *Mountain Men*. Although Fisher's work on Eilley Orrum Bowers is not considered his best, it is a wonderful piece of writing. Fisher's first wife had committed suicide, and in *City of Illusion* we find in his version of Bowers, a simple little Scottish woman, who, when confronted with unfathomable sorrows, is transcendent. Bowers was a larger-than-life figure even in her day; in the twentieth century she has served as a foil to Julia Bulette and is the subject of count-less books and articles.[30] Comstock writers of the 1940s liked Eilley Bow-ers. As Beebe said, there is joy to her story; one relishes its humor while one winces at the brutality of her foes.[31]

Of the 1940s women novelists to write of the Comstock and develop its myths, Zola Ross became the Danielle Steele of Nevada, producing three Nevada titles.[32] Ross drew from her visits as a child to Virginia City for her *Bonanza Queen*. While her fiction is solidly in the style of Margaret Mitchell, with all her breathless, ruined, antebellum finery, Ross's Nevada characterizations are engaging and oddly convincing. Ross was the first fic-tion writer to apply post–World War II sensibilities to Comstock women. Her heroine, Rene Courtelot, is the sixteen-year-old daughter of the dis-tinguished, southern, down-at-the-heel newspaperman, Tempest Courte-lot, who is writing for the *Territorial Enterprise*. The action is staged before the Great Fire of 1875, when Virginia City was at the very epicenter of the universe. The heroine, Rene, having been raised by her widowed father, is worldly and educated yet still innocent about the crasser side of life. She is a southerner from the same red earth that gave up Margaret Mitchell's Scarlett. The novel finds her trading barbs with Dan De Quille and speak-ing French. She is liberated, sophisticated, and falls for the rakish, hand-some, Rhett Butleresque millionaire. Although Ross's work is genre, it is, like Fisher's writings, good genre.

Ross's male counterpart is the pulp writer Louis L'Amour. Even in pulp fiction, men are judged differently from women. L'Amour's protagonist, Val Trevallion, a miner from Cornwall, is based loosely on Nevada's U.S. Senator John Percival Jones, although Trevallion came to the Comstock as a loner with a mysterious, possibly violent past. L'Amour described his heroine, Grata Radii, as radiantly beautiful, innocent, and true. Her work as an ingenue allowed her to break from the confines of Victorian behavior but to remain pure of heart and soul. There is no haunting past for Grata to reconcile; instead there is an unfulfilled need.

The far western fictional (i.e., male) stereotype, according to Bernard DeVoto, was based upon "a pattern of platitudes and convention" that had not been broken since Bret Harte embodied them in 1869.[33] Louis L'Amour exploited those same conventions quite successfully. L'Amour, who had come briefly to Virginia City during the Depression to find work in the mines, returned as a vigorous man of seventy-two to promote his novel, *Comstock Lode*.[34]

In the *Silver Platter*, written by Ellin Mackay Berlin and published in 1957, we find a unique, intimate portrait of the author's grandmother, Louise Mackay of Virginia City. Berlin, while not a resident, was well known to the Comstock literati. Numerous popular twentieth-century accounts, including those authored by Beebe, refer to Marie-Louise Hungerford Bryant Mackay, but they are often unfavorable.[35] Her granddaughter's long-awaited biography took ten years to complete. Marie-Louise Hungerford was born in New York to poor parents who emigrated to Downieville, California. There, at fifteen, she married a young physician named Edmond Bryant. Unable to resist the magnetic tug of the Comstock, the young couple soon found themselves in Virginia City. By 1863 Dr. Bryant had become hopelessly addicted to morphine and had cast his family, including two daughters born in Virginia City, into abject poverty. Louise Bryant took in sewing and taught at the Daughters of Charity school until, as a widow, she met and married the kind, Dublin-born, mining millionaire, John Mackay, who promised her wealth and delivered a life on a silver platter.

As a writer, Ellin Berlin possessed a natural flair, although there is a provincial quality to her work. She was a flapper who broke into print in 1925 with an article on speakeasies for Harold Ross's fledgling *New Yorker* magazine. Beebe's 1957 review of her book is scathing: "Ellin Berlin's long awaited biography of Mrs. John Mackay," wrote Beebe, had become a matter of controversy among her readers "not over content but whether she should have written the book at all or if it should have been done by a professional in the field of western or American social history."[36]

Ellin Berlin's biography is flawed by her omissions. Like her male counterparts, she fails to recognize many important influences, particularly those Comstock role models who gave Louise Mackay the courage to endure. Berlin barely examines the influence of Catholicism or the importance of Sister Frederica S. S. Xavier and the intercessory role the Daughters of Charity played on the Comstock. Mackay's beliefs may have kept her from despair, but the sisters gave her employment, a more tangible gift. In 1875 both Louise Mackay and Theresa Fair made the building of the sister's Mary Louise Hospital their most important charity.

Perhaps the most serious flaw in Berlin's book is her own lack of awareness. One wonders whether women's inferior position ever occurred to Louise Mackay. While the struggles of the 1860s—especially the issue of slavery—seem far more fundamental than did the later women's suffrage movement, Berlin's biography fails to mention what Louise Mackay thought about any of those nineteenth-century "isms" played out in Comstock lecture halls and stages: unionism, feminism, abolitionism, Fenianism, and spiritualism. Indeed, there is no question that Berlin also took part in the myth-making of the Comstock by placing her grandmother in the rarified atmosphere of a sanctified heroine. Using her grandmother's scrapbook, Berlin's book does, however, afford readers poignant insight into the bleakness widowed women of the Comstock endured.

One of the most interesting people to appear on the vast Comstock nineteenth-century stage was Mary Ellen Pleasant, a woman of color who was born into slavery, earning her freedom through marriage and becoming a *gens de couleur*. After her husband's death, she returned to the New Orleans area, where she became an active participant in the Underground Railroad. Fleeing West in fear for her life, she settled in San Francisco, where she operated boardinghouses and earned distinction for her culinary arts. She was a frequent visitor of and an early investor in the Ophir, the famous Comstock mine.

Pleasant was also intimately connected with the Comstock Lode through Thomas Bell, a director of the powerful Bank of California, the Virginia and Truckee Railroad, and ultimately the Union Mill and Mining Company. It is said that one stock tip from Bonanza king James Fair earned her 30 million dollars. Pleasant was a practitioner of voodoo, having studied with Marie LaVeau while in New Orleans; her religion was a jambalaya containing voodoo, African ancestor reverence, Native American pantheism, and a goodly dash of Catholicism and occultism. It was her plan to make money on the Comstock through stocks to aid her people. A letter from her promising more funds was found on the personage of John Brown upon his capture. She was an active participant in Sarah Althea Hill's protracted divorce from Nevada Senator and Comstock baron William Sharon.

"Mammy Pleasant," as she was known, has been the subject of many books. Paperback novelist Helen Holdredge made her the subject of two books, published in 1953 and in 1954. Working with enviable primary source materials, some belonging to Pleasant, Holdredge succeeds in producing sensationalized accounts, breathlessly telling of Pleasant's supposed sexual doings. A careful telling of her life awaits an author: she would make a remarkable subject, whether of a novel or of a nonfiction work. It is a tale that needs telling without the taint of racism or sexism and without Holdredge's snide humor. Mary Ellen Pleasant was truly a remarkable bird of plumage who passed through the Comstock. She was remarkable because she fully participated in commerce and was neither diminutive nor dutiful, as women of color were supposed to be.

An area of writing that evolved significantly during the 1940s was the memoirists' school, which drew upon the experiences of those who lived on the Comstock. It became a special section in the *Territorial Enterprise*, and these accounts are an incredible source of information on women, minorities, and children of the nineteenth century. These recollections belong more appropriately to the Real West school, because they attempt to capture the history of the place realistically and in a balanced fashion. Two of the most important of these books were Harry M. Gorham's work and John Taylor Waldorf's stories of growing up on the Comstock. Gorham's *My Memories of the Comstock* is a wonderful coming-of-age account of a young man of eighteen. He came to the mining district in 1877 to work for his uncles, Senator John P. Jones and Sam Jones, and their extensive mining interests. Likewise, John Taylor Waldorf's stories of growing up in the shadows of Saint Mary in the Mountains, *A Kid on the Comstock: Reminiscences of a Virginia City Childhood,* is another important Comstock work that was published through the Friends of the Bancroft. Miriam Michelson, who was born on A Street and became an important journalist and author, also wrote of her childhood recollections. Small Nevada presses and foundations have been responsible for a number of these publications as well as for reprints of articles, abstracts, and booklets about nineteenth-century Comstock life. David Basso's Falcon Hill Press, for instance, reprinted David Belasco's 1914 *Scribner's* magazine piece about the playwright's days at Piper's Opera House in Virginia City.[37]

Comstock memoirists—Gorham, Waldorf, and certainly Belasco—paid eloquent tribute to the women of the Comstock, and there is little bias found in their writing. Of course, in the case of Waldorf, as a child, the most central figure in his life was his mother. Only Belasco refers to the tenderloin, the red-light district, which stood cheek by jowl with the theater district on D Street in the 1870s.

The Michelson, Gorham, Belasco, and Waldorf memoirs are but one facet of a large body of writings by former residents. Uniformly, these writers speak of Virginia City as a metropolis with manners, a town elegant and refined. Still, they also refer to a family town, where schools and churches dominated the skyline. Taken together, these various memoirs embody the

Real West school that was lost amidst the popular culture mythologizing of the Wild and Woolly West authors.

After the Wild and Woolly school had its way with the Comstock during the 1940s and 1950s, certain mythic elements became integrated into the image of the past, regardless of their veracity. Julia Bulette, for example, became a prominent part of Virginia City's history after her death, though she had not figured significantly in the town during her life.

The sad fact is that much of the truth of the Comstock story was lost because of the 1940s writers' good intentions. As they sought to create a mythic Comstock in literature, they spawned a new reality that shaped the evolution of the actual place. Writing in the Wild and Woolly West School, they created an appetite among visitors for a wild and woolly western place. While the Real West quality of the Comstock attracted the writers, their fascination with the Wild and Woolly West and their subsequent publications changed the reality of the setting. The tension between the two schools is healthy, yet the true history and the great Comstock novel awaits an author. While much of the original place survives, unpainted cedar façades testify to the time when business owners attempted to market the television Cartwrights of *Bonanza* more than they did the real Comstock. The real Virginia City prided itself on being a place of refinement and distinguished tastes, one that would not have tolerated such a roughshod presence on its main street.

The 1940s writers realized too late what they had done when the television show *Bonanza* was visited upon them in 1959, together with all those "rutting cowboys with criminal sideburns," as Beebe remarked indignantly.[38] Yet Beebe and Clegg were part of the process of fictionalization. The television writers fabricated women as token backdrops having as little to do with the reality of the place as did the cedar siding that grew up to hide historic brick. Only too late did writers like Beebe and Emrich realize how quickly the dandelion seeds of myth multiply, becoming something unruly. A foreign, opportune guest, the dandelion is not strong enough to choke out the hearty natives, the bachelor buttons, daisies, Indian paintbrush, lupin, and yellow roses that flourish on the mountainside. Like the flora, the two realities of the Comstock, the original Real West and the Wild and Woolly West, now grow side by side and give testimony to the 1940s, yet another fabulous period in the development of one of the world's most remarkable mining districts.

❧ 14 ❧

Gender and Archaeology on the Comstock

Donald L. Hardesty

Introduction

Images of the women who lived on the Comstock come not only from written accounts and oral testimony but also from the material things contained in the architecture of surviving buildings, the archaeological record, and museum collections. Thus in his book, *The Past Is a Foreign Country*, David Lowenthal argues that access to the past is gained by traveling along the routes of history, memory, and relics. But the routes, he observes, are best traversed in combination: "Each route requires the others for the journey to be significant and credible. Relics trigger recollection, which history affirms and extends backward in time. History in isolation is barren and lifeless; relics mean only what history and memory convey."[1] Taking such material expressions into account, therefore, offers a richer, more comprehensive portrayal of Comstock women. Here, I explore "relics" and archaeology as a pathway to understanding the lives of women on the Comstock in the 1860s and 1870s. Rather than describing the results of archaeological studies, for very little has been done on the Comstock, I focus upon what archaeology might do to offer insights into how gender structured Comstock society and culture.

Toward an Engendered Archaeology of the Comstock

The importance of using research strategies that employ theories of gender to interpret the archaeological record is increasingly recognized.[2] Such research strategies ask specific questions about the role of gender in the social and cultural organization of human groups; they consider, for example, how the principles of gender organize work, domestic households, and landscape. Joan Gero and Margaret Conkey argue that: "By adopting gender as an explicit conceptual and analytical category, by applying gender concepts and categories to familiar and original sets of archaeological data, women are brought into view as active producers, innovators, and contextualizers of the very material world by which we know the past."[3]

The physical remains of mining camps provide an opportunity to explore the use of archaeological record in constructing an engendered history of what traditionally has been thought of as a mostly male society and culture.[4] Thus, a research strategy that uses concepts of gender to document and interpret the archaeological record of the Comstock promises to reveal much that is otherwise invisible about the lives of women and men.

The development of such a strategy, however, requires combining relics, history, and memory as sources of information about the past. James Deetz, Robert Schuyler, and others argue that they should be used together within an "interactive" framework.[5] In this approach, a beginning model or first approximation of a material world for the Comstock that reflects gender is developed from preliminary documentary or archaeological data or oral testimony. The model then is used to identify hypotheses that can be tested with new information collected by additional archaeological and documentary research. Finally, the beginning model is modified to take into account the new data, leading to yet other hypotheses, a second generation, that can be tested with more research. Thus, the construction of an engendered history of the Comstock from the three sources of information is cyclical and continually evolving.

The interpretation of the archaeological record requires the development of a social and cultural context for material things used on the Comstock that is gender sensitive. Rather than being examined in isolation, artifacts and other objects are understood within the context of sex/gender systems. Within this context, material things acquire meaning, uses, and functions not otherwise apparent. Indeed, material things, far from being passive, often actively motivate the actions of individuals and groups and are used intentionally by them for social purposes.

French anthropologist Pierre Bourdieu, for example, shows that tastes in consumer goods helped to create and maintain the power structure among social classes in twentieth-century France.[6] Thus, he found significant differences in the style of tableware used by women in working-class and bourgeois families while entertaining visitors in their homes. Women in working-class families focused upon the substance of the food served at the meal and used common tableware. The meal symbolized the *contrast* between the home and the public worlds. Working-class families entertained only friends who could be treated like family and who they felt "at home with." Women in bourgeois families, on the other hand, focused upon the form of the service used for the meal and used special-purpose tableware. The meal symbolized the *continuity* between the home and the public worlds. Bourgeois families often entertained professional and business acquaintances toward whom they felt social obligations and who required an appropriate ritual performance.

In addition to developing an engendered context for material things, the approach used to study the material expression of sex/gender systems on the Comstock should also be comparative. Archaeologist James Deetz, for example, argues that: "It is proper, then, that we look at everything we do in a comparative, international perspective. . . . One can hold both culture (form), for instance, English society, and time constant and observe how a national European culture at a particular time manifested itself in two different places. At this point, the archaeologist would ask how English culture was different, and how it was alike in these two locations."[7] In the same vein, sex/gender systems in mining camps and their material expression should be compared on regional, national, and global scales for differences and similarities. How and why they differ are equally pressing questions.

The Material Expression of Sex/Gender Systems

The material expression of sex/gender systems on the Comstock includes both *symbolic* and *ecological* dimensions. Material things often serve as symbols of socially and culturally defined categories such as gender identities and carry strong meaning.[8] The symbols may be a large Italianate-style house, a style of tableware, designed gardens, a parlor for entertaining, or the rodeos and all-important cowboy boots, hats, and clothing of male cattle ranchers.[9] Sex and gender identities involve the acquisition, maintenance, and control of material symbols communicating a distinct message. Consumer behavior, therefore, is an important clue to sex and gender categories.

Understanding the material expression of Comstock sex/gender systems begins with the emergence of national markets, industrialization, and a middle class in nineteenth-century America. For the first time, national advertising and the pursuit of middle-class status created a large demand for material goods. As Daniel Sutherland points out, "this desire, increasingly realized through the sale of mass-produced imitations of upper-class houses, furniture, clothing, and art, had, by the 1870s, begun to transform the appearance, even the character, of everyday life."[10] Without question, the consumer revolution also transformed everyday life on the Comstock during the 1860s and 1870s. Not only did the vast wealth coming from the silver mines greatly increase the purchasing power of the community, but the completion of the transcontinental railroad by 1869 opened up markets and made many more commodities available. During the Victorian period between 1875 and World War One the demand for material goods intensified even more, and America was rapidly transformed into a mass consumption society.[11] The period was marked by the emergence of an ideology of conspicuous consumption and by the homogenization of material life through mass factory production of commodities, large retail outlets, mass advertising, and mail order marketing.

At the same time, it is clear that the shift to a mass consumer society and culture did not mean simply purchases of more commodities but the reinterpretation of new goods within the contexts of evolving sex/gender systems. The role of women and men as active agents in using material things to bring about change in the construction of sex/gender systems should not be overlooked. Consider, for example, Mary Ryan's study of Oneida County, New York, during its capitalist transformation between 1790 and 1865.[12] In contrast to Marxist-feminist arguments portraying women as passive victims of structural inequalities in a new capitalistic social order, she found that women as well as men in the county actively took advantage of new opportunities not only to change the family but also to create a middle class in the community. Out of the changes came the new gender identities defined by a widespread Victorian ideology that assigned middle-class men and women dramatically different gender identities by creating two separate worlds or spheres: the public/commercial sphere of men and the private/domestic sphere of women.[13]

Paula Petrik's study of Helena, Montana, shows the active role of women in a western mining town during the same time period as the Comstock.[14] Middle-class women in early Helena between 1865 and the 1880s, for example, pursued a much wider range of economic activities both inside and outside the home than was typical elsewhere in the United States. At least part of the reason for this lies in high divorce rates, large age differences between marriage partners, long business trips taken by husbands, shortages of women in the community, and poorly developed schools and churches (the key power base of American women at the time), making it possible for and even encouraging or forcing the women to search for new economic and social opportunities. In addition, Petrik documents the role of many of Helena's single, young, underclass women during this period as "proprietor-prostitutes," actively working not only as successful capitalists but also as social entrepreneurs in the Helena community.[15]

How women actively used material things such as house architecture and tableware to "socially reproduce" the domestic sphere is illustrated by Diane Wall's study of mid-nineteenth-century New York City women living in middle-class families.[16] On the one hand, they used artifacts as "boundary markers" to set off the home from the "heartless world of the marketplace." On the other hand, they also used material things to "display" and promote "their home's and the family's image of refinement, gentility, and fashion among friends and acquaintances" as a way of "negotiating the family's position in the perilous class structure." New York City women living in middle-class families used both the architecture and organization of the house and the style of interior furnishings to do this. The widespread use of domestic Italianate architecture by the middle class, for example, lent itself easily to being used both as boundary markers and for social displays. Its potential for elegance and rich architectural detail allowed "householders to make statements about wealth, fashion, and taste on an unprecedented scale" and "emphasized the role of the women in the house as social negotiators in making a visible, public statement of the family's class position."

New York City women living in middle-class families used the layout of the house and its furnishings to symbolize and fulfill their dual roles as "moral guardian" of the family and as "social negotiator" of its class

Virginia City was urban and industrialized with structures close together and smoke stacks everywhere. (Courtesy of Comstock Historic District Commission)

Virginia City's businesses included the same wide variety of services one would expect in any refined community of the nineteenth century. (Courtesy of Nevada Historical Society)

position in the community or in the outside world. Consider, for example, the dramatic contrast between the front parlor and the back dining room in the typical middle-class family residence in nineteenth-century New York City.[17] The family used the front parlor for public purposes as a display to negotiate and to reproduce social status. Here, women entertained friends and acquaintances with tea parties and other ceremonial or ritualized activities and used teaware reflecting the special meaning of the occasion. Thus, Diane Wall found documentary and archaeological evidence that Eliza Robson and her family used gilded and pedestaled European porcelain for this purpose in the mid-nineteenth century.[18] The style, according to a contemporary house-plan book, "partakes of the gay spirit of the drawing-room and social life."[19] In contrast, the family used the back dining room for private family meals. Here, the wife and mother "exercised her almost sacred role as moral guardian of family members" and used tableware reflecting the meaning of the event.[20] Thus, Eliza Robson used dishes in the Gothic style to serve family meals, which were meant to reflect "the quiet, domestic feeling of the library and the family circle."[21]

The New York example may be a good model of the material things used for social displays by women living in middle-class households of the Comstock. Certainly one can assume that Victorian culture, so pervasive in the urban centers of Europe and the United States played an important role in the organization of Comstock society. The evidence for or against such an assumption comes not only from written accounts but also from the surviving standing buildings and archaeological remains dating back to the nineteenth century. Without question, archaeology on the Comstock makes it possible to find the traces of designed gardens, floor plans, and the material life of middle-class households.

There is no reason, however, to assume that all Comstock women, whether they lived in middle-class families or not, can be understood within the context of this sex/gender system. Lee Virginia Chambers-Schiller, for example, documents an alternative gender identity for Victorian women that focused upon a divine "calling" that was higher than the worldly pleasures of marriage and domesticity. Toward that end, its followers pursued careers in the public sphere as a means of personal fulfillment.[22] Much like Chambers-Schiller, Frances Cogan documents another gender identity available to Victorian women, particularly after 1890.[23] According to Cogan, women sought fulfillment in "intelligence, physical fitness and health, self-sufficiency, economic self-reliance, and careful marriage."[24] The followers of this gender identity valued work, including domestic work, as a means by which to develop independence and moral character.

Suzanne Spencer-Wood shows that domestic reformers in the nineteenth and early twentieth centuries carrying such alternative gender ideologies used material things to bring about change.[25] They, therefore, created distinctive and new material expressions of gender identities for women. The domestic reformers focused especially upon the household, with the purpose of making women dominant in the domestic sphere. They applied "scientific-industrial principles and technology . . . to recreate housework as a female profession."[26] Among the most popular ideas for reforming housework were those advanced in manuals written by Catherine E. Beecher, Harriet Beecher Stowe, Ellen Swallow Richards, and Christine Fredericks. Spencer-Wood observes that:

> These manuals exemplify how materialistic domestic reformers created rational spatial arrangements of innovative equipment in order to efficiently organize housework processes. They designed house interiors and work sequences to minimize the distance and time required for spatial movements among equipment used in related sets of tasks. . . . The reformers promoted the application of the latest scientific technology to create efficient house designs and housekeeping machinery that would decrease the amount of work and time required by housework. The Beechers and Richards, as well as a number of other domestic reformers, also proposed dress reform to eliminate tight corsets and heavy petticoats that physically crippled women, making housework more difficult to perform.[27]

Archaeology provides a way to determine the extent to which the housekeeping machinery and architectural innovations described in the manuals were actually adopted by individual households as part of the reforms.

The domestic reformers also advocated a number of changes in the public sphere. Perhaps the best known are the household cooperatives in utopian communities, neighborhood housekeeping cooperatives, cooperative tenements and apartment houses or hotels, and working women's cooperative homes.[28] But the reformers also created new institutions that provided domestic services to the working class. These included day-care and infant-education facilities, industrial schools for girls, and public kitchens. All of these have material expressions that may be revealed in the archaeological record. In Boston, for example, Suzanne Spencer-Wood found that "starting in 1877, Pauline Agassiz Shaw founded thirty-one racially integrated charitable kindergartens that used both ideal Froebelian equipment and a variety of other toys" as part of the domestic reform movement. Froebel's toys, created to "develop children's mental, spiritual, and manual abilities" are quite distinctive and show up in the archaeological remains of the playgrounds associated with schools instituting the reforms.[29] Whether similar public reforms occurred on the Comstock is an interesting question that can be answered by combining archaeology and the methods of documentary history.

Not all gender identities on the Comstock, however, are as well defined and subject to material expression as are the ones discussed above. The reminiscences of Mary McNair Mathews, for example, who lived in Virginia City and neighboring towns in the 1870s, illustrates what may have been a common gender identity for middle- and working-class women on the Comstock. A widow from Buffalo, New York, Mathews came to the Comstock in 1869 to investigate her brother's murder and stayed for nearly ten years. While there, she worked as a seamstress, schoolteacher, nurse, laundress, lodging-house operator, and businesswoman and entrepreneur, investing in mining stocks and real estate. She was a proud, independent, and bigoted "survivalist," who held strong beliefs about temperance and child abuse. The many roles played by Mathews suggest a gender identity with a quite complex material expression that potentially varies greatly from one individual to another.[30]

The archaeological study of sex/gender systems on the Comstock also must take into account the multi-ethnic composition of the population, which greatly increased the diversity of gender identities in the community. Chinese women on the Comstock, for example, participated in several household types, engaged in many occupations, belonged to different social classes, and lived under a wide variety of conditions.[31] A few lived in seclusion as traditional wives in wealthy households of physicians or merchants. Some worked as restaurant cooks, laundresses, and domestic servants, occasionally even participating in the running of Euro-American middle-class households. Others worked as prostitutes and lived in brothels, sometimes under horrendous conditions. Chinese working-class households on the Comstock seldom included women, who often had to be purchased from one of the Chinese tongs. All of this suggests that the potential for creating gender categories within the Comstock Chinese community is enormous.

Perhaps the key problem with using archaeological strategies to study sex/gender systems on the Comstock is identifying house sites or other archaeological features that have been formed by households that included women. In most cases, finding Comstock women in the archaeological record demands good grounding in documentary history or oral testimony. Sex or gender markers are not as obvious as one might expect, for example, and the artifact assemblages of all-female and all-male households are remarkably similar in many cases.

Some differences, however, can often be found. Using well-documented archaeological sites throughout the American West, for example, Catherine Blee found distinctive artifact profiles for different household types: family, all male, all-female brothels, and military.[32] The differences, however, are quantitative (e.g., differences in the percentage of some artifact categories such as decorated tableware or canning jars) rather than qualitative. Ethnicity introduces yet another issue. We know little, for example, about the material expressions of Chinese women on the Comstock or the households in which they participated. Some studies from elsewhere in the American West, however, suggest that such things as fan handles, jade rings, marriage cups, traditional clothing, jewelry, grooming items (e.g., bone hair picks), and cosmetics may be good artifact markers.[33] Even here, however, class confounds the material expressions of ethnicity. Only Chinese women in traditional upper-class households, for example, wore shoes for bound feet.

The Archaeology of Soiled Doves

The lifestyles of Comstock women take place within a social and cultural context that includes, among other things, gender, class, and ethnicity. Certainly there is a relationship between the gender identities available to women and their lifestyles. Comstock women working as prostitutes are a good example of such a relationship. Alexy Simmons studied the material expression of prostitutes as a gender identity from several mining towns, including the Comstock's Virginia City, from oral testimony and from documentary sources such as probate records, biographies, newspapers, photographs, and fire records. The material expression includes personal belongings, household furnishings, and architecture. She defines three patterns, which vary according to the social status of the prostitutes and their clients: better brothels and parlors, mid-status houses, and low-status establishments.[34]

High-class prostitutes typically occupied houses with more than one story that had cooking and dining facilities. In 1866, for example, Cad Thompson managed a two-story parlor house on D Street and Sutton Avenue. The "working girls" lived in the basement, and the upstairs housed a parlor with a piano. Jessie Lester's probate inventory provides detail of how a high-class establishment was furnished.[35] The D Street brothel housed two parlors downstairs furnished with chandeliers, lace curtains, carpets, antimacassars, two sofas, and ten chairs, all mostly unmatched.

Five upstairs bedrooms contained spittoons, basins, water pitchers, linen window shades, carpets, and beds with spring and wool or horsehair-and-wool mattresses. Plaster busts, clocks, bottles, vases, and other items also furnished some of the bedrooms. In addition, Jessie Lester furnished her bedroom with a matched mahogany bedroom set.

The personal belongings found in the better parlors and brothels include those associated with commercial sexuality and entertainment such as birth control devices, female tonics, perfumes and cosmetics, musical instruments, board and card games, paraphernalia for alcohol and opium use, clothing, and personal adornment. Toys and other children's artifacts appear often. In addition, high-class houses often kept male grooming items, such as razors, for overnight clients. In contrast to the situation in lower-status establishments, in high-class establishments personal belongings abundant, varied, and relatively expensive.

Mid-status prostitutes usually occupied less luxurious houses but gave the appearance of being relatively well off. Julia Bulette's establishment illustrates the material image of a mid-status courtesan in Virginia City.[36] Perhaps the most famous of the prostitutes on the Comstock, she was murdered in 1867, much to the sorrow of the miners and to the delight of married women in Virginia City, who served cookies to the jailed murderer. She was a self-employed prostitute who lived in a small two-room house on D Street, which she rented but furnished on her own. The house contained only a parlor and a bedroom. In the parlor was a carved black walnut set made up of a sofa, two rocking chairs, four matching chairs, several unmatched cane-seated chairs arranged around a stove, lace curtains, hanging gas lamps, a carpet, and a spittoon. The bedroom was furnished with a small wood box stove, a mahogany bedstead, two wash basins, two spittoons, coal oil lamps, damask curtains, window shades, fancy bedspreads, white lace curtains, and rugs. Clothing included several silk, linen, and cotton dresses; skirts; a riding habit; chemises; drawers; women's shoes and hats, corsets, and stockings. Other personal belongings included a box of hair, a parasol, and several pieces of jewelry. The personal belongings in general are less abundant, diverse, and costly than were those found at the better brothels and parlors.

The material image of low-status prostitutes, often housed in rooms over saloons, brothels, cribs, cottages, and shacks, is typically submerged in the remains of other activities and is difficult to find. Certainly the personal belongings are typically sparse, not very diverse, and cheap. In Virginia City the newspapers described Nellie Sayer's Bawdy Saloon and the Corcoran Union Saloon as "low one storey doggeries," with barrooms in the front and bedrooms in the back. What they looked like inside is suggested by the following description of the interior of one of Denver's Market Street cribs:

> Through the door of a crib a man entered a small reception room furnished with a chair and perhaps a small couch. Awaiting him was a woman clad only in a nightshirt or kimono. Once money had changed hands, she would lead him through another door or heavy curtain to the other of the crib's two rooms, the boudoir or "workshop." In it the customer would see a washstand holding a pitcher, basin, and bottle of carbolic acid, a clothes trunk, and a decrepit iron bed with a bright but dirty spread.[37]

The artifacts excavated from the archaeological remains of brothels provide another source of information about the material expression of prostitution that is independent of documents. Catherine Blee, for example, studied the artifact assemblages excavated from the remains of two well-documented brothels in western mining towns. She found that liquor-related items and bottle stoppers/caps dominated the assemblages. Together they made up almost half of the total number of artifacts found.[38] Certainly the consumption of alcoholic beverages is one of the principal activities that takes place in brothels, and in this way their artifact assemblages are quite similar to those of saloons. Female-specific artifacts are relatively common, greatly outnumbering male-specific artifacts and nearly three times as abundant in the assemblages of brothels as in family household assemblages. Undecorated dishes are also common, making up nearly 14 percent of the assemblage. Decorated tableware makes up less than one percent of the total. In this way, the assemblages of brothels are similar to those of restaurants, hotels, and boardinghouses.

Donna Seifert's study of artifact assemblages excavated from turn-of-the-century brothels in Washington, D.C., is another example. She found that the assemblages of brothels and other working-class households in the same neighborhood were quite similar. She concludes, however, that although both contained a low percentage of kitchen artifacts, brothel assemblages had a "relatively high percentages of clothing, personal, tobacco, and activities (e.g., toys, heating and lighting devices, hand tools) artifacts."[39]

The Archaeology of Women's Living Conditions

The ecological dimension of sex/gender systems on the Comstock focuses upon the material conditions of existence. Certainly the ecological niches of gender identities such as prostitute or domestic reformer vary in diet, housing, crowding, disease, violence, and the like. Archaeology provides not only a vast repository of engendered information about Comstock ecology but also sometimes a quite different view of the material conditions of existence from the ones appearing in written accounts. Thus, William Rathje and Cullen Murphy's book, *Rubbish!*, shows that quite different conclusions about American food habits can be reached by contrasting the archaeological evidence left behind in garbage cans of households with the information that the same households leave in response to written questionnaires and oral interviews.[40] The three sources of information are independent and often quite different.

In another example, Mary Beaudry combined archaeology and history to construct the sanitation, hygiene, and nutrition of nineteenth-century women working in the Boott cotton mill in Lowell, Massachusetts. She focused upon the interaction between ideology and practice and found that: "despite the penetration of elements of 19th-century domestic ideology (e.g., notions of economy and scientific housekeeping) into the design, operation, and maintenance of the Lowell boardinghouses, workers and keepers retained traditional notions of health and nutrition."[41] Her analysis of the diet of workers living in the company boardinghouses, at which she arrived through studying written accounts and archaeological remains, for example, shows that foods high in fatty meats, starch, and carbohydrates were typical. The diet not only was contrary to the new ideology of health and nutrition in the late nineteenth century but also shows that traditional food habits were retained in the face of corporate ideology. In fact, documents show that the workers were quite happy and satisfied with a traditional diet, even though it was known to be unhealthy and clashed with the image of workers' well-being that the cotton mill company tried to convey to the public. It is also possible that the retention of the traditional diet may be related to people's high caloric needs as a result of the work they performed.

In a similar way, archaeological remains, written records, and oral testimony can be combined for the purpose of interpreting the living conditions of Comstock women. Animal remains, plant remains, soil chemistry profiles such as lead content, and other archaeological data on nutrition, sanitation, and health can be combined with documentary and ethnographic accounts of not only what was eaten but also what women thought about diet and health conditions generally in the two kinds of towns. Pollen, phytoliths, and macro fossils surviving in the archaeological record can be used to reconstruct vegetation in the back lots of houses in which women lived. Another dimension of women's living conditions is housing. The material remains and written accounts of women's housing, as well as oral testimony on the subject, can be used to reconstruct not only variability in the sizes of buildings and their layout but also the number of occupants in each building or room.

Engendering the Comstock Landscape

The landscape conveys yet another material image of sex/gender systems. Archaeologist Stephen Mrozowski, for example, shows that:

> The mixed residential/commercial landscape of eighteenth century Boston was replaced by a spatial order characterized by the separation of work from the home. One effect was the eventual loss of the household's productive function. More effort was now placed on social reproduction among the upper and middle classes. Part of this process involved new roles for both men and women in the household: with production replaced by the factory, the space surrounding the dwelling which had, in traditional agrarian society, been the domain of the women, now was an extension of male status in a new social order. The urban landscape of present-day America is a product of this process.[42]

Landscape architect Robert Melnick identifies three processes and seven components of landscapes. The landscape processes include land uses and activities, patterns of landscape spatial organization, and responses to natural environments. Landscape components include circulation networks, boundary demarcation, vegetation related to land use, cluster arrangements, structures, and small-scale elements.[43] Engendering the Comstock landscape through archaeology involves looking at each of these as a possible material expression of sex/gender systems. Ethnicity, of course, plays a significant role. Chinese and Native American women, for example, viewed and used the Comstock landscape quite differently from the way that Irish,

Cornish, and American-born women living in the same community did. How the principles of gender structure settlement patterns is an obvious research focus for the archaeology of the Comstock landscape; another is the use of different landscape resources by men and women. Yet another research focus is the extent to which Comstock landforms symbolize gender, such as the myth and practice of mines as a male domain.

Consider, for example, Comstock settlement patterns. The settlement patterns of some mining landscapes strongly reflect sex/gender systems through the practice of spatially segregating some gender identities such as prostitutes or social formations such as middle-class families. Compare, for example, the similarities and differences between the Comstock and the Bodie mining district in northeastern California. Both districts were organized in 1859, but, unlike the Comstock, the Bodie district did not attract a significant number of people until a rich ore body at the Bunker Hill (Standard) Mine was discovered in 1877. Not much archaeological or documentary evidence of engendered space could be found during the early years of the Bodie district between 1859 and 1877.[44] That the population was mostly male is illustrated by mining journalist J. Ross Browne's humorous but chauvinistic observation of the "joys" of domestic life without women in Bodie during his visit in 1864.[45] Still, there is documentary evidence that at least three families moved to Bodie from the neighboring mining camp of Aurora in 1863.[46] That brothels were clustered together in the same neighborhood is suggested by Browne's reference to a "Maiden Lane" in the vicinity of what later became the town of Bodie, which implies the existence of a red-light district as early as 1864.

After the discovery of the rich ore body in 1877 the Bodie district boomed, and its population reached nearly 7,000. In the boom period between 1877 and 1882, most of the people lived in the town of Bodie, but several large outlying or satellite settlements grew up around the major mines and mills in the district, especially at High Peak, the Tioga Mine, the Red Cloud Mine, and the Noonday Mine. Documentary data on the town of Bodie suggest that during that period women tended to be clustered in a few neighborhoods. Green Street, for example, was the most common location for women living in middle-class family households; Bonanza Street was the most common location for women living in brothel households;[47] and women living in working-class family households lived in other parts of town. Archaeological data primarily provide information about the gender structure of the surrounding satellite settlements, made up mostly of all-male households but with a single family household at the Tioga Mine at an arrastra site just outside the town of Bodie.

Most of the Bodie mines closed in 1882, bringing about a rapid depopulation of the district. Within five years, the number of people living in Bodie had fallen as low as 500. During the period between 1882 and 1895, the Bodie community lost much of its single adult male population to other mining districts, but miners with families tended to stay behind, hoping for better times. Middle-class households continued to be clustered on Green Street; most of the brothels remained on Bonanza Street, but only a few prostitutes stayed behind, one remaining until her death in 1900.[48] In 1895 the then-new cyanide leaching technology was introduced into the Bodie district. The cyanide revival changed gender patterns in the district but did not radically reverse the trend toward abandonment of the district. Most of the population lived in the town of Bodie, but a few people still lived in the surrounding satellite settlements. Almost half of the 202 households listed in the 1910 census were families, including women, men, children, and boarders; the others were single adult males or small co-residential groups of adult-male partners or siblings. At least one brothel with seven or eight prostitutes is also listed. Most of the satellite settlements had been abandoned by this time, but archaeological remains document all-male households at the Jupiter Mine and both all-male and all-female households at the Bodie and Benton railroad depot.

The evolution of settlement patterns on the Comstock, however, does not show the same practice of sexual and gender segregation and probably reflects a quite different population history. The 1880 population census, for example, the first to include street addresses, shows that Comstock women were more or less evenly distributed geographically rather than clustering in a few family neighborhoods or red-light districts. Although it is without street addresses, the 1860 census also suggests that even then women were well integrated into the spatial structure of the Comstock community. The pattern probably reflects, at least in part, the early migration of working-class families, such as those of teamsters, to the Comstock. In contrast, the first women moving to many other mining towns came as prostitutes or with middle-class, professional families such as those of merchants, doctors, and lawyers.

Still, Alexy Simmons gives some evidence, mostly from newspaper accounts, suggesting that at least one gender identity, that of prostitute, played a role in structuring the evolution of Comstock settlement patterns.[49] Virginia City, for example, does not appear to have limited where prostitutes worked and lived during the first few years after the Comstock strike, but most prostitutes seem to have gravitated towards the central business/commercial district. On August 13, 1863, the city passed an ordinance defining a "red light district" by making it "unlawful to open or maintain any house of ill-repute or brothel in the district of this city west of D Street, or south of Sutton Avenue or north of Mill Street."[50] Several later ordinances reestablished the district. Notwithstanding the ordinance, however, many prostitutes in Virginia City probably continued to work and live outside the official red-light district.

Fires played a more significant role. The great fire of October 26, 1875, for example, changed the boundaries of the red-light district by destroying most of the buildings. As a result, many prostitutes took up residence in the "Barbary Coast" area of C Street, and others left the city, probably moving to nearby towns such as Gold Hill, Silver City, Dayton, Carson City, and Reno. The 1875 fire also changed the housing situation in the red-light district. Brothels in larger buildings, operated by commercial enterprises, replaced the ubiquitous small cottages of the small entrepreneur.

Population mobility, or the way that men and women moved around, is another dimension of settlement patterns that may reflect gender in the Comstock landscape. Where women were on the Comstock is a critical question. Margaret Purser documents the importance of mobility in understanding how men and women were distributed over the landscape in Paradise Valley, north of Winnemucca. "Visiting," for example, was a common practice that moved women from one settlement to the next or even further away for several days at a time.[51] Mary McNair Mathews's reminiscences of her life on the Comstock during the 1870s documents a similar pattern. She moved around frequently, often spending several days at other places in Nevada and California.[52] Another aspect of this issue is the mobility of prostitutes, mentioned above. The sex and gender-specific mobility of the Comstock population adds another dimension to the issue of engendering the landscape.

Finally, the two interrelated questions of engendered land uses and landscape symbolism should be considered. How gender affected the use of resources on the Comstock landscape in part reflects differences in occupations. Males, for example, typically monopolized mining—an occupation that excluded women—on the Comstock.[53] Still, William White and Ronald M. James found a piece of possible archaeological evidence—what appears to be a woman's basque (or close-fitting undergarment), discovered in a small underground mine—that women may have worked in the Comstock mines on occasion.[54] The general pattern of occupational exclusion of women from the Comstock mines, however, and the myth that women in mines bring bad luck, create a social and cultural context for resource use.

Conclusion

In conclusion, then, archaeology offers a new avenue by which to explore the lives of women on the Comstock that is both independent of and complementary to written accounts and oral testimony. The material world of the Comstock potentially provides evidence not only of women's living conditions but also of the symbolic system in which gender identities were immersed. Archaeological studies of Comstock households that are well documented in written accounts are likely to yield the most fruit, making possible the use of multiple lines of inquiry. Ultimately, however, comparative studies in archaeology and history are needed to understand how the lives of women on the nineteenth-century Comstock differed from and were similar to those of women in other mining communities and other urban settlements.

Statistical Profile of Women on the Comstock

The tables in this section are part of the available statistical profile that describes Comstock females. While the data immediately suggests some conclusions, the information presented here and in the following appendices may serve as the basis for further research.

APPENDIX I.I Population Size for Storey County by Sex, 1860–1990

Census	Males (*N*)	Males (%)	Females (*N*)	Females (%)	Total (*N*)
1860	2,857	95	159	5	3,016
1862	3,843	85	655	˙15	4,498
1870	7,814	69	3,505	31	11,319
1875	13,415	69	6,113	31	19,528
1880	9,221	58	6,783	42	16,004
1890	5,144	58	3,662	42	8,806
1900	1,874	53	1,686	47	3,560
1910	1,748	59	1,228	41	2,976
1920	803	55	666	45	1,469
1930	378	57	289	43	667
1940	709	58	507	42	1,216
1950	354	53	317	47	671
1960	295	52	273	48	568
1970	343	49	352	51	695
1980	767	51	736	49	1,503
1990	1,250	49	1,276	51	2,526

The statistics for 1860 summarize the communities of Gold Hill and Virginia City, Utah Territory, before the creation of Storey County in 1861. Figures for 1870, 1880, 1900, and 1910 are based on the census manuscripts, because they are more accurate than the census reports.

Sources: U.S. census manuscripts and reports, 1860–1990; the Nevada territorial census of 1861 and the Nevada state census of 1875.

APPENDIX I.2   Occupations of Virginia City's Women by Marital
Status, 1880

Occupation	Married (%)	Widowed (%)	Divorced (%)	Single (%)	Total (N)	Total (%)
Keeping house	89.0	6.5	0.6	3.9	1,422	63.7
Servant	5.1	1.0	1.3	92.6	156	7.0
Seamstress	4.4	20.9	9.9	64.8	91	4.0
Prostitution	16.9	7.8	7.8	67.5	77	3.4
Lodging house	37.5	42.9	6.3	13.3	64	2.9
Teacher	20.6	5.9	2.9	70.6	34	1.5
Nun	—	—	—	100.0	16	0.7
Health care	12.5	43.8	—	43.0	16	0.7
Laundry	33.3	60.0	6.7	—	15	0.7
Milliner	30.1	7.7	—	62.2	13	0.6
Restaurant work	23.1	15.4	—	61.5	13	0.6
Other	42.9	47.6	4.8	4.7	21	1.1
None	27.9	9.9	0.7	61.5	294	13.1
Total					2,232	100.0
Averages (Total Female pop.)	64.7	10.2	1.5	23.6		

Women are defined as females seventeen or older with the exception of those at school (who are not included here) and of those who were younger but working (who are included here). "Keeping House" signifies a woman with an occupation listed as such or a woman with no occupation listed but who is the wife of the head of household or who is herself head of household with no occupation listed. "None" signifies the other women seventeen years or older with no occupation listed and not attending school. "Single" is identified by the census and indicates women who had never been married. These statistics are for Virginia City only; other locations in Storey County lacked addresses to evaluate occupation in the context of prostitution.
Numbers for each marital status group listed are as follows: Married = 1,443; Widowed = 227; Divorced = 34; and Single = 528.
Source: Tenth U.S. manuscript census of Virginia City, Nevada, 1880.

APPENDIX 1.3 Occupations of Virginia City's Women by Age and
Household Size, 1880

Occupation	Age (Years)	Household Size (N)	Unemployed (Months)	Family (N)	Total (N)
Other	40	9	0	2	21
Lodging house	39	9	0	3	64
Laundry	39	9	0	3	15
Health care	38	7	1	2	16
Keeping house	36	7	0	4	1,422
Sister of Charity	36	14	0	1	16
Seamstress	29	7	0	2	91
Milliner	29	9	0	4	13
None	28	12	0	4	294
Servant	28	10	1	10	156
Prostitute	28	6	1	6	77
Restaurant work	28	14	3	14	13
Teacher	25	10	0	3	34
Total					2,232
Averages (total for women)	33	8	0	4	—
Total men in workforce					3,954
Averages (total for men)	37	10	2	2	—

Women are defined as in appendix 1.2. The "months unemployed" column is the total from the previous twelve months and as the census enumerator recorded it.
Source: Tenth U.S. manuscript census of 1880 for Virginia City, Nevada.

APPENDIX 1.4 Virginia City Women's Occupation by Ethnicity, 1880

Occupation	U.S.	Irish	British	German	Canadian	Chinese	Hispanic	French	Afro–Am	Swiss	Jewish	Other	Total
Keeping house	530	448	165	113	47	19	15	19	11	17	10	28	1,422
Servant	51	78	6	2	8	—	—	—	6	1	—	4	156
Seamstress	56	6	11	3	8	—	—	2	3	1	—	1	91
Prostitute	31	3	3	4	2	15	12	4	3	—	—	—	77
Lodging house	24	25	5	5	1	2	—	1	—	1	1	—	64
Teacher	28	2	1	—	2	—	—	1	—	—	1	—	34
Sister of Charity	8	6	1	—	—	—	—	—	—	—	—	—	16
Health care	7	6	1	2	—	—	—	1	—	—	—	—	16
Laundry	2	9	—	2	1	—	—	1	1	—	—	—	15
Milliner	6	3	—	—	—	—	—	—	—	—	2	1	13
Restaurant	2	8	—	1	2	—	—	—	—	—	—	—	13
Other	3	6	3	4	1	1	—	—	1	—	—	2	21
None	217	29	16	8	13	—	4	2	1	1	2	1	294
Total	965	629	212	144	85	37	31	30	26	21	15	37	2,232

Ethnicity equals nativity, with the following exceptions: African American equals racial designation by the census enumerator as black or mulatto; Jews are of diverse nativity and are identified by cemetery records and other primary sources dealing with Jewish organizations; U.S denotes Euro-Americans born in the United States and includes children of European immigrants; British includes those of English, Scottish, Welsh, Cornish, or Manx nativity. All Asians are Chinese born in China.

Sources: Tenth U.S. manuscript census of 1880 for Virginia City, Storey County, Nevada.

Birthplace of Females in Storey County, 1860–1910

The following is a list of the places of birth for females in Storey County (or its 1860 equivalent) from 1860 through 1910. It is a contribution of Kenneth Fliess and represents some of the preliminary data available through the Nevada Census Project. Beginning in 1880, the place of birth of each individual's father and mother was recorded, and this information is also presented here. The 1860 census does not contain any information about parents, and the 1870 census only reports whether the individual's father and mother were foreign born or native born. At the end of this list are the percentages of Storey County females born in the United States and of those who were foreign born. In general, approximately 60 percent or more were native born, a figure that grew to more than 75 percent by 1910.

Birthplace	1860 N	1860 (%)	1870 N	1870 (%)	1880 N	1880 (%)	1900 N	1900 (%)	1910 N	1910 (%)
Alabama	1	0.68	7	0.20	10	0.15	—	—	—	—
Alaska	—	—	—	—	—	—	—	—	1	0.08
Arizona	—	—	—	—	1	0.01	1	0.06	—	—
Arkansas	—	—	1	0.03	1	0.01	1	0.06	—	—
California	14	9.52	475	13.58	816	12.03	167	9.91	138	11.25
Colorado	—	—	—	—	7	0.10	2	0.12	6	0.49
Connecticut	1	0.68	19	0.54	25	0.37	6	0.36	1	0.08
Dakota	—	—	1	0.03	—	—	—	—	—	—
Delaware	—	—	3	0.09	2	0.03	1	0.06	1	0.08
District of Columbia	—	—	8	0.23	10	0.15	2	0.12	—	—
Florida	—	—	1	0.03	2	0.03	—	—	—	—
Georgia	1	0.68	—	—	1	0.01	—	—	1	0.08
Idaho	—	—	—	—	4	0.06	1	0.06	6	0.49
Illinois	10	6.80	57	1.63	88	1.30	10	0.59	5	0.41
Indiana	3	2.04	20	0.57	25	0.37	1	0.06	4	0.33
Indian Territory	—	—	—	—	1	0.01	—	—	—	—
Iowa	3	2.04	31	0.89	43	0.63	8	0.47	7	0.57
Kansas	—	—	—	—	14	0.21	1	0.06	4	0.33
Kentucky	1	0.68	28	0.80	46	0.68	6	0.36	2	0.16
Louisiana	—	—	13	0.37	37	0.55	7	0.42	4	0.33
Maine	2	1.36	49	1.40	91	1.34	14	0.83	6	0.49
Maryland	1	0.68	20	0.57	27	0.40	6	0.36	2	0.16
Massachusetts	7	4.76	110	3.14	140	2.06	27	1.60	14	1.14

Michigan	2	1.36	27	0.77	73	1.08	10	0.59	14	1.14
Minnesota	—	—	3	0.09	11	0.16	1	0.06	2	0.16
Mississippi	1	0.68	1	0.03	3	0.04	—	—	—	—
Missouri	3	2.04	39	1.11	89	1.31	18	1.07	11	0.90
Montana	—	—	—	—	1	0.01	1	0.06	1	0.08
Nebraska	—	—	8	0.23	2	0.03	1	0.06	1	0.08
Nevada	—	—	652	18.64	1,893	27.91	790	46.88	639	52.08
New Hampshire	1	0.68	16	0.46	22	0.32	4	0.24	3	0.24
New Jersey	1	0.68	25	0.71	49	0.72	8	0.47	2	0.16
New York	17	11.56	200	5.72	321	4.73	49	2.91	25	2.04
North Carolina	—	—	—	—	1	0.01	—	—	—	—
Ohio	3	2.04	70	2.00	118	1.74	14	0.83	10	0.81
Oregon	—	—	3	0.09	8	0.12	—	—	5	0.41
Pennsylvania	4	2.72	91	2.60	170	2.51	23	1.36	14	1.14
Rhode Island	—	—	2	0.06	9	0.13	1	0.06	—	—
South Carolina	—	—	3	0.09	—	—	—	—	—	—
Tennessee	1	0.68	4	0.11	10	0.15	1	0.06	2	0.16
Texas	4	2.72	3	0.09	2	0.03	1	0.06	2	0.16
Utah	9	6.12	7	0.20	18	0.27	1	0.06	12	0.98
Vermont	2	1.36	17	0.49	29	0.43	3	0.18	3	0.24
Virginia	2	1.36	25	0.71	20	0.29	1	0.06	—	—
West Virginia	—	—	—	—	3	0.04	2	0.12	1	0.08
Wisconsin	7	4.76	35	1.00	53	0.78	6	0.36	4	0.33
Wyoming	—	—	—	—	1	0.01	—	—	1	0.08
America	—	—	—	—	1	0.01	—	—	—	—
United States	—	—	—	—	2	0.03	—	—	—	—
At home	—	—	—	—	1	0.01	—	—	—	—
Mississippi River	—	—	—	—	1	0.01	—	—	—	—

Birthplace	1860 N	(%)	1870 N	(%)	1880 N	(%)	1900 N	(%)	1910 N	(%)
At sea	1	0.68	—	—	4	0.06	—	—	1	0.08
Australia	—	—	11	0.31	14	0.21	2	0.12	—	—
Austria	—	—	4	0.11	10	0.15	5	0.30	3	0.24
Azores	—	—	—	—	2	0.03	—	—	—	—
Belgium	—	—	—	—	4	0.06	—	—	1	0.08
Bermuda	—	—	—	—	1	0.01	—	—	—	—
Brazil	—	—	—	—	1	0.01	—	—	—	—
British Guiana	—	—	—	—	1	0.01	—	—	—	—
Canada	4	2.72	69	1.97	247	6.64	50	2.97	36	2.93
Chile	—	—	5	0.14	1	0.01	2	0.12	—	—
China	—	—	102	2.92	43	0.63	3	0.18	1	0.08
Colombia	—	—	—	—	2	0.03	—	—	—	—
Denmark	1	0.68	—	—	7	0.10	3	0.18	3	0.24
East Indies	1	0.68	—	—	—	—	—	—	—	—
Ecuador	—	—	1	0.03	1	0.01	—	—	—	—
England	7	4.76	219	6.26	473	6.97	101	5.99	50	4.07
Europe	—	—	2	0.06	1	0.01	—	—	—	—
Fayal (Azores)	—	—	—	—	1	0.01	—	—	—	—
France	1	0.68	30	0.86	48	0.71	3	0.18	4	0.33
Germany	9	6.12	148	4.23	206	3.04	44	2.61	20	1.63
Guatemala	—	—	—	—	1	0.01	—	—	—	—
Holland	—	—	—	—	1	0.01	—	—	—	—
Hungary	—	—	—	—	1	0.01	—	—	—	—

India	10	6.80	—	—	—	—	1	0.06	—	—
Ireland	—	—	732	20.93	1,159	17.09	218	12.94	110	8.96
Isle of Man	—	—	2	0.06	3	0.04	—	—	—	—
Italy	—	—	8	0.23	16	0.24	11	0.65	17	1.39
Jamaica	—	—	—	—	1	0.01	—	—	—	—
Mexico	6	4.08	36	1.03	24	0.35	3	0.18	1	0.08
New Gourdland	—	—	—	—	1	0.01	—	—	—	—
New Zealand	—	—	2	0.06	1	0.01	—	—	—	—
Norway	—	—	—	—	9	0.13	2	0.12	2	0.16
Panama	—	—	4	0.11	6	0.09	1	0.06	—	—
Poland	—	—	—	—	6	0.09	2	0.12	—	—
Portugal	—	—	—	—	16	0.24	—	—	1	0.08
Russia	—	—	—	—	3	0.04	1	0.06	1	0.08
Scotland	4	2.72	42	1.20	66	0.97	10	0.59	5	0.41
South America	1	0.68	—	—	—	—	—	—	—	—
Spain	—	—	—	—	3	0.04	—	—	—	—
Sweden	1	0.68	—	—	20	0.29	5	0.30	6	0.49
Switzerland	—	—	—	—	40	0.59	10	0.59	4	0.33
Tobago Island	—	—	—	—	1	0.01	—	—	—	—
Wales	—	—	7	0.20	29	0.43	8	0.47	4	0.33
West Indies	—	—	—	—	3	0.04	—	—	—	—
Unknown	—	—	—	—	1	0.01	2	0.12	1	0.08
Left blank	—	—	—	—	2	0.03	2	0.12	2	0.16
United States	101	68.71	2,074	59.29	4,302	63.43	1,196	70.98	954	77.75
Other countries	46	31.29	1,424	40.71	2,477	36.52	485	28.78	270	22.00
Other	—	—	—	—	3	0.04	4	0.24	3	0.24
Total	147		3,498		6,782		1,685		1,227	

APPENDIX II.2   Parents' Birthplaces, 1880–1910

Birthplace	1880 Father	1880 Mother	1900 Father	1900 Mother	1910 Father	1910 Mother
Alabama	2	1	2	—	1	—
Arizona	—	—	—	—	1	—
Arkansas	—	—	—	1	—	1
California	1	66	54	128	64	102
Colorado	—	—	—	2	1	—
Connecticut	45	38	17	10	8	2
Delaware	—	2	1	2	1	1
District of Columbia	5	5	3	1	—	1
Florida	1	—	—	—	—	—
Georgia	7	5	1	—	1	1
Idaho	—	—	—	—	2	2
Illinois	42	79	14	9	13	5
Indiana	18	33	4	2	5	3
Iowa	6	22	5	12	11	10
Kansas	2	—	—	—	1	3
Kentucky	95	88	12	15	10	13
Louisiana	15	25	2	6	4	4
Maine	117	109	25	26	19	15
Maryland	30	34	11	14	4	6
Massachusetts	106	132	26	41	23	25
Michigan	32	33	14	13	13	14
Minnesota	—	1	—	3	—	2
Mississippi	4	4	14	—	1	—
Missouri	38	52	18	21	14	14
Nebraska	—	—	—	—	—	4
Nevada	66	65	41	88	117·	194
New Hampshire	39	32	5	9	7	7
New Jersey	33	25	3	8	5	5
New York	317	328	55	61	51	43
North Carolina	8	7	2	1	—	—
Ohio	110	119	25	14	12	12
Oregon	1	—	—	—	2	2
Pennsylvania	160	176	40	44	22	35
Rhode Island	7	5	1	—	—	—
South Carolina	8	5	1	—	1	—
Tennessee	25	22	5	1	4	5
Texas	3	7	—	—	—	—
Utah	—	6	—	—	2	10
Vermont	65	51	10	10	8	4
Virginia	49	54	8	7	3	1
Washington	—	—	—	—	1	—
West Virginia	1	1	2	1	1	—

Birthplace	1880 Father	1880 Mother	1900 Father	1900 Mother	1910 Father	1910 Mother
Wisconsin	11	16	9	4	8	4
America	4	4	—	—	—	—
United States	15	7	—	—	7	5
At home	1	1	—	—	—	—
At sea	1	1	—	—	3	—
Australia	5	10	1	1	1	—
Austria	28	16	6	6	8	5
Azores	3	3	—	—	—	—
Belgium	9	6	—	—	2	1
Bermuda	2	1	—	—	—	—
Brazil	—	3	—	—	—	—
British Guiana	—	1	—	—	—	—
Canada	301	288	98	69	66	51
Chile	1	1	1	1	—	—
China	43	43	3	3	1	1
Colombia	1	1	—	—	—	—
Denmark	17	15	8	7	13	10
Ecuador	2	2	—	—	—	—
England	1,017	899	268	213	165	113
Europe	1	1	—	—	—	—
Fayal (Azores)	1	1	—	—	—	—
France	78	88	10	6	10	8
Germany	532	486	146	113	81	61
Greece	3	1	—	—	—	—
Guatemala	3	3	—	—	—	—
Holland	8	4	—	—	—	—
Hungary	1	1	—	—	—	—
India	—	—	1	1	—	—
Ireland	2,688	2,745	571	567	319	318
Isle of Man	12	9	—	—	—	—
Italy	27	23	30	26	40	36
Luxembourg	2	—	—	—	—	—
Mexico	27	34	2	3	—	—
Norway	15	18	3	6	6	6
Panama	6	12	40	1	—	—
Poland	14	10	2	7	—	—
Portugal	33	27	1	—	2	1
Russia	9	7	1	2	—	2
Scotland	205	187	52	36	—	20
Spain	7	5	1	3	2	2
Sweden	33	32	11	8	11	12
Switzerland	68	66	19	24	13	13

APPENDIX II.2 *continued*

Birthplace	1880		1900		1910	
	Father	Mother	Father	Mother	Father	Mother
Tiparary Island	1	—	—	—	—	—
Tobago Island	—	—	—	1	—	—
Wales	56	47	18	21	7	8
West Indies	1	3	—	—	—	—
Unknown	12	9	10	7	8	3
Left blank	16	10	—	—	2	3

Occupations of Women on the Comstock, 1860–1910

The following are lists of occupations recorded for women in the censuses from 1860 to 1910. Like Appendix II, this is a product of Kenneth Fliess's Nevada Census Project. While some occupations are listed for individuals younger than age fifteen, the censuses are designed to record these data for individuals at age fifteen and older. No attempt has been made to collapse categories. In 1910 the census added "industry" as a new type of data recorded. This allows for a better understanding of women in the workplace because of the added specificity.

APPENDIX III.1   Female Professions, Occupations, and Trades for Age Fifteen or Older, 1860

Profession	N	Profession	N
None listed	88	Saloon keeper	1
Bar keeper	1	School teacher	1
Boardinghouse	1	Sewing	1
Housekeeper	5	Theatrical	1
Machine sewing	1	Washing	1
Milliner	2		

APPENDIX III.2   Female Professions, Occupations, and Trades for Age Fifteen or Older, 1870

Profession	N	Profession	N
None listed	156	Laundry	2
Actress	7	Laundrywoman	1
At home	22	Lodgers	1
Boarding	15	Lodging house	8
Boarding home	1	Melodeon keeper	1
Boardinghouse	7	Milliner	13
Boarding; Housekeeping	1	Millinery store	8
Chambermaid	2	Nurse	2
Circus performer	1	Prostitute	109
Cook	2	Restaurant	1
Dressmaker	39	Saloon keeper	3
Harlot	48	School	11
Hotel de refreshment	1	School marm	1
Hotel keeper	4	School teacher	6
House holder	1	Seamstress	17
Housekeeper	59	Servant	46
Housekeeping	122	Serving the public	3
House servant	1	Shop woman	1
Housemaid	1	Nun at St. Mary's School	8
Intelligence office	1	Student at St. Mary's	9
Keeping hotel	1	Theatrical	2
Keeping house	766	Upholsterer	1
Keeps house	652	Waiter	4
Keeps boarders	3	Wash house	1
Keeps dairy	1	Wash woman	3
Keeps lodgers	4	Washerwoman	7
Keeps store	1	Wife	1
Laundress	1		

APPENDIX III.3 Female Professions, Occupations, and Trades for Age Fifteen or Older, 1880

Profession	N	Profession	N
None listed	223	House work	2
Assistant Sister of Charity	3	House of prostitution	1
At home	319	Janitor in public school	1
At school	108	Keeping hotel	1
Attending school	1	Keeping house	2273
Authoress	1	Keeping lodging house	5
Blacksmith	1	Keeping lodgers	2
Boarder	28	Keeping toll house	1
Boarding	32	Keeps boarders	2
Boarding and lodging	1	Keeps boardinghouse	2
Boardinghouse	10	Keeps house	348
Boarding housekeeper	1	Keeps lodge house	1
Boarding house partner	1	Keeps lodgers	6
Book folder	1	Keeps lodging house	25
Book keeper	1	Keep restaurant	2
Candy store	2	Keeps saloon	1
Chamber maid	2	Ladies hair dresser	1
Clerk in drug store	1	Ladies nurse	1
Confectioner	1	Laundress	1
Cook	8	Lodging and boardinghouse	1
Daughter	1	Liquor dealer	1
Deceased	1	Liquor saloon	1
Doctress	1	Lodging	27
Does housework	2	Lodging house	26
Domestic	6	Matron, cares for hotel	1
Dressmaker	95	Millinery	1
Dressmaking	4	Milliner	21
Dressmaker's apprentice	1	Mistress, keeps house	2
Fancy dry goods store	1	Mother-in-law	3
Fancy goods store	1	Music teacher	12
Fancy store	2	Niece	1
Fortuneteller	1	No occupation	18
French teacher	1	None	20
General work	1	Nurse	28
Hair dresser	2	Nurse at hospital	1
Help mother	1	Nurses	1
Helper in laundry	1	Opium den	1
Helping in restaurant	1	Pauper	1
Helps in house	1	Peddler	1
House girl	1	Proprietor of boardinghouse	1
Housekeeper	30	Property owner	1
Housekeeping	5	Prostitute	42
House servant	1	Public school teacher	3

APPENDIX III.3 *continued*

Profession	N	Profession	N
Rents house	1	Teacher of French	1
Restaurant	1	Teacher in public school	1
Restaurant keeper	2	Teaching	2
Roomer	11	Teaching school	1
Rooming	2	Upholsterer	1
Saloon keeper	1	Visiting	2
School teacher	1	Visitor	4
Seamstress	26	Waiter	9
Seamstress and keeps house	1	Waitress	6
Servant	183	Wash and iron	1
Sewer	1	Washerwoman	6
Sewing	1	Washing	1
Sewing girl	1	Washing and ironing	5
Sewing woman	3	Washwoman	8
Sister of Charity	12	Wife	3
Superior sister	1	With daughter	1
Tailoress	1	Works in lodging house	1
Teacher	8	Works in private family	1
Teacher of elocution	1		

APPENDIX III.4   Female Professions, Occupations, and Trades for Age
Fifteen or Older, 1900

Profession	N	Profession	N
None listed	945	Millinery store	1
Artist (sewing)	1	Millinery	2
At home	1	Mining secretary	1
At school	97	Music teacher	6
Boardinghouse	1	Nurse	16
Boarding and lodging house	1	Paper route	1
Bookkeeper	1	Peddler	1
Clerk	3	Peddling	1
Clerk in store	2	Post mistress	1
Courtesan	8	Post office clerk	2
Drawing teacher	2	Restaurant	1
Dressmaker	29	Restaurant keeper	1
Dressmaking	3	Sales lady	1
Fancy-goods store	1	Saloon keeper	1
Farmer	1	School teacher	30
Governess	1	Servant	25
Grocer	2	Sewing	1
Housekeeper	5	Tea store clerk	1
Janitress	1	Teacher	9
Keeping lodgers	1	Telegraph operator	1
Lace maker	1	Typewriter	1
Laundress	5	Waiter	3
Lawyer	1	Waitress	2
Lodging housekeeper	4	Washing and sewing	1
Lodging home	2	Washing and ironing	1
Lodging house	11	Washwoman	1
Lodging home keeper	1		

APPENDIX III.5   Female Professions, Occupations, and Trades for Age Fifteen or Older, 1910

Profession	Industry	N
—	Livery Stable	1
Accountant	Butcher shop	1
Agent	Books	1
Assistant matron	County hospital	1
Bookkeeper	Dry goods store	1
Bookkeeper	General work	1
Bookkeeper	Wholesale house	1
Care taker	Private house	1
Cashier	Dry goods store	1
Chamber maid	Lodging house	2
Clerk	Post office	1
Cook	Boardinghouse	2
Cook	Farm	2
Cook	Father's family	1
Cook	Private family	1
Cook	Private house	1
Deputy recorder	Storey County	1
Dressmaker	—	3
Dressmaker	At home	2
Farmhand	Vegetables	1
Hairdresser	—	1
House help	Father's family	1
Housekeeper	At home	2
Housekeeper	Private family	1
Housekeeper	Private home	3
Keeper	Boardinghouse	7
Keeper	Lodging house	4
Keeper	Rooming house	1
Launderer	Own house	1
Laundress	At home	1
Laundress	Own home	1
Laundress	Own house	1
Laundress	Private family	1
Laundress	Public school	1
Laundress	Working out	1
Librarian	State library	1
Lodging house	Lodging house	4
Manage	Opera house	1
Manicure	House	1
Matron	County hospital	1
Milliner	At home	1

Profession	Industry	*N*
Milliner	House	1
Milliner	Own shop	1
Milliner	Shop	1
Music teacher	—	1
Musician	—	1
None	—	721
Nurse	—	1
Nurse	Confinement	2
Nurse	Hospital	3
Nurse	Odd jobs	1
Nurse	Private family	3
Operator	Telegraph	1
Operator	Telephone	1
Operator	Telephone exchange	1
Own income	—	26
Portrait artist	—	1
Postmistress	—	1
Retail merchant	—	1
Retail merchant	Dry goods	1
Retail merchant	Dry goods store	1
Retail merchant	Fancy goods	1
Retail merchant	Own store	1
Retail merchant	Stationery	1
Saleslady	Clothing	2
Saleslady	Dry goods	2
Saleswoman	Bakery	1
Saleswoman	Clothing store	1
School teacher	Public school	5
Seamstress	Working out	1
Servant	Hotel	1
Servant	Lodging house	1
Servant	Private family	11
Servant	Private home	6
Servant	Working out	1
Stenographer	Courthouse	1
Stenographer	Office	2
Student	University	1
Teacher	High school	1
Teacher	Music	3
Teacher	Odd jobs	3
Teacher	Public school	18
Teacher	Stenography	1
Waitress	Restaurant	2
Washwoman	Private family	1

1. "I Am Afraid We Will Lose All We Have Made"

1. The most detailed contemporary history of the Comstock is Eliot Lord, *Comstock Mining and Miners* (1883; reprint, Berkeley: Howell-North, 1959). See also Rodman Wilson Paul, *Mining Frontiers of the Far West, 1848–1880* (1963; reprint, Albuquerque: University of New Mexico Press, 1974). For Comstock tourism, see Wilbur S. Shepperson with Ann Harvey, *Mirage-Land: Images of Nevada* (Reno: University of Nevada Press, 1992). The opening quotations are from Louise Palmer, "How We Live in Nevada," *Overland Monthly* (May 1869): 457–62, reprinted in part in *So Much to Be Done: Women Settlers on the Mining and Ranching Frontier,* ed. Ruth B. Moynihan, Susan Armitage, and Christiane Fischer Dichamp (Lincoln: University of Nebraska Press, 1990), 110–18; and Georgina Joseph Letters, Manuscript Collections, Nevada State Historical Society, Reno, Nevada. The title of the essay comes from Georgina Joseph, June 1, 1870.

2. Mary McNair Mathews, *Ten Years in Nevada, or Life on the Pacific Coast* (1880; reprint, Lincoln: University of Nebraska Press, 1985); Mark Twain, *Roughing It* (1871; reprint, New York: Harper and Brothers, Publishers, 1913); J. Ross Browne, *A Peep at Washoe and Washoe Revisited* (1863; 1864; 1869; reprint, Balboa Island, Calif.: Paisano Press, 1959); Walter van Tilberg Clark, ed., *The Journals of Alfred Doten, 1829–1903,* 3 vols. (Reno: University of Nevada Press, 1973).

3. See chapter two by Ronald James and Kenneth Fliess for more information on the Nevada Census Project and compilations of data relating to Virginia City during the boom period. Eventually the project will encompass the entire state.

4. Elliott West, "Beyond Baby Doe: Child Rearing on the Mining Frontier," in *The Women's West,* ed. Susan Armitage and Elizabeth Jameson (Norman: University of Oklahoma Press, 1987), 182.

5. For the evolution of the parlor in other regions of the United States, and for its significance, see Karen Haltunnen, *Confidence Men and Painted Ladies* (New Haven: Yale University Press, 1986); Sallie McMurry, *Families and Farmhouses in Nineteenth-Century America: Vernacular Design and Social Change* (New York: Oxford University Press, 1988); and Jane C. Nylander, *Our Own Snug Fireside* (New Haven: Yale University Press, 1992). The Palmer quote is found on page 113 of the version in *So Much To Be Done.*

6. Marion S. Goldman, *Gold Diggers and Silver Miners: Prostitution and Social Life on the Comstock Lode* (Ann Arbor: University of Michigan Press, 1981). See also the chapter in this volume by Ronald James and Kenneth Fliess, "Women of the Mining West: Virginia City Revisited."

7. For a suggestive exploration of the role of chance in men's lives and class distinctions in their responses, see Gunther Peck, "Manly Gambles: The Politics of Risk on the Comstock Lode, 1860–1880," *Journal of Social History* 26 (summer 1993): 701–23.

8. For other western examples see Peggy Pascoe, *Relations of Rescue: The Search for Female Moral Authority in the American West, 1874–1939* (New York: Oxford University Press, 1990); and Paula Petrik, *No Step Backward: Women and Family on the Rocky Mountain Mining Frontier, Helena, Montana 1865–1900* (Helena: Montana Historical Society Press, 1987).

9. Susan Armitage and Elizabeth Jameson, eds., *The Women's West* (Norman: University of Oklahoma Press, 1987); Lillian Schlissel, Vicki L. Ruiz, and Janice Monk, eds., *Western Women: Their Land, Their Lives* (Albuquerque: University of New Mexico Press, 1988).

10. See especially John Mack Faragher, *Women and Men on the Overland Trail* (New Haven: Yale University Press, 1979); Julie Roy Jeffrey, *Frontier Women: The Trans-Mississippi West, 1840–1880* (New York: Hill and Wang, 1979); Sandra L. Myres, *Westering Women and the Frontier Experience, 1800–1915* (Albuquerque: University of New Mexico Press, 1982); and Lillian Schlissel, *Women's Diaries of the Westward Journey* (New York: Schocken Books, 1982).

11. See, for example, the 1992 special issue of *Pacific Historical Review* (vol. 61) devoted to Western women's history, as well as the classic review essay, "The Gentle Tamers Revisited: New Approaches to the History of Women in the American West," by Joan M. Jensen and Darlis A. Miller, *Pacific Historical Review* 49 (1980): 173–213. In "Western Women at the Cultural Crossroads," Peggy Pascoe summarizes and cites much of this literature in Patricia Nelson Limerick, Clyde A. Milner II, and Charles E. Rankin, eds., *Trails Toward a New Western History* (Lawrence: University Press of Kansas, 1991), 40–58.

12. For suggestive explorations of the implications of using gender, rather than just women's experience, as a category of historical analysis in the West, see Susan Lee Johnson, "'A Memory Sweet to Soldiers': The Significance of Gender in the History of the 'American West,'" *Western Historical Quarterly* 24 (1993): 495–517; and Katherine G. Morrissey, "Engendering the West," in *Under an Open Sky: Rethinking America's Western Past,* ed, William Cronon, George Miles, and Jay Gitlin (New York: W. W. Norton, 1992), 132–44.

13. Notable exceptions are West and Petrik, both cited earlier. See also forthcoming works by Mary Murphy, *Recreating Butte: Gender, Work and Leisure in a Western Mining Camp, 1914–41* (Champaign: University of Illinois Press, 1996); and Elizabeth Jameson, *All That Glitters: Class, Culture and Community in Cripple Creek* (Champaign: University of Illinois Press, in press). Also important is Andrea Yvette Huginnie, "'Strikitos': Race, Class, and Work in the Arizona Copper Industry, 1870–1920" (Ph.D. diss., Yale University, 1991). For women in California mining see Joann Levy, *They Saw the Elephant: Women in the California Gold Rush* (Hamden, Conn.: Archon Books, 1990). For a twentieth-century interior mining camp that was much shorter lived than Virginia City, see Sally Zanjani, *Goldfield: The Last Gold Rush on the Western Frontier* (Athens: Ohio University Press, 1992). Mary Lou Locke, "Out of the Shadows and Into the Western

Sun: Working Women of the Late Nineteenth-Century Urban Far West," *Journal of Urban History* 16 (1988): 175–204, makes the important point that region may be as important as class in analyzing women's lives.

14. Glenda Riley, *The Female Frontier* (Lawrence: University Press of Kansas, 1988), 4. "My intention in this study is to demonstrate not only that women did play highly significant and multifaceted roles in the development of the American West but also that their lives as settlers displayed fairly consistent patterns which transcended geographic sections of the frontier . . . [and] constituted a 'female frontier.'" (2).

15. Twain, *Roughing It,* 220–21.

16. Though Frank Joseph was only a miner and was frequently so ill he couldn't work, the family invested in stocks. On August 18, 1869, Georgina wrote: "We have had very back luck. All the money we saved since we have been here we put in the mines here . . . they took fire last April . . . We have had to pay assessments . . . and we are in debt."

17. For miners' expensive taste in food, see Joseph Conlin, *Bacon, Beans, and Galantines* (Reno: University of Nevada Press, 1986).

18. The quotation is from Franklin Buck, a New England man in Gold Rush California, quoted in Levy, *They Saw the Elephant,* 108. In her study of Helena, Petrik discusses how the peculiar circumstances of a hard-rock mining community affected the lives of the women who lived in it.

2. *Women of the Mining West*

1. Grant Smith, *The History of the Comstock Lode: 1850–1920* (Reno: Nevada Bureau of Mines, 1943), 292–93. See also Eliot Lord, *Comstock Mining and Miners* (1883; reprint, Berkeley: Howell-North, 1959) for an excellent overview of Comstock history.

2. See, for example, Rodman W. Paul, *Mining Frontiers of the Far West, 1848–1880* (New York: Holt, Rinehart and Winston, 1963); Otis E. Young, *Western Mining* (Norman: University of Oklahoma Press, 1970); and Donald L. Hardesty, *The Archaeology of Mining and Miners: A View from the Silver State* (The Society for Historical Archaeology: Special Publication Series, No. 6, 1988).

3. For two excellent sources from the nineteenth century, see Mary McNair Mathews, *Ten Years in Nevada: Or Life on the Pacific Coast* (1880; reprint, Lincoln: University of Nebraska Press, 1985) and John Waldorf, *Kid on the Comstock: Reminiscences of a Virginia City Childhood* (1970; reprint, Reno: University of Nevada Press, 1991). For a bibliography of Comstock sources, see Ronald M. James, "A Plan for the Archeological Investigation of the Virginia City Landmark District," addendum to the *Nevada Comprehensive Preservation Plan: Addendum* (Carson City: Nevada State Historic Preservation Office, 1992). Among the more useful publications are Susan Armitage and Elizabeth Jameson, *The Women's West* (Norman: University of Oklahoma Press, 1987); Paula Petrik, *No Step Backward: Women and Family on the Rocky Mountain Mining Frontier, Helena, Montana 1865–1900* (Helena: Montana Historical Society Press, 1987); and Mary Murphy, "Bootlegging Mothers and Drinking Daughters: Gender and Prohibition in Butte, Montana," *American Quarterly* 46, no. 2 (June 1994): 174–94.

4. The purpose of this project is ultimately to computerize the complete census manuscripts for the entire State of Nevada between 1860 and 1920. The data are being preserved in easily accessible computerized data files, using FoxBase 2.0, that will be available to scholars and other interested individuals in Nevada and around the United

States sometime in the future. The territorial censuses of 1861 and 1863 and the state census of 1875 are excluded from all analyses, because the quality and coverage of these documents is in dispute. Also, the federal censuses are used, since they will allow for future comparisons to other areas and communities because of format, data quality, and coverage. The five federal censuses considered here do not all ask the same questions, and, as a result, some information, such as marital status, is not available in 1860 and 1870.

5. Nevada is an ideal location for this type of project, because its small population between 1860 and 1920 eliminates the need for sophisticated sampling procedures. The entire population for each census year will ultimately be available for use in all analyses. Additionally, once the census data entry and verification process is complete, Nevada can serve as a model for comparison with other frontier populations in the western United States. The manuscripts for the 1890 census burned in a warehouse fire in the 1920s before they were microfilmed and therefore are not available. While the 1920 census manuscripts have recently been released, these data are not used in the analyses presented here.

6. Elizabeth Jameson, "Women as Workers, Women as Civilizers: True Womanhood in the American West," in *The Women's West*, ed. Susan Armitage and Elizabeth Jameson, 158.

7. In 1860 the entire western section of the Utah Territory bordering on California was Carson County. It ran from Oregon on the north to the Arizona Territory on the south. Consequently, for 1860 this analysis only uses data from Gold Hill, Virginia City, and the Virginia Mining District. Analysis of subsequent censuses draws on data from Storey County, created in 1861, which encompasses these communities.

8. On demographic analysis, see, for example, Vivian Z. Klaff, *DEM-LAB: Teaching Demography Through Computers* (Englewood Cliffs, N.J.: Prentice Hall, 1992); Charles B. Nam and Susan O. Gustavus, *Population: The Dynamics of Demographic Change* (Boston: Houghton Mifflin, 1976); James A. Palmore and Robert W. Gardner, *Measuring Mortality, Fertility and Natural Increase: A Self-Teaching Guide to Elementary Measures* (Honolulu: East-West Center, 1983); Henry S. Shryock, Jacob Siegel and Associates, *The Methods and Materials of Demography* (New York: Academic Press, 1976, condensed edition); John R. Weeks (1994) *Population: An Introduction to Concepts and Issues* (Belmont, Calif.: Wadsworth, 1994); Robert H. Weller and Leon F. Bouvier, *Population: Demography and Policy* (New York: St. Martin's Press, 1981); and David Yaukey, *Demography: The Study of Human Population* (New York: St. Martin's Press, 1985).

9. One can only surmise location from the 1860 eighth U.S. manuscript census, placing sites on the assumption that the enumerator went from house to house, a pattern probably not followed consistently. Adult women are defined here as those age sixteen or older. For a discussion of some of this, see Jameson, "Women as Workers," Armitage and Jameson, eds., *The Women's West*, 146–47.

10. Sally Zanjani, *Goldfield: The Last Gold Rush on the Western Frontier* (Athens, Ohio: Ohio University Press, Swallow Press, 1992), 102–3, 104–8.

11. J. S. Holliday, *The World Rushed In: The California Gold Rush Experience* (New York: Simon and Schuster, 1981), 354. Jacqueline Baker Barnhart seems to verify this in her *The Fair but Frail: Prostitution in San Francisco, 1849–1900* (Reno: University of Nevada Press, 1986). Ralph Mann, *After the Gold Rush: Society in Grass Valley and Nevada City, California, 1849–1870* (Palo Alto: Stanford University Press, 1982), 201, discusses the scarcity of women in California during the Gold Rush.

12. See JoAnn Levy, *They Saw the Elephant: Women in the California Gold Rush* (Norman: University of Oklahoma Press, 1992).

13. For additional observations about the 1860 census, see Ronald M. James, "Women of the Mining West: Virginia City Revisited," *Nevada Historical Society Quarterly* 36, no. 3 (fall 1993): 153–77; see especially 157.

14. See Michael S. Coray, "African-Americans in Nevada," *Nevada Historical Society Quarterly* 35, no. 4 (winter 1992): 239–57 and "Influences on Black Family Household Organization in The West, 1850–1860," *Nevada Historical Society Quarterly* 31, no. 1 (spring 1933): 1–31; and see Elmer R. Rusco, *"Good Times Coming?": Black Nevadans in the Nineteenth Century* (Westport, Conn.: Greenwood Press, 1975).

15. See Mary Ellen Glass, "Nevada's Census Taker: A Vignette," *Nevada Historical Society Quarterly* 19, no. 4 (winter 1966).

16. Association with a particular class is not possible to determine accurately from the census.

17. Marion Goldman, *Gold Diggers and Silver Miners: Prostitution and Social Life on the Comstock Lode* (Ann Arbor: University of Michigan Press, 1981). Goldman focuses on the 1880, tenth U.S. manuscript census because it includes street addresses. See also George Blackburn and Sherman L. Ricards, Jr., "The Prostitutes and Gamblers of Virginia City, Nevada, 1870," *Pacific Historical Review* 48, no. 2 (May 1979): 235–59.

18. For a full treatment of Goldman's approach, see James, "Women of the Mining West." Mathews, *Ten Years,* 165, describes the limits of the red-light district in her contemporary overview of life in Virginia City.

19. Goldman, *Gold Diggers,* 172.

20. The ambiguity of the line dividing prostitution and other occupations is addressed by Anne M. Butler, *Daughters of Joy, Sisters of Misery: Prostitutes in the American West, 1865–90* (Urbana: University of Illinois Press, 1985), and by Elliott West in *The Saloon on the Rocky Mountain Mining Frontier* (Lincoln: University of Nebraska Press, 1979), 49. See also Jameson, "Women as Workers."

21. See James, "Women of the Mining West," for an overview of these problems.

22. The data to verify this conclusion are on file at the State Historic Preservation Office, Carson City, Nevada. Use of the 1890 Sanborn Perris Fire Insurance Map to assess associations with the red-light district more precisely would have benefited Goldman's study. Her problematic approach to geography indicates that she did not use this important document, particularly in on-site work, for which it can do a great deal toward clearing up misunderstandings of geographic relationships. Goldman's assertions about supposed prostitutes would also have been more convincing had she reinforced her identifications with citations from local court records, but she appears not to have used this rich source of information in such a way.

23. Goldman claims that prostitution was the most common occupation for women. *Gold Diggers,* 159.

24. These generalizations contradict Goldman's statistical overview, which is based on her flawed data. *Gold Diggers,* 71. The 1870 census provides limited confirmation of this overview, because without street addresses or a record of marital status, household and individual profiles are difficult to ascertain. For an overview of regional prostitution, see Butler, *Daughters of Joy.*

25. The term "Hispanic" denotes representatives from throughout the Spanish-speaking world. Although immigrants from Mexico constituted the majority of the

group, the Comstock also served as home to Central and South Americans and to people from Spain.

26. See Goldman, *Gold Diggers,* 69–70, 177–78.

27. Ibid., 67ff. See also Sucheng Chan, *Asian Americans: An Interpretive History* (Boston: Twayne, 1991); Shih-shan Henry Tsai, *The Chinese Experience in America* (Bloomington: Indiana University Press, 1986); Judy Yung, *Chinese Women of America: A Political History* (San Francisco: Chinese Cultural Foundation of San Francisco, 1986); and Levy, *They Saw the Elephant,* 151–54.

28. Goldman, *Gold Diggers,* 67ff.

29. Ibid., 67–69.

30. Goldman claims to have found three African American prostitutes in the 1880 census. She missed the Panamanian altogether and apparently overlooked the mulatto designation for two other women whom she includes as white prostitutes. There is no reason to conclude that her three African Americans designated as prostitutes were in fact employed as such. The forty-two-year-old African American servant at the prestigious Bow Windows brothel was the only servant in an otherwise all Euro-American brothel, and it is reasonable to assume that her occupation was the one the enumerator recorded. The other two women lived on C Street, far from the red-light district, and had no obvious association with prostitution. See *Gold Diggers,* 70.

31. Levy, *They Saw the Elephant,* 91–103. See also the chapter in this volume on lodging houses by Julie Nicoletta.

32. See, for example, Elliott West, *Growing up with the Country: Childhood on the Far-Western Frontier* (Albuquerque: University of New Mexico, 1989); *idem,* and "Beyond Baby Doe: Child Rearing on the Mining Frontier," in *Women's West,* 179–92. See also Waldorf, *Kid on the Comstock.* Again, it is not possible to draw conclusions about marital status in 1870.

33. Georgina Joseph letters, 2 June 1867. Print File, Nevada Historical Society, Reno, Nevada.

34. Eighth U.S. manuscript census (1860). See also J. Wells Kelly, *First Directory of the Nevada Territory* (1862; reprint, Los Gatos, California: The Talisman Press, 1962), 162, for a rare citation of a servant in the directory two years later.

35. J. Ross Browne, *A Peep at Washoe and Washoe Revisited* (1863; 1864; 1869; reprint, Balboa Island, Calif.: Paisano Press, 1959), 105–6. See also Mark Twain, *Roughing It* (1871; reprint, New York: Harper and Brothers, Publishers, 1913), 106.

36. See Mathews, *Ten Years,* 251–56.

37. Ibid., 252.

38. See *Territorial Enterprise,* 28 April 1876, 2:1; 24 June 1876, 2:3. And see Mathews, *Ten Years,* 132, 135.

39. Domestic service discouraged courting and postponed marriage, making it an undesirable occupation for most women. Many Irish women may have regarded this as a positive attribute of the profession, however, according to Hasia R. Diner; see her *Erin's Daughters in America: Irish Immigrant Women in the Nineteenth Century* (Baltimore: John Hopkins University Press, 1982), 20–22.

40. Jews on the Comstock are not identified by the 1880 tenth U.S. manuscript census. Identification has been possible only through Storey County records (including those related to the Jewish cemetery) and citations from local newspapers that identify Jewish families. Many Jews are missed in this process, however.

41. Mathews, *Ten Years,* 36, 39ff., 130.

42. Ronald M. James, Richard D. Adkins, and Rachel J. Hartigan, "Competition and Coexistence in the Laundry: A View of the Comstock." *Western Historical Quarterly* 25, no. 2 (summer 1994). For a California perspective, see Levy, *They Saw the Elephant,* 103–6.

43. Mathews, *Ten Years,* 258; and see also 92, 252–58, 287.

44. Ibid., 98–99, 104, 110, 137. Mathews gave up her school for a while to take up nursing. See 117ff.

45. One woman involved in health care is listed as a "doctress," but for statistical purposes, she is included here. At least one of the women listed in the census as a nurse was a midwife who probably treated patients independently for a wide variety of illnesses, according to Dr. Morris Gallagher of Elko, Nevada, one of her descendants. For parallels in the California Gold Rush, see Levy, *They Saw the Elephant,* 115–20.

46. *Virginia Evening Bulletin,* 7 July 1863, 2:4.

47. *Territorial Enterprise,* 6 April 1866, 3:2, 3:5.

48. Cedric E. Gregory, *A Concise History of Mining* (New York: Pergamon Press, 1980), 221–22.

49. *Territorial Enterprise,* 8 April 1871, 3:2; 14 April 1871, 3:2.

50. William G. White and Ronald M. James, "Little Rathole on the Big Bonanza: Historical and Archeological Assessment of an Underground Resource" (unpublished report; Carson City: Nevada State Historic Preservation Office, 1991).Thanks to Jan Loverin of the Nevada State Museum for identification of the clothing. See also, Charles Howard Shinn, *The Story of the Mine* (1896; reprint, Reno: University of Nevada Press, 1992), 2–3, for a reference to men and women working in remote mines of the West.

51. Alice B. Addenbrooke, *Mistress of the Mansion* (Palo Alto: Pacific Books, 1950); Mathews, *Ten Years,* 121. For a treatment of women in mining in California see Levy, *They Saw the Elephant,* 109–14.

52. Joseph letters; see throughout.

53. David Belasco, *Gala Days of Piper's Opera House and the California Theater* (Sparks, Nev.: Falcon Hill Press, 1991), 30.

54. Georgina Joseph died on December 11, 1870. Her newborn baby died three months later. See the transcripts of the records at the Nevada Historical Society, Reno, Nevada, page 28, or see the original document at the Recorder's Office, Storey County Courthouse, Virginia City, Nevada.

55. Susan Armitage, "Through Women's Eyes: A New View of the West," in *The Women's West,* 14.

3. Redefining Domesticity

1. See Bernadette Francke, "The Neighborhood and Nineteenth-Century Photographs: A Call to Locate Undocumented Historic Photographs of the Comstock Region," *Nevada Historical Society Quarterly* 35, no. 4 (winter 1992): 258–69.

2. The records for Storey County have been put into a computerized database from which much of the data for this chapter is taken.

3. Lawrence B. De Graaf, "Race, Sex, and Region: Black Women in the American West, 1850–1920," *Pacific Historical Review* 49, no. 2 (May 1980): 286.

4. Dan De Quille (William Wright), *The Big Bonanza* (1876; reprint, New York: Alfred A. Knopf, 1947), 20.

5. Myron Angel, ed. *History of Nevada* (1881; reprint, Berkeley: Howell-North, 1958), 571.

6. Eliot Lord, *Comstock Mining and Miners* (1883; reprint, Berkeley: Howell-North, 1959), 66–67.

7. J. Ross Browne, *A Peep at Washoe and Washoe Revisited* (1863; 1864; 1869; reprint, Balboa, Calif.: Paisano Press, 1959), 70.

8. Ibid., 205.

9. Ibid., 181.

10. Ibid.

11. For a good survey of styles in American houses see Virginia and Lee McAlester, *A Field Guide to American Houses* (New York: Alfred A. Knopf, 1990).

12. Contemporary descriptions in newspapers and books, historic photographs, and extant buildings provide much information concerning the appearance of nineteenth-century streetscapes and houses.

13. Browne, *A Peep at Washoe,* 200.

14. Elizabeth Jameson, "Women as Workers, Women as Civilizers: True Woman-hood in the American West," in *The Women's West,* ed. Susan Armitage and Elizabeth Jameson (Norman: University of Oklahoma Press, 1987), 150.

15. *Daily Union,* 14 January 1867, 1.

16. Ibid.

17. For contemporary accounts of Virginia City that helped promote its reputation as a coarse mining town, see Browne, *A Peep at Washoe;* De Quille, *Big Bonanza;* and Mark Twain, *Roughing It* (Hartford: American Publishing Company, 1872). For a rebuttal to some of these views, see Louise Palmer, "How we live in Nevada," *Overland Monthly* (May 1869): 457–62; reprinted as "Of course we move in the best society," in *So Much to Be Done: Women Settlers on the Mining and Ranching Frontier,* ed. Ruth B. Moynihan, Susan Armitage, and Christiane Fischer Dichamp (Lincoln: University of Nebraska Press, 1990), 110–18.

18. For more information on lower-class and middle-class housing of the nineteenth century, see Lizbeth A. Cohen, "Embellishing a Life of Labor: An Interpretation of the Material Culture of American Working-Class Homes, 1885–1915," in *Material Culture Studies in America,* ed. Thomas J. Schlereth (Nashville: American Association of State and Local History, 1982): 289–305; and Sally McMurry, *Families and Farmhouses in Nineteenth Century America* (New York: Oxford University Press, 1989).

19. Historian Sandra L. Myres notes that boardinghouse keeping was a town or city occupation for women, but even with the demand for accommodations in these areas, most boarding establishments were small operations. See Myres, *Westering Women and the Frontier Experience, 1800–1915* (Albuquerque: University of New Mexico Press, 1982), 243.

20. Sutro *Independent,* 6 November 1875.

21. *Mining Reporter,* 27 April 1876.

22. Lord, *Comstock Mining and Miners,* 373.

23. Mary McNair Mathews, *Ten Years in Nevada: Or Life on the Pacific Coast* (Lincoln: University of Nebraska Press, 1985), 35.

24. Ibid., 36.

25. Ibid., 37.

26. For an overview of late nineteenth- and early twentieth-century interiors, see

William Seale, *The Tasteful Interlude: American Interiors through the Camera's Eye, 1860–1917,* 2nd ed. (Nashville: American Association for State and Local History, 1981).

27. Mathews, *Ten Years,* 91.

28. Cohen, "Embellishing a Life," 300–1.

29. Mathews, *Ten Years,* 119–20.

30. Ibid., 122.

31. Ibid., 138–40.

32. Paula Petrik, *No Step Backward: Women and Family on the Rocky Mountain Mining Frontier, Helena, Montana, 1865–1900* (Helena: Montana Historical Society Press, 1987), 12.

33. Women who moved to California during the Gold Rush discovered that domestic work could bring in a substantial income. See JoAnn Levy, *They Saw the Elephant: Women in the California Gold Rush* (Norman: University of Oklahoma Press, 1992), 92–95.

34. Mathews, *Ten Years,* 329.

35. Petrik, *No Step Backward,* 155. Helena reached its population peak in the 1880s.

36. Mathews, *Ten Years,* 308–18 and Lord, *Comstock Mining,* 219–20, 269–77.

37. *Bishop's Directory of Virginia City, Gold Hill, Silver City, Carson City and Reno. . . . 1878–79* (San Francisco: B.C. Vandall, 1878), 527.

38. Walter Van Tilburg Clark, ed., *The Journals of Alfred Doten, 1849–1903,* 3 vols. (Reno: University of Nevada Press, 1973), 2:1403.

39. Assuming this sample holds true for the entire Storey County population of 1880, about 260 married women living with their husbands took in lodgers.

40. Ronald M. James, Richard D. Adkins, and Rachel J. Hartigan, "Competition and Coexistence in the Laundry: A View of the Comstock," *Western Historical Quarterly* 25, no. 2 (summer 1994): 165–84.

41. *Territorial Enterprise,* 11 May 1871, 3:1.

42. Petrik, *No Step Backward,* 60.

43. Ibid.

44. Myres, *Westering Women,* 181.

45. Moynihan, Armitage, and Fischer, *So Much to Be Done,* 4.

46. For the first important delineation of the Cult of True Womanhood, see Barbara Welter, "The Cult of True Womanhood: 1820–1860," *American Quarterly* 18, no. 2 (summer 1966): 151–74. Elizabeth Jameson discusses the difficulties of applying the idea of the cult to the history of western women in her article "Women as Workers, Women as Civilizers: True Womanhood in the American West," in *The Women's West.*

47. Marion S. Goldman, *Gold Diggers and Silver Miners: Prostitution and Social Life on the Comstock Lode* (Ann Arbor: University of Michigan Press, 1981), 28–29.

48. Glenda Riley, *A Place to Grow: Women in the American West* (Arlington, Ill.: Harlan Davidson, 1992), 51–52.

49. For a historiographical overview of women of color in the West, see Antonia I. Castañeda, "Women of Color and the Rewriting of Western History: The Discourse, Politics, and Decolonization of History," *Pacific Historical Review* 61, no. 4 (November 1992): 501–33.

50. Much work has been done on families in urban communities. For an excellent examination of families who took in lodgers but did not run lodging houses in the northeastern United States from the eighteenth to the early twentieth centuries, see John Modell and Tamara K. Hareven, "Urbanization and the Malleable Household: An

Examination of Boarding and Lodging in American Families," in *Family and Kin in Urban Communities, 1700–1930*, ed. Tamara K. Hareven (New York: New Viewpoints, 1977), 164–86.

51. Ibid., 177.

4. *"They Are Doing So to a Liberal Extent Here Now"*

1. Myron Angel, ed., *History of Nevada with Illustrations and Biographical Sketches of its Prominent Men and Pioneers* (Oakland: Thompson and West, 1881), 35.

2. Mary McNair Mathews, *Ten Years in Nevada: Or Life on the Pacific Coast* (Buffalo: Baker, Jones, 1880), 132. Alarm over soaring divorce rates in the United States in the nineteenth century resulted in a national study of marriage and divorce supervised by Commissioner of Labor, Carroll D. Wright. Wright's report, *Marriage and Divorce in the United States, 1867 to 1886* (Washington, D.C.: Government Printing Office, 1897), is a valuable source of divorce statistics. Wright, *Marriage and Divorce,* 140, 352–53.

3. Glenda Riley, *Divorce: An American Tradition* (New York: Oxford University Press, 1991), 5, 79; Paula Petrik, "Not a Love Story: Bordeaux v. Bordeaux," *Montana* 41 (1991): 32.

4. Wright, *Marriage and Divorce,* 547.

5. Although these records must be used with caution, bearing always in mind that their purpose was to convince a judge or jury to terminate a marriage, studies of divorce in other western states have demonstrated that court documents can be used effectively to illuminate women's lives. Robert Griswold in "Apart But Not Adrift: Wives, Divorce, and Independence in California, 1850–1890," *Pacific Historical Review* (1980): 265–83 and *Family and Divorce in California, 1850–1890: Victorian Illusions and Everyday Realities* (Albany: State University of New York Press, 1982) examines divorce documents from rural San Mateo and Santa Clara counties in California. For the urban areas of San Diego and Sacramento counties in California, see Bonnie L. Ford's dissertation, "Women, Marriage, and Divorce in California, 1849–1872" (University of California, Davis, 1985) and Susan Gonda, "Not a Matter of Choice: San Diego Women and Divorce, 1850–1880," *Journal of San Diego History* 37, no. 3 (1991): 194–213. Paula Petrik has studied the urban mining and transportation frontier of the Rocky Mountains in "If She Be Content: The Development of Montana Divorce Law," *The Western Historical Quarterly* 18, no. 3 (1987): 261–92; *No Step Backward: Women and Family on the Rocky Mountain Mining Frontier, Helena, Montana 1865–1900* (Helena: Montana Historical Society Press, 1987); "Not a Love Story: Bordeaux v. Bordeaux," *Montana* 41, no. 2 (1991): 32–46; and "'Send the Bird and Cage': The Development of Divorce Law in Wyoming, 1868–1900," *Western Legal History* 6 (summer/fall 1993): 153–82.

6. Assessment of social status is based on Griswold's adaptations in *Family and Divorce* of Michael B. Katz, *The People of Hamilton, Canada West: Family and Class in a Mid-Nineteenth Century City* (Cambridge: Harvard University Press, 1975), and Petrik's adaptations in *No Step Backward* of Stephen B. Thernstrom, *The Other Bostonians: Poverty and Progress in the American Metropolis, 1880–1970* (Cambridge: Harvard University Press, 1973).

7. Wilbur S. Shepperson, *Restless Strangers: Nevada's Immigrants and their Interpreters* (Reno: University of Nevada Press, 1970), 13. The African American population of Storey County numbered only 95 in 1870 and only 108 in 1880 according to Elmer R.

Rusco, *"Good Times Coming?": Black Nevadans in the Nineteenth Century* (Westport, Conn.: Greenwood Press, 1975).

8. Mathews, *Ten Years,* 133. Louise M. Palmer, "How We Live in Nevada," *Overland Monthly* 2 (1869): 459–60.

9. *Allers v. Allers,* Probate Court Utah Territory, 1861. Angel, *History of Nevada,* 572. The fate of Theresa's marriage to Leonard Dirks is unknown; according to the 1860 federal census, Theresa was an unmarried woman with three children operating a boardinghouse in Virginia City. Eighth U.S. manuscript census (1860).

10. *Allers v. Allers.*

11. *Dean v. Dean,* First District Court, 1872, 2430. *Charles v. Charles,* First District Court, 1872, 2447. Robert Charles and Theresa Dirks marriage contract, 1872.

12. *Walsh v. Walsh,* First District Court, 1870, 2381. Ninth U.S. manuscript census (1870).

13. Tenth U.S. manuscript census (1880).

14. *Compiled Laws of Utah,* 1876, 375–77. For an overview of divorce and divorce law, see Glenda Riley, *Divorce: An American Tradition* (New York: Oxford University Press, 1991); and Roderick Phillips, *Putting Asunder: A History of Divorce in Western Society* (Cambridge: Cambridge University Press, 1988). Other useful surveys are *Divorce Reform: Changing Legal and Social Perspectives* (New York: Free Press, 1980) by Lynne Halem, and Nelson Blake's *The Road to Reno: A History of Divorce in the United States* (New York: Macmillan, 1962).

15. *Zottman v. Zottman,* Probate Court Utah Territory, 1861. Even though Nevada had become a separate territory in March 1861, the Utah Territory probate court continued to hear divorce cases until October 1864. Discussions of Utah's statute can be found in Carol Cornwall Madsen, "'At Their Peril': Utah Law and the Case of Plural Wives, 1850–1900," *The Western Historical Quarterly* 21, no. 4 (1990): 425–43 and in Leonard J. Arrington and Davis Bitton, *The Mormon Experience: A History of the Latter-day Saints* (New York: Alfred A. Knopf, 1979).

16. *Wilkins v. Wilkins,* Probate Court Utah Territory, 1860. *Hayse v. Hayse,* Probate Court Utah Territory, 1860. *Bristol v. Bristol,* Probate Court Utah Territory, 1859.

17. *Cowan v. Cowan,* Carson County Probate Court. Vardis Fisher, *City of Illusion* (Caldwell, Ida.: The Caxton Printers, Ltd., 1941). Swift Paine, *Eilley Orrum: Queen of the Comstock* (Palo Alto: Pacific Books, 1949).

18. *Hilsaa v. Hilsaa,* Probate Court Utah Territory, 1860.

19. *Laws of the Territory of Nevada, passed at the First Regular Session of the Legislative Assembly, begun the First Day of October and Ended on the Twenty-Ninth Day of November, 1861, at Carson City* (San Francisco: Valentine & Co., 1862), 94–99.

20. *Statutes of the State of Nevada passed at the First Session of the Legislature, 1864–5* (Carson City: State Printing Office, 1865), 99, 116. Nevada's divorce law was amended only once between 1864 and 1880. In 1875 the waiting period for desertion or neglect was reduced to one year. *Statutes of the State of Nevada, passed at the Seventh Session of the Legislature, 1875* (Carson City: State Printing Office, 1875), 63.

21. Susan Gonda, "Not a Matter of Choice: San Diego Women and Divorce, 1850–1880," *Journal of San Diego History* 37, no. 3 (1991): 194–213.

22. *Wood v. Wood,* First District Court, 1876, 2947. *Robinson v. Robinson,* First District Court Nevada Territory, 1863, 79.

23. *Hodges v. Hodges,* First District Court, 1877, 3021. *Goffin v. Goffin,* First District

Court Nevada Territory, 1863, 80. *Hall v. Hall*, First District Court Nevada Territory, 1863, 75.

24. *O'Byrne v. O'Byrne*, First District Court, 1876, 2949.

25. Andrew J. Marsh, Official Reporter, *Official Report of the Debates and Proceedings in the Constitutional Convention of the State of Nevada, Assembled at Carson City, July 4th, 1864, to Form a Constitution and State Government* (San Francisco: Frank Eastman, Printer, 1866), 154, 277.

26. Gonda, Petrik, and Ford all discuss the issue of married women's employment. B. Ford, *Women, Marriage, and Divorce in California*, 113.

27. Henry C. Cutting, compiler, *The Compiled Laws of Nevada In Force From 1861 to 1900 (Inclusive)* (Carson City: State Printing Office, 1900), 118–23.

28. James Dean and Theresa Dirks marriage contract. Theresa Dirks inventory of separate property. *Dean v. Dean*. Robert Charles and Theresa Dirks marriage contract.

29. *Ogden v. Ogden*, First District Court, 1871, 2375.

30. *Reyes v. Reyes*, First District Court, 1872, 2445.

31. Wright, *Marriage and Divorce*, 544–47. On the national level as well cruelty was a leading cause of divorce. For a discussion of the evolution of this ground see Robert L. Griswold's "Law, Sex, Cruelty, and Divorce in Victorian America, 1840–1900," *American Quarterly* 38, no. 5 (1986): 721–45; and "Sexual Cruelty and the Case for Divorce in Victorian America," *Signs: Journal of Women in Culture and Society* 11, no. 3 (1986): 529–41.

32. See Robert L. Griswold, "The Evolution of the Doctrine of Mental Cruelty in Victorian American Divorce, 1790–1900," *Journal of Social History* 20, no. 1 (1986): 127–48.

33. *Zimmerman v. Zimmerman*, First District Court Nevada Territory, 1864, 157.

34. Thomas P. Hawley, ed. *Reports of Decisions of the Supreme Court of the State of Nevada*, vols. 3 and 4 (San Francisco: A. L. Bancroft and Company, 1877), 836–41.

35. *LeGuen v. LeGuen*, First District Court, 1871, 2356. Ninth U.S. manuscript census (1870). Thomas LeGuen ran regular advertisements for his services in the *Territorial Enterprise* in 1871.

36. *Schenadeke v. Schenadeke*, First District Court, 1870, 2264.

37. Ninth U.S. manuscript census (1870). *Booth v. Booth*, First District Court, 1872, 2448.

38. *Booth v. Booth*, First District Court, 1873, 2552. The Booths also illustrate some of the difficulties inherent in using federal manuscript census returns for research. In the 1870 census the Booth family was enumerated, as noted above, on July 1, 1870. On July 25 it was counted again at a different location in Gold Hill. In this entry, Joseph is thirty-six, Margaret, twenty-seven, and the youngest child, Susan, had aged a year (or had celebrated a birthday) and was born in California, not Nevada.

39. Cruelty charges against a wife were rare. Wright's researchers found only thirty-two instances in Nevada records where cruelty alone or in combination with other causes was cited by male plaintiffs. Wright, *Marriage and Divorce*, 544–47; *Kruttschmitt v. Kruttschmitt*, First District Court, 1873, 2547.

40. Wright, *Marriage and Divorce*, 544–47.

41. *Jewett v. Jewett*, First District Court, 1876, 2906.

42. *Charles v. Charles. Porteous v. Porteous*, First District Court, 1871, 2379.

43. *Dunn v. Dunn*, First District Court, 1871, 2405. *Rogers v. Rogers*, First District Court, 1876, 2959.

44. *Megger v. Megger,* First District Court Nevada Territory, 1863, 65. *Legg v. Legg,* First District Court, 1869, 2222.

45. *Gomez v. Gomez,* First District Court, 1871, 2393.

46. *Green v. Green,* First District Court, 1873, 2559.

47. *Rothenbucher v. Rothenbucher,* First District Court, 1876, 2872. Ninth U.S. manuscript census (1870). Census of the Inhabitants of the State of Nevada 1875. See Lowe, "'The Secret Friend': Opium in Comstock Society, 1860–1887" for details of Sue Rothenbucher's fate.

48. *Glander v. Glander,* First District Court Nevada Territory, 1863, 66. Ninth U.S. manuscript census (1870). *Daley v. Daley,* First District Court, 1871, 2352. *Dill v. Dill,* First District Court Nevada Territory, 1863, 87.

49. *Burns v. Burns,* First District Court, 1873, 2450.

50. *Allers v. Allers. Garhart v. Garhart,* First District Court, 1872, 2465.

51. Wright, *Marriage and Divorce,* 544–47. Petrik, "'Send the Bird and Cage,'" 156–57.

52. *Snyder v. Snyder,* First District Court, 1870, 2295.

53. *Adams v. Adams,* First District Court Nevada Territory, 1863, 149.

54. *Gardner v. Gardner,* First District Court, 1876, 2825.

55. Riley, *Divorce,* 83.

56. Riley, *Divorce,* 92. *Randolph v. Randolph,* First District Court, 1870, 2244. *Walsh v. Walsh.*

57. *LeGuen v. LeGuen.*

58. *Derby v. Derby,* First District Court, 1876, 2820.

59. *Porteous v. Porteous,* First District Court, 1871, 2379.

60. *O'Neil v. O'Neil,* First District Court, 1872, 2419.

61. *Gelzter v. Gelzter,* First District Court, 1870, 2286.

62. *Legg v. Legg,* First District Court, 1869, 2222.

63. Several of the essays in Ruth B. Moynihan, Susan Armitage, and Christiane Fischer Dichamp, eds., *So Much to Be Done: Women Settlers on the Mining and Ranching Frontier* (Lincoln: University of Nebraska Press, 1990) and in Susan Armitage and Elizabeth Jameson, eds., *The Women's West* (Norman: University of Oklahoma Press 1987) consider the work of women in the West. For the Comstock, see Ronald M. James, Richard D. Adkins, and Rachel J. Hartigan, "Competition and Coexistence in the Laundry: A View of the Comstock," *The Western Historical Quarterly* 25, no. 2 (1994): 173. *Martha and the Doctor: A Frontier Family in Central Nevada* (Reno: University of Nevada Press, 1977) by Marvin Lewis provides a view of one woman's experiences in a similar mining environment. Mathews's *Ten Years* illustrates many of the economic opportunities for women on the Comstock, and *The Journals of Alfred Doten, 1849–1903,* 3 vols. (Reno: University of Nevada Press, 1973) edited by Walter Van Tilburg Clark chronicles the crucial contributions of Doten's wife, Mary.

64. Wright, *Marriage and Divorce,* 546–47. Tenth U.S. manuscript census (1880).

65. *Lacy v. Lacy.*

66. *Goffin v. Goffin.* Ninth U.S. manuscript census (1870). *Green v. Green.* The 1870 U.S. manuscript census lists C. C. Green's property as $1,000 in real estate and $10,000 in personal property. *Miller v. Miller,* First District Court, 1876, 2840.

67. James, Adkins, and Hartigan, "Competition and Coexistence in the Laundry," 173.

68. Tenth U.S. manuscript census (1880).

69. *Cleveland v. Cleveland,* First District Court, 1870, 2278. Ninth U.S. manuscript census (1870). *Snyder v. Snyder,* First District Court, 1870, 2297. *O'Byrne v. O'Byrne.* Tenth U.S. manuscript census (1880).

70. *Hodges v. Hodges.* Tenth U.S. manuscript census (1880). *Meza v. Meza,* First District Court, 1869, 2148. Census of the Inhabitants of the State of Nevada 1875. *O'Byrne v. O'Byrne.* The Nevada state census of 1875 presents additional difficulties, because only surnames and first initials are included, and neither marital status nor household numbers are provided.

71. *Bonafous v. Bonafous,* First District Court, 1877, 3009. Ninth U.S. manuscript census (1870).

72. W. H. Sheridan and Kate Bonafous marriage license.

73. *Muckle v. Muckle,* First District Court, 1876, 2866. Census of the Inhabitants of the State of Nevada 1875. Thomas Muckle and Bessie Gallagher marriage license.

74. *Cowan v. Cowan. Thompson v. Thompson,* 1876, 2862. *Quigley v. Quigley,* First District Court, 1876, 2805. E. H. Smith and Annie Lowden marriage license. David Estey and Catherine Holland marriage license.

75. For a comparison of the Nevada environment with that of other states and territories in the West, see the works of Griswold, Petrik, and Riley previously cited.

5. The "Secret Friend"

1. *Territorial Enterprise,* 24 September 1873.

2. *Nevada Appeal,* 1 February 1987.

3. Rodman W. Paul, *Mining Frontiers of the Far West 1848–1880* (New York: Holt, Rinehart and Winston, 1963), 72.

4. Paul G. Barth, *Bitter Strength* (Cambridge: Harvard University Press, 1964), 1–108.

5. Russell Elliott, *History of Nevada* (Lincoln: University of Nebraska Press, 1973), 112, 166.

6. Mary McNair Mathews, *Ten Years in Nevada: Or Life on the Pacific Coast* (Lincoln: University of Nebraska Press, 1985), 250.

7. Ibid., 259–60.

8. David Courtwright, "Opiate Addiction in the American West, 1850–1920," *Journal of the West* 21, no. 3 (1982): 25.

9. Ibid., 260.

10. Marion S. Goldman, *Gold Diggers and Silver Miners: Prostitution and Social Life on the Comstock Lode* (Ann Arbor: University of Michigan Press, 1981), 30.

11. Courtwright, "Opiate Addiction," 27.

12. Ibid.

13. David Allan Johnson, *Founding the Far West: California, Oregon, and Nevada, 1840–1890* (Berkeley: University of California Press 1992), 75.

14. Ibid., 75.

15. Harry Hubbell Kane, *Opium-Smoking in America and China* (New York: G. P. Putnam's Sons, 1881), 17.

16. Ibid., 2.

17. Ibid., 71.

18. Goldman, *Gold Diggers,* 133.

19. Storey County District Court Indictment Records suggest that there were "several arrests" made for violations of opium laws as reflected in the General Statutes of the State of Nevada from 1861–1885. 4726 Sec. 4 states that: "It shall not be lawful for any person to resort to any house, room, apartment, or other place kept for any of the purposes forbidden by this Act, for the purpose of indulging in the use of opium, or any preparation containing opium, by smoking or otherwise, and any person who shall violate the provisions of this section, shall on legal conviction thereof, be punished." In the case of the State of Nevada vs. Miss Becky, March 1, 1880, the Grand Jury of Storey County indicted her for the crime of "unlawfully resorting to a room or apartment used as a place of resort—by persons for the purpose of indulging in the use of opium." One could assume that because Miss Becky was in the home of Hop Sing and because the connection between opium and Chinese trafficking had given rise to a backlash of anti-Chinese sentiment, Miss Becky was prosecuted as a warning to other marginal members of society who might consider breaking the law.

20. Walter Van Tilburg Clark, ed. *The Journals of Alfred Doten,* 3 vols. (Reno: University of Nevada Press, 1973), 2:976.

21. George Gould, *Illustrated Dictionary of Medicine, Biology and Allied Sciences* (Philadelphia: P. Blakiston, 1894), 1214; Elliot Lord, *Comstock Mining and Miners* (1883; reprint, Berkeley: Howell-North, 1959), appendix, 436.

22. Clark, *Journals of Alfred Doten,* 2:1172.

23. A fire in 1875 destroyed many of the records at the Storey County Court House in Virginia City. After this date, evidence in the coroner's reports indicates that the cause of death was an overdose of opium and/or its derivatives. Whether or not these cases were all planned suicides is a matter of conjecture. One thirty-five-year-old woman who appears to have been a prostitute was said to have died of an overdose of laudanum, yet the coroner's report is inconclusive as to whether the death was planned by the victim or was accidental. *Coroner's Certificate of death,* Rose Barker's death certificate, December 17, 1884. Storey County Court Records State of Nevada, Box 1.

24. Goldman, *Gold Diggers,* 134.

25. Clark, *Journals of Alfred Doten,* 2:867.

26. Ibid., 1:296.

27. Virginia City *Evening Chronicle,* 9 July 1877.

28. Ibid.

29. Ibid.

30. Ibid.

31. Kane, *Opium Smoking in America and China,* 3.

32. San Francisco *Chronicle,* 25 July 1881.

33. Alfred Lindesmith, *Opiate Addiction* (Evanston: Principia Press of Illinois, 1957), 87. In 1888 in Deadwood, opium smoking, which had been introduced by the Chinese, spread to the white inhabitants. Legislation to make opium smoking illegal was being considered at that time. In other cities, such as New York, similar patterns could be observed.

34. Courtwright, "Opiate Addiction," 28.

35. Appendix to the Senate and Assembly—eighth Session, "Recapitulation of Inhabitants of Storey County, State of Nevada," vol. 3, Carson City, 1876.

36. The Statutes of the State of Nevada, eighth Session of 1877, Section 1, redefine "dispose" in Nevada Revised Statutes, Volume 27, Section 453, 056. "Dispense" means

the furnishing of a controlled substance in any amount greater than that which is necessary for the present and immediate needs of the ultimate user. This term does not include the furnishing of a controlled substance by a hospital pharmacy for inpatients.

37. Virginia City *Evening Chronicle,* 30 July 1877.

38. Courtwright, "Opiate Addiction," 29.

39. Goldman, *Gold Diggers,* 133.

40. Phillip Earl, Nevada *Appeal,* 1 February 1987.

41. David Musto, *The American Disease: Origins of Narcotic Control* (New Haven: Yale University Press, 1973), 28–35.

42. Courtwright, "Opiate Addiction," 25.

43. Cynthia C. Palmer, *Shaman Woman, Mainline Lady* (New York: Quill, 1982), 46.

44. William Rothstein, *American Physicians in the 19th Century* (Baltimore: Johns Hopkins University Press, 1985), 191.

45. Ibid., 191.

46. Charles E. Terry and Mildred Pellens, *The Opium Problem* (New York: N.p., 1928), 46.

47. Storey County Certificate of Death, Virginia City, Nevada, January 9, 1879.

48. Palmer, *Shaman Woman,* 47.

49. Ibid.

50. Nevada State Hospital Commitment Records, Row 32, Sec. 2, 1873–1894. See the chapter by Totton, "Women and Divorce on the Comstock, 1859–1880," in this volume for a discussion of Rothenbucher's earlier difficulties.

51. Records of the Nevada State Orphans Home, 1870–1892, Nevada State Archive Microfilm Series.

52. Storey County Certificate of Death, Virginia City, Nevada, June 20, 1879.

53. Nevada State Hospital Commitment Records, Row 32, Sec. 2, 1873–1894.

54. Carroll Smith-Rosenberg, "The Hysterical Woman: Roles and Role Conflict in Nineteenth Century America," *Social Research* (winter 1972): 657–58.

55. Palmer, *Shaman Woman,* 49.

56. Reno *Evening Gazette,* 7 March 1881.

57. Storey County Certificate of Death, Virginia City, Nevada, December 11, 1879.

58. Ibid.

59. Wayne Morgan, *Drugs in America a Social History, 1800–1980* (New York: Syracuse University Press, 1981), 38.

60. A. T. Schertzer, "Excessive Opium Eating," *Boston Medical and Surgical Journal* N.S. 5 (January 20, 1890): 56, and "The Medical Abuse of Opium," *Medical Age* 12 (October 25, 1884): 17.

61. Morgan, *Drugs in America,* 39.

62. H. P. Wilson, "The Indiscriminate Use of Opiades in the Pelvic Diseases of Women," *North American Practitioner* (January 1881): 9–13.

63. "Stimulants and Narcotics," New York *Times,* 6 January 1878, 19.

64. Morgan, *Drugs in America;* John S. Haller and Robin M. Haller, *The Physician and Sexuality in Victorian America* (Urbana: University of Illinois Press, 1974), 25.

65. Mathews, *Ten Years,* 180.

66. D. C. Bishop, "Hypodermic Medication," *Southern Medical Record* 4 (June 1874): 315–33, as seen in Morgan, *Drugs in America,* 40.

67. Courtwright, "Opiate Addiction," 24.

68. *Warner and Munkton Papers,* no. 20, Ms-nc 254-1-263 (1862–63).

69. Dan De Quille, *The Big Bonanza* (Reno: University of Nevada Press, 1974), 297. Census report as seen in Bibliography—Storey County, State of Nevada, 1880.

70. Ibid.

71. Ibid.

72. As a result of muckraking that publicized conditions in slaughterhouses, harmful ingredients in food preservatives, poison in drugs, and addictive properties of patent medicines were exposed. The Pure Food and Drug Act was passed in 1906, and it restricted the manufacture, sale, and transportation of adulterated, misbranded, or harmful food and drugs. The law did not fully protect the public, but it did attack the worst abuses of those industries, including patent medicines.

73. See C. W. Earle's statistical summary in the appendix, as seen in J. Haller and S. Haller's work. This evidence reflects the fact that women who worked in the home as a group were the largest users of opium in Chicago in 1880. Earle interviewed fifty druggists and found that of 235 identified addicts, 169 were women, among whom prostitutes made up the largest group (56), followed by women who worked in the home (45).

6. Creating a Fashionable Society

1. Louise M. Palmer, for example, wrote an article that defends the society of the Comstock, claiming that it could be quite distinguished. See "How We Live in Nevada," *Overland Monthly*, May 1869.

2. Lois Banner, *American Beauty* (New York: A. Knopf, 1983), 70, 73–74; and Marilyn J. Horn and Lois M. Gurel, *The Second Skin: An Interdisciplinary Study of Clothing* (Boston: Houghton Mifflin, 1981), 34. Thorstein Veblen, *The Theory of the Leisure Class* (New York: Modern Library, 1931) was the first to identify clothing, behavior, and its relationship to economic consumption and personal display.

3. Phyllis Tortora and Keith Eubanks, *Survey of Historic Costume: A History of Western Dress* (New York: Fairchild, 1994), 255–56 discusses morality and values in the nineteenth century and refers to the Victorian period as "marked by a code of conduct which was determined by rigid notions about the right ordering of society and individual behavior." Also see W. J. Reader, *Life in Victorian England* (New York: Putnam, 1964), 6. Mary Ellen Roach and Kathleen Musa, *New Perspectives on the History of Western Dress* (New York: Nutriguides, 1981), 11, define dress as, "The total arrangement of all outwardly detectable modifications of the body itself and all materials objects added to it."

4. Gold Hill *News*, 26 January 1865, 2; *Territorial Enterprise*, 14 September 1870, 3, states that there is an ordinance against women wearing trousers.

5. For more on wearing bloomers while crossing the prairie, see S. L. Myres, *Westering Women and the Frontier Experience 1800–1915* (Albuquerque: University of New Mexico Press, 1982), 310 n. 31; and Shelly Foote, "Bloomers," *Dress* 7 (1981): 1–12. For more on children's dress see Sally Helvenston, "Advice to American Mothers on the Subject of Children's Dress 1800–1920," *Dress* 7 (1981): 40; Stella Blum, *Victorian Fashions and Costumes from Harper's Bazaar: 1867–1898* (New York: Dover Publications, 1974), vii; and Marilyn J. Horn and Lois M. Gurel, *The Second Skin: An Interdisciplinary Study of Clothing* (Boston: Houghton Mifflin, 1968), 202–6. For more on specific Victorian toilettes see Blum, *Victorian Fashions*.

6. Claudia Kidwell and Margaret Christman, *Suiting Everyone: The Democratization of Clothing in America* (Washington, D.C.: Smithsonian, 1974), 47–75; Eliot Lord, *Comstock Mining and Miners* (1883; reprint, Berkeley: Howell-North, 1959), 372.

7. Sally Helvenston, "Fashion on the Frontier," *Dress* 15 (1990): 146. Helvenston only addresses dressmakers.

8. Kidwell and Christman, *Suiting Everyone,* 13.

9. Elizabeth W. Barber, *Women's Work: the First 20,000 Years* (New York: W. W. Norton, 1994), 24, 30. Barber examines the idea of perishability associated with weaving and the early production of cloth. Her study focuses on the Stone Age through the Iron Age (approximately 20,000 B.C. to 500 B.C.). Also see Wendy Gamber, "Reduced to Science: Gender, Power and Technology in the American Dressmaking Trade, 1860–1910," *Technology and Culture* 36, no. 3 (July 1995): 456. For more on women and housework, see Susan Strasser, *Never Done: A History of American Housework* (New York: Pantheon, 1982), 125–44; and Ruth Schwartz Cowan, *More Work for Mother* (New York: Basic Books, 1983), 16–68.

10. *Territorial Enterprise,* 31 January 1882, 3, describes the dress Mrs. John Mackay wore to the opening of the Paris Grand Opera House. See also Helvenston, "Fashion on the Frontier," 145–47. *Territorial Enterprise,* 20 July 1861, 2. S. F. A. Caulfield and Blanche Saward, *Encyclopedia of Victorian Needlework* (1887; reprint, New York: Dover, 1972), 2:342, define merino as a "thin, woolen, twilled cloth, made of wool of the Spanish Merino sheep."

11. *Compendium of the Tenth Census* (June 1, 1880) Part II, Revised Edition, Washington Government Printing Office 1888; and Patricia Trautman, "Personal Clothiers: A Demographic Study of Dressmakers, Seamstresses and Tailors 1880–1920," *Dress* 5 (1979). Trautman (74) coined the phrase "personal clothiers," which includes dressmakers, tailoresses, and tailors.

12. Helvenston, "Fashion on the Frontier," and Brenda Brandt, "Arizona Clothing: A Frontier Perspective," *Dress* 15 (1989): 69–76, both state that dress performed a social function in the interaction of women, and although there may have been some adaptations and more freedom in personal attire, both communities followed fashionable norms. Christie Dailey, "A Woman's Concern: Millinery on Center Iowa, 1870–1880," *Journal of the West* 21 (1982): 26–31 notes the significance of millinery shops to social interaction and as producers of fashionable headgear.

13. Sessions Wheeler with William Bliss, *Tahoe Heritage, The Bliss Family of Glenbrook, Nevada* (Reno: University of Nevada Press, 1992), 11. For an understanding of point and thread lace, see Pat Earnshaw, *Bobbin and Needle Laces: Identification and Care* (London: Batsford, 1983), 28, 83, 95, 134.

14. L. Palmer, "We Live in Nevada," 459.

15. Mary McNair Mathews, *Ten Years in Nevada: Or Life on the Pacific Coast* (Lincoln: University of Nebraska Press, 1985), 37; Charlotte Calasibetta, *Fairchild's Dictionary of Fashion* (New York: Fairchild, 1985), 426, defines (426) a wrapper as "1. Any robe made in wrap-around style. 2. Early nineteenth century term for women's housedress." Also see Helvenston, "Fashion on the Frontier," 147–52.

16. *Territorial Enterprise,* 6 September 1862, 2.

17. J. Wells Kelly, *Second Directory of Nevada Territory, 1863* (San Francisco: Valentine, 1863), Business Directory of Advertisers, 250.

18. *Daily Silver Age,* 20 October 1861.

19. Mathews, *Ten Years,* 115, 120. *Territorial Enterprise,* 20 July 1861, 2.

20. Jane Farrell-Beck, "Nineteenth-Century Construction Techniques: Practice and Purpose," *Dress* 13 (1987): 18; Farrell-Beck further explains (15, 16) that in many cases women's ensembles had machine-stitched bodices and hand-sewn skirts. Several things

account for this: 1) skirt seams tended to pucker when sewn by machine, especially after washing; 2) removing hand-sewn stitches was easier when the skirt had to be "turned." ("When the outside was worn or soiled, it was turned to the inside and the fresh surface was shown.") Another reason for hand-stitched skirts was that they could be permanently dismantled, and the fabric could be used for something else.

21. *Territorial Enterprise*, 13 March 1877, 1.

22. Tailoresses are not addressed in this study. Ella A. Davis is listed as a tailoress in the Storey County 1870 census. Gamber, "Reduced to Science," 458, identifies tailoresses [see previous query] and seamstresses as needlewomen who "typically stitched together garments that had been cut from cloth by male tailors." See also Claudia Kidwell, *Cutting a Fashionable Fit* (Washington, D.C.: Smithsonian, 1979), 3.

23. The opposite is true, as well. The rapid decline also parallels the bust cycle of the mining industry. By 1900 (the 1890 census burned) there were only three milliners and thirty-two dressmakers left.

24. The Ninth Census—Vol 1, *The Statistics of the Population of the United States* (Washington, D.C.: Government Printing Office, 1872) does not list any hoopmakers in Nevada. Hoopmakers made crinolines, an undergarment of graduated hoops with a waistband. *The Statistics of the Population of the United States at the Tenth Census* (Washington, D.C.: Government Printing Office, 1883) does not list any galloonmakers or gimpmakers in Nevada. Galloonmakers made binding for women's dresses. Gimpmakers made an openwork trimming for women's dresses.

25. See James and Fliess in this work. This sample size is small and limiting. The 1860 and 1870 censuses do not record marital status; therefore statistics analyzing age are only from the 1880 census and are so cited. Also see Dailey, "A Woman's Concern," 27, who states that millinery was the third most popular employment option for women in Iowa from 1870 to 1880.

26. The censuses provides identification of occupation, while the city directories identify employee/employer relationships. For a detailed discussion of the business nature of milliners and dressmakers, see Wendy Gamber, "A Precarious Independence: Milliners and Dressmakers in Boston, 1860–1890," *Journal of Women's History* 4, no. 1 (1992): 60–88.

27. The eighth manuscript census (1860) does not list any dressmakers. Lord, *Comstock Mines and Miners*, 94, lists one dressmaker shop and four machine sewing rooms as being present on October 13, 1860.

28. Caulfield, *Encyclopedia of Victorian Needlework*, 2:346.

29. Mathews, *Ten Years*, 132. In this case Mathews is actually quoting Brother McGrath. Also see Wendy Gamber, "The Female Economy: The Millinery and Dressmaking Trades, 1860–1930" (Ph.D. diss., Brandeis University, 1990), 260; and Dailey, "A Woman's Concern," 29.

30. Caulfield, *Encyclopedia of Victorian Needlework*, 2:346.

31. *Territorial Enterprise*, 14 May 1868, 2. Also see Gamber, "Female Economy," 104.

32. The 1870 census lists Elizabeth and Annie Vincent (aged forty and twenty-four respectively) as working at a "millinery store." The 1870 census does not list addresses. The 1880 census lists Fanny and Fanny Mayer (aged forty-one and eighteen respectively) living at 26 North A Street and Louisa and Jennie Jackson (aged forty-nine and twenty-one respectively) living at 330 South B Street.

33. *Territorial Enterprise*, 20 July 1861, 2.

34. Kelly, *Second Directory of Nevada Territory*, 135.

35. *Territorial Enterprise,* 17 November 1861, 2. Mrs. Loryea's dry goods included: fine dress silks, French merinos, poplins, paramettas, ruffling, ladies' fine shawls, French flowers, chantilly lace, woolen goods, ladies' and children's boots and shoes, ladies' white kid gaiters, and slippers.

36. Kelly, *First Directory of Nevada Territory,* 135–36.

37. Gamber, "Female Economy," 2, 110, 115.

38. According to the 1870 manuscript census, Mr. Jackson was from Prussia and Mrs. Jackson from Russia.

39. *Territorial Enterprise,* 8 May 1863, 2; 1 January 1867, 2; 26 May 1867, 3. According to Caulfield, *Victorian Needlework,* 1:203, "fancies" are varieties of fabrics with unusual or different patterns, weaving or color combinations, as compared to basic functional dry goods. The Jacksons' stock also included "curls," whose definition is not listed in any major fashion history texts or dictionaries, but more than likely are hairpieces.

40. Ninth U.S. manuscript census (1870). Gamber notes (45) that approximately 10.3 percent of the businesses in Boston were husband-and-wife teams, but in analyzing the Comstock, no other such arrangements were found.

41. County Directory Publishing Company, *Storey, Ormsby, Washoe and Lyon Counties Directory for 1871–2* (Sacramento: H. S. Crocker, 1872), 153.

42. *Territorial Enterprise,* 1 January 1867, 3. Also see Dailey, "A Women's Concern," 29, who states that Iowa millinery shops were usually located in prime mercantile locations.

43. The ninth U.S. manuscript census (1870). A four-year-old African American child also appears directly under Elizabeth Vincent's name, suggesting that she was the mother; however it is possible that her twenty-four-year-old daughter, Annie, was the mother. The 1870 census does not record marital status.

44. *Pioneer Nevada* (Reno: Harold Club, 1956), 106.

45. William R. Gillis, *The Nevada Directory for 1868–9* (San Francisco: M. D. Carr, 1868).

46. *Territorial Enterprise,* 14 May 1868, 2; 23 October 1868, 1; 14 August 1870, 2.

47. *Territorial Enterprise,* 31 March 1868, 2; Gamber, "Female Economy," 57.

48. Ibid, 115.

49. Mathews, *Ten Years,* 287; *Territorial Enterprise,* 12 June 1871, 3.

50. Mathews, *Ten Years,* 131.

51. *Territorial Enterprise,* 24 June 1874, 4. The firm of Isaac Berck and Co. advertised "Ladies Underwear, Dresses, Cloaks, etc." See also Helvenston, "Fashion on the Frontier," 146.

52. Kidwell, *Cutting a Fashionable Fit,* 3.

53. Ibid., 11–13. The term "mantua maker" was derived from the mantua dress, a popular eighteenth-century woman's gown.

54. *Territorial Enterprise,* 12 May 1872, 2.

55. Gamber, "Reduced to Science," 455–82, provides a comparative look at dressmaking techniques and their relationship to gender.

56. Kidwell, *Cutting a Fashionable Fit,* 20–74. Also see Patricia Trautman, *Clothing America: A Bibliography and Location Index of Nineteenth Century American Pattern Drafting Systems* (Costume Society of America, Region II, 1987).

57. Ibid., 25.

58. Gamber, "Female Economy," 321. Also see Linda Jonason, "Dressmaking in

North Dakota between 1890 and 1920: Equipment, Supplies and Methods" (M.A. thesis, North Dakota State University, 1977) 19, 23.

59. *Territorial Enterprise,* June 27, 1878; 2, 15 August 1878. Also see Gamber, "Female Economy," 323.

60. Kidwell, *Cutting a Fashionable Fit,* 16.

61. *Bishop's Directory of Virginia City, Gold Hill, Silver City, Carson City and Reno, 1878–9* (San Francisco: B.C. Vandall, 1879), Also see Nancy Page Fernandez, "Pattern Diagrams and Fashion Periodicals 1840–1900," *Dress* 13 (1987): 5–10; and Nancy Page Fernandez, "'If a Woman Had Taste . . .': Home Sewing and the Making of Fashion 1850–1910" (Ph.D. diss., University of California, Irvine, 1987), 161–200. Gamber, "Reduced to Science," 479, suggests that paper patterns may have been used by professional dressmakers, but, more than likely, they were preferred by those who lived in a rural environment. Also see Margaret Walsh, "The Democratization of Fashion: The Emergence of the Women's Dress Pattern Industry," *The Journal of American History* 66, no. 2 (September 1979): 299–313. For a history of the Demorest family and their fashion business, see Ishbel Ross, *Crusades and Crinolines: The Life and Times of Ellen Curtis Demorest and William Jennings Demorest* (New York: Harper and Row, 1963).

62. *Territorial Enterprise,* 3 December 1868, 2; 24 June 1874, 2; Mathews, *Ten Years,* 129.

63. Mathews, *Ten Years,* 130. See also Norah Waugh, *The Cut of Women's Clothes 1600–1930* (London: Faber and Faber, 1968), 183–90.

64. Goldman, *Gold Diggers,* 89 (Papers Concerning the Estate of Julia Bulette, 1867–68 Storey County Courthouse).

65. *Territorial Enterprise,* 25 May 1867, 3; 29 March 1868, 3. The term "dress pattern" is equivalent to the French phrase *en disposition,* according to personal communications (at the Costume Society of American Symposium, May 1991) with Claudia Kidwell, Curator of Costumes, Smithsonian Institution and Madeleine Ginsburg, retired Curator of Costumes, Victoria and Albert Museum, London. See also Blanche Payne, *History of Costume* (New York: Harper and Row, 1965), 509, who states that, "Materials were often woven or printed 'en disposition,' that is, with designs planned for definite parts of the costume." See also Elizabeth Coleman, *The Opulent Era: Fashions of Worth, Doucet and Pingat* (London: Brooklyn Museum in association with Thames and Hudson, 1990), 11; Waugh, *The Cut of Women's Clothes,* 141; and Otto Thieme, Elizabeth Coleman, Michele Oberly, and Patricia Cunningham et al., *With Grace and Flavor: Victorian and Edwardian Fashion in America* (Cincinnati: Cincinnati Art Museum, 1993), 35.

66. The Nevada *State Journal,* 9 September 1881, 4, stated that a "dress pattern" with a value of $100 was given as first prize for the Ladies' Grand Tournament in the Nevada State Fair. The Nevada State Museum owns the "Margaret Ormsby Gown," which was worn to the Territorial Legislative Ball in 1861. This peach silk taffeta dress is *en disposition,* with a supplemental weft weave, which has been attributed to Lyon, France, by Virginia Vogel, University of Nevada, Reno.

67. Gamber, "The Female Economy," 47, 105, 110.

68. Trautman, "Personal Clothiers," 90. This study is limited, since Trautman identifies dressmakers listed only in city directories, not in the census.

69. Gamber, "Female Economy," 51.

70. *Territorial Enterprise,* 13 March 1877, 1; 9 April 1878, 2; also see Gamber, "Female Economy," 51.

71. Presumably Mrs. George A. Gray is Mary Gray, the forty-six-year-old African American dressmaker listed in the 1870 census. The listings in the city directories refer to her only as Mrs. George A. Gray.

72. County Directory Publishing, *Storey, Ormsby, Washoe and Lyon Counties Directory for 1871–72*. This directory lists Ellen Reudenbough as a seamstress who is employed by Miss Macguigan. Yet the 1870 census identifies Ellen Reudenbough as a dressmaker. Macguigan is more than likely Ellen Reudenbough's employer and is probably doing seamstressing, since Macguigan lists $500 in her personal estate and Ellen Reudenbough lists nothing. Verification of her actual occupation is not possible.

73. *Territorial Enterprise*, 8 January 1875, 2.

74. Ibid., 3 December 1868, 2; 12 May 1872, 2.

75. Ibid., 9 April 1878, 2; Gamber, "Female Economy," 346.

76. *D. M. Bishop's Directory of Virginia City, Gold Hill, Silver City, Carson City, and Reno* (San Francisco: B.C. Vandall, 1879).

77. *Territorial Enterprise*, 27 June 1878, 2; and *Bishop's Directory of Virginia City, Gold Hill, Silver City, Carson City and Reno, 1878–79*.

78. The tenth U.S. manuscript census (1880) lists the following sister dressmaking teams: Mary and Jane McLaughlin, Mary and Nora Sheeley, Isabel and Christine Rose; Charles Collins, *Mercantile Guide and Directory for Virginia City, Gold Hill, Silver City and American City* (San Francisco: Agnew and Deffebach, 1864), lists Misses E. and C. O'Connell; Gillis, *The Nevada Directory for 1868–69* lists Susan and H. E. Hoyd; *Storey, Ormsby, Washoe and Lyon Counties Directory for 1871–72* lists Katie and Josie Shanahan.

79. The tenth U.S manuscript census (1880) lists the following people as having the same address: Sarah Bartell and Darah Reed, and Alice Johnson and Mrs. Johnson.

80. The tenth U.S. census (1880) lists Mary Main and Martha Hosking at the same address. Gamber, "A Precarious Existence," 65, notes the frequency of sisters and/or unrelated women joining forces in an often hostile commercial world.

81. Kidwell, *Cutting a Fashionable Fit*, 139, explains that "suit" was another term for a dress; *Territorial Enterprise*, 20 August 1878, 2.

82. *Territorial Enterprise*, 10 May 1870, 2.

83. Ibid.; Gamber, "Female Economy," 40.

84. *Territorial Enterprise*, 6 April 1866, 3; 11 June 1870, 1. The 1870 manuscript census lists two Irish women, twenty-eight-year-old Mary Conway and twenty-seven-year-old Mary Dowd. It does not list Susan Carroll.

85. Kidwell, *Cutting a Fashionable Fit*, 25.

86. Caufield and Saward, *The Encyclopedia of Victorian Needlework*, 2:442.

87. Ibid., 394–95; Gamber, "Female Economy," 290; Fernandez, "'If a Woman Had Taste . . . ,'" 108–9.

88. *Storey, Ormsby, Washoe and Lyon Counties Directory for 1871–72*, 218.

89. Christina Walkley, *The Ghost in the Looking Glass: The Victorian Seamstress* (London: Peter Owen, 1981), 1.

90. "'If a Woman Had Taste . . . ,'" 109–10 and Walkley, *The Ghost in the Looking Glass*, 2.

91. Trautman, "Personal Clothiers," 74–75; Ellin Berlin, *Silver Platter* (New York: Doubleday, 1957), 115; Fernandez, "'If a Woman Had Taste . . . ,'" 109.

92. Mathews, *Ten Years*, 21.

93. Ibid., 41. Mrs. Hungerford traded Mathews flour, oatmeal, sugar, tea, milk, and

other groceries worth two dollars and paid her the remaining one dollar in cash.

94. Ibid., 66–67. Berlin, *Silver Platter,* 128. In this case, Mrs. Bryant, recently widowed, describes the woman making her mourning garments: "The kind woman pinned and sewed and fashioned her mourning garments." Gamber, "Reduced to Science," indicates that the pin-to-form method was preferred in urban areas.

95. Mathews, *Ten Years,* 38.

96. Ibid., 131–32.

97. Walkley, *Ghost in the Looking Glass,* 1–31.

98. Mathews, *Ten Years,* 104.

99. Ibid., 41, 130.

100. Ibid., 92.

101. Ibid., 173; The *Carson Morning Appeal,* 15 May 1889, stated that a local milliner filed a civil suit to collect $113.50 from a delinquent customer.

102. Berlin, *Silver Platter,* 79, 113, 133, 137, 141, 153.

103. Several nineteenth-century Nevada dresses, now in the collection of the Nevada State Museum, are finished using these techniques: 1) CM-2866-G-1, worn by Mary Kennedy when she married W. C. Ross in Carson City, 1873; 2) CM-4191-G-4, worn by Eliza Mott as part of her wedding trousseau in 1894.

104. Trautman, "Personal Clothiers," 86.

105. Ibid. This study is limited and may not be accurate, as she identified seamstresses only through city directories, not through census materials.

106. Mathews, *Ten Years,* 115.

107. Ibid., preface, 5, 6.

108. Gamber, "Female Economy," 52. Twenty-eight percent of Bostonian milliners and dressmakers combined were foreign born, compared to 31 percent of Comstock milliners and dressmakers. For more information on Nevada immigrants, see Wilbur S. Shepperson, *Restless: Nevada's Immigrants and Their Interpreters* (Reno: University of Nevada Press, 1970).

109. See Helvenston, "Fashion on the Frontier," 141, 150–2 for a complete discussion of the importance of fashion to women in a rural frontier environment.

110. Kidwell, *Cutting a Fashionable Fit,* 2.

111. Nevada *State Journal,* 27 January 1881, 3.

112. Gamber, "Reduced to Science," 456, states that dressmakers worked in a predominantly female world, one that provided highly skilled work, creative labor, relatively high wages, and the very real possibility of someday opening one's own business.

113. Eleventh U.S. census report (1890) for Storey County, Nevada.

114. For a complete study of the transition from custom clothiers to mass production, see Kidwell and Christman, *Suiting Everyone.*

115. Gamber, "Female Economy," 325–29, 420.

116. Gamber states that, "Women lost not only skill but authority. The typical millinery or dressmaking shop was operated by a woman; the typical department store employed women but was controlled by men." Ibid., 188.

117. Ibid., 346.

7. Mission in the Mountains

1. Thomas Gorman, *Seventy-Five Years of Catholic Life in Nevada* (Reno: Journal Publishing, 1935), 119.

2. Ibid., 127–29.

3. Ibid., 41–53.

4. Sister Ann Frederick Hehr, O.P., "History of the Catholic Church in Virginia City, Nevada: St. Mary's In the Mountains, 1860–1967" (M.A. thesis, University of San Francisco, 1969), 49–50, 64, 70.

5. Three articles that incorporate the two elements of western and religious history and exclude the Sisters of Charity are: Francis P. Weisenburger, "God and Man in a Secular City: The Church in Virginia City, Nevada," *Nevada Historical Society Quarterly* 14 (summer 1971): 3–23; John Bernard McGloin, S.J., "Patrick Manogue, Gold Miner and Bishop," *Nevada Historical Society Quarterly* 15 (summer 1972): 25–31; and Vincent A. Lapomarda, S.J., "Saint Mary's in the Mountains: The Cradle of Catholicism in Western Nevada," in Notes and Documents, *Nevada Historical Society Quarterly* 35 (spring 1992): 58–62.

6. There is a rich literature that addresses western womanhood. Among others, see: Glenda Riley, *A Place to Grow* (Arlington Heights, Ill.: Harlan Davidson, 1992), and *The Female Frontier: A Comparative View of Women on the Prairies and the Plains* (Lawrence: University Press of Kansas, 1988); Sandra Myres, *Westering Women* (Albuquerque: University of New Mexico Press, 1982); Susan Armitage and Elizabeth Jameson, eds. *The Women's West* (Norman: University of Oklahoma Press, 1987); and Paula Petrik, *No Step Backward: Women and Family on the Rocky Mountain Mining Frontier, Helena, Montana, 1865–1900* (Helena: Montana Historical Society Press, 1987). For Catholic sisters, see Susan Peterson and Courtney Vaughn-Roberson, *Women of Vision: The Presentation Sisters of South Dakota* (Urbana: University of Illinois Press, 1991); Eileen Mary Brewer, *Nuns and the Education of American Catholic Women, 1860–1920* (Chicago: Loyola University Press, 1987); Mary Ewens, *The Role of the Nun in Nineteenth-Century America* (New York: Arno Press, 1978); and Ursula Stepsis, C.S.A. and Delores Liptak, R.S.M., eds., *Pioneer Healers: The History of Women Religious in American Health Care* (New York: Crossroad, 1978).

7. It is inappropriate to use either the word "congregation" or "nun" when referring to the Daughters of Charity. Rather, they formed a community of sisters, distinguishable by the taking of simple vows and a lack of enclosure. Sister Daniel Hannefin, *Daughters of the Church: A Popular History of the Daughters of Charity in the United States, 1809–1987* (Brooklyn: New City Press, 1989), x–xi. The canonical distinction between nuns and sisters rests on detailed points of church law, generally of little importance to the public at large. This accounts for the wide boulevard use of the word "nun" to refer to any Roman Catholic women living in religion. Occasionally, the word appears in the text of this article, but it is used in a generic sense and does not directly apply to the Daughters of Charity.

8. The mining town life later spawned a host of memoirs and reminiscences written by those who took part in the epic. Examples of such publications for Virginia City include: Emmett L. Arnold, *Gold Camp Drifter: 1906–1910* (Reno: University of Nevada Press, Bristlecone Paperback, 1973); Henry Ernst Dosch, *Vigilante Days at Virginia City: A Personal Narrative*, Facsimile Reproduction (Seattle: Shorey Book Store, 1967); Wells Drury, *An Editor on the Comstock Lode* (1936; reprint, Reno: University of Nevada Press, 1984); and John Taylor Waldorf, *A Kid on the Comstock: Reminiscences of a Virginia City Childhood* (Palo Alto: American West Publishing, 1970; reprint, Reno: University of Nevada Press, 1991).

9. Gorman, *Seventy-Five Years*, 42, and Hannefin, *Daughters of the Church*, 102–3.

10. Annals, Sisters of Charity of San Francisco, 134. The annals of a religious congregation contain a great mix of materials—some letters and documents of the community or a particular mission, unmarked newspaper articles, sisters' reminiscences, or transcripts of all such items. This makes community annals both rich and unreliable research tools. Most notably, they are usually edited for materials the community feels may cast a shadow on matters of spiritual confidentiality. The annals quoted in this paper are copies of the materials from the Daughters of Charity of St. Vincent de Paul, Seton Provincialate, Los Altos Hills, California, and are held by Carolyn Beaupré, Archivist, St. Mary in the Mountains Church, Virginia City, Nevada. Annals citations hereafter are referred to as MS, copy, SMMA (St. Mary in the Mountains Archives).

11. Annals, "Unidentified Sister's Reminiscence of Sister Frederica and the Journey from Maryland to Nevada," MS, copy, SMMA.

12. "Remarks on Sister Alice McGrath, Who Died April 18th, 1913, at Mount St. Joseph's Infant Asylum, San Francisco, California, 88 Years of Age, 60 of Vocation," Annals, copy, MS, SMMA, Virginia City.

13. Ninth U.S. manuscript census (1870), Storey County, Nevada, 85. Many communities only accepted white women for membership throughout the nineteenth century. African American women usually joined one of two African American congregations: the Sisters of the Holy Family in New Orleans, Louisiana, or the Oblates Sisters of Providence in Baltimore, Maryland. Hispanic women entered convents in Mexico but also joined Anglo communities, such as the Sisters of Divine Providence in Texas. Individual community histories show occasional exceptions to general practices of racial exclusion, as well as changing trends over time.

14. Tenth U.S. manuscript census (1880), Storey County, Nevada. Computer search data provided by Ronald James, State Historic Preservation Office, Carson City, Nevada. The manuscript for 1870 shows Sister Frederica's age as forty, but in 1880 she is listed as being fifty-five years old. Information in the Daughters of Charity Annals indicates Sister Frederica was thirty-nine when she came to Virginia City. That would have made her forty-five at the time of the 1870 census and fifty-five in 1880.

15. Douglas McDonald, *Virginia City: The Silver Region of the Comstock Lode* (Las Vegas: Nevada Publications, 1982), 74.

16. Waldorf, *Kid on the Comstock,* 40–53.

17. James Hulse, *The Nevada Adventure,* 6th ed. (Reno: University of Nevada Press, 1990), 112; Russell R. Elliott, *History of Nevada* (Lincoln: University of Nebraska Press, 1973), 144–46.

18. Drury, *Editor on the Comstock,* 13–14. For other examples, see Waldorf, *Kid on the Comstock,* especially 80–86, 90–93, 130–33; McDonald, *Virginia City: The Silver Region;* J. Ross Browne, *A Peep at Washoe and Washoe Revisited* (1863; 1864; 1869; reprint, Balboa Island, Calif.: First Paisano Press, 1959) and from "Our Artist Correspondent at Large," *Harper's Weekly* (February 1863); Walter van Tilburg Clark, ed., *The Journals of Alfred Doten: 1849–90,* 3 vols. (Reno: University of Nevada Press, 1973).

19. The call for stricter rules of enclosure for women religious emerged as a common theme in the evolution of monastic life in the twelfth and thirteenth centuries. Nuns who moved in secular society were often highly criticized. C. H. Lawrence, *Medieval Monasticism: Forms of Religious Life in Western Europe in the Middle Ages,* 2nd ed. (London: Longman, 1989), 221, 229. The first Irish Sisters of Charity, established in Dublin in 1815, took simple vows and had no enclosure rule. Caitriona Clear, "The Limits of Female Autonomy: Nuns in Nineteenth Century Ireland," in *Women Surviv-*

ing: Studies in Irish Women's History in the 19th and 20th Centuries, ed. Maria Luddy and Cliona Murphy (Dublin: Poolberg, 1989), 15–50. The debates about enclosure intensified when various European congregations extended to the United States. See Anne M. Butler, "Adapting the Vision: Caroline in America," in Virgina Geiger, SSND and McLaughlin, SSND, eds., *One Vision, Many Voices* (Lanham, Md.: University Press of America, 1993), 37–51.

20. William Jarvis, "Mother Seton's Sisters of Charity" (Ph.D. diss., Columbia University, 1984), 74–83, 173–204.

21. A number of American Sisters of Charity communities trace their roots to the organizational work of Mother Elizabeth Seton. Her biography and philosophy of religious life are treated in Annabelle M. Melville, *Elizabeth Beyley Seton, 1774–1821* (New York: Charles Scribner's Sons, 1951).

22. See especially, Drury, *Editor on the Comstock,* and Elliott, *History of Nevada,* 145–47.

23. Annals of St. Mary in the Mountains, Annals, MS, copy, SMMA.

24. Robert P. Maloney, *The Way of Vincent de Paul: A Contemporary Spirituality in the Service of the Poor* (Brooklyn: New City Press, 1992), 14–15, 23, 25–27.

25. Unidentified Sister, "Extracts from Another Sister's Account of Nevada," Annals, MS, copy, SMMA.

26. Ibid.

27. Siftings—Virginia City, Nevada, Rev. Asmuth, CM to Father Burlando, July 5, 1864; Rev. Asmuth, CM to unknown, August 15, 1864; Sister Frederica, DC to Father Burlando, December 16, 18__ (date unclear, perhaps 1864), Annals, MS, copy, SMMA. For more on the philosophy of Vincent de Paul that guided the organization of the Daughters of Charity, see James R. Cain, *The Influence of the Cloister on the Apostolate of Congregations of Religious Women* (Rome: Pontifical Lateran University, 1965), 28–37.

28. Bishop O'Connell to Father Burlando, December 6, 1866, Annals, MS, copy, SMMA. John T. Dwyer, *Condemned to the Mines: The Life of Eugene O'Connell, 1815–1891, Pioneer Bishop of Northern California and Nevada* (New York: Vantage Press, 1976), 80–100.

29. Hannefin, *Daughters of the Church,* 102–3.

30. Sister Frederica to Father Burlando, January 6, 1865, July 7, 1865, Annals, Virginia City, Nevada, Annals, MS, copy, SMMA.

31. Ibid.

32. James G. Scrugham, ed., *Nevada: A Narrative of the Conquest of a Frontier Land,* 3 vols. (Chicago: American Historical Society, 1935), 1:357.

33. Act of Incorporation, Sister Frederica to Father Burlando, May 16, 1867, Annals, MS, copy, SMMA; *Territorial Enterprise,* March 3, 1867; Gorman, *Seventy-Five Years,* 45.

34. *Territorial Enterprise,* 3 March 1867.

35. Ibid., 3 March 1867.

36. Hannefin, *Daughters of the Church,* 81, 142–43, 157. This study indicates that in the nineteenth century the Daughters of Charity worked with African Americans in both integrated and segregated situations. As such, their work represents the increasing shift in the nineteenth-century American church to a concern for social justice coupled with political action.

37. *Territorial Enterprise,* 5 March 1867.

38. Annals of St. Mary's School and Asylum and Nevada Orphan Asylum, Virginia

City, State of Nevada, n.p., n.d., Annals, MS, copy, SMMA; *Territorial Enterprise,* 5 February 1873.

39. *Territorial Enterprise,* 21 February 1873.

40. Cain, *The Influence of the Cloister,* 36.; Annals of St. Mary's School and Asylum and Nevada Orphan Asylum, Virginia City, State of Nevada, n.p., n.d., Annals, MS, copy, SMMA.

41. *Territorial Enterprise,* 21 February 1873.

42. Ibid.

43. Annals of St. Mary in the Mountains, Annals, MS, copy, SMMA.

44. Clark, *Journals of Alfred Doten,* vol. 3.

45. Father Asmuth to Father Burlando, June 5, 1864, and Sister Frederica to Father Burlando, July 7, 1866 Annals, MS, copy, SMMA; Drury, *Editor on the Comstock,* 30; McGloin, "Gold Miner and Bishop and his 'Cathedral on the Comstock,'" *Nevada Historical Society Quarterly* 24 (summer 1972): 27–28; and Gorman, *Seventy-Five Years,* 42.

46. Gold Hill *News,* 31 May 1865.

47. Clark, *Journals of Alfred Doten,* 2:817.

48. Drury, *Editor on the Comstock,* 58–59; Margaret G. Watson, *Silver Theatre: Amusements of Nevada's Mining Frontier, 1850–1864* (Glendale, Calif.: Arthur H. Clark, 1964).

49. Annals of St. Mary in the Mountains, Annals, MS, copy, SMMA. See also the chapter in this volume by Watson et al.

50. See *Territorial Enterprise,* 15 August 1868; 17 September 1868; 3, 8, 22 October 1868.

51. Virginia City Reminiscence, Unidentified Sister, n.p., n.d., Annals, MS, copy, SMMA. In addition, these relationships complied with Vincent de Paul's establishment of the Ladies of Charity, lay women dedicated to good works. Maloney, *The Way of Vincent de Paul,* 12.

52. For examples see *Territorial Enterprise,* 1 January 1867, 1 March 1867, 12 February 1867, 27 February 1867, 27 December 1870, 3 January 1873.

53. For examples of the violence and death in Virginia City, see the appendix, "From a Comstock Editor's Scrapbook," in Drury, *Editor on the Comstock,* 297–330.

54. Elliott, *History of Nevada,* 373.

55. Gorman, *Seventy-Five Years,* 45.

56. Annals of St. Mary's School and Asylum and Nevada Orphan Asylum, Virginia City, State of Nevada, Annals, MS, copy, SMMA.

57. Weisenburger, "God and Man in a Secular City," 8–9.

58. Annals of St. Mary in the Mountains, Annals, MS, copy, SMMA. Both the annals and the local newspaper referred to women prominent in civic work by a marital designation only. No first names were given.

59. Ibid.

60. Annals of St. Mary's School and Asylum and Nevada Orphan Asylum, Annals, MS, copy, SMMA.

61. There are countless tales about Mackay, some certainly grounded in legend. See Oscar Lewis, *Silver Kings* (New York: Alfred A. Knopf, 1947), 58–60.

62. Reminiscences, Unidentified Sister, n.d., n.p., Annals, MS, copy, SMMA.

63. *Territorial Enterprise,* 12 February 1867.

64. Reminiscence, Unidentified Sister, n.d., n.p., Annals, MS, copy, SMMA.

65. Gorman, *Seventy-Five Years,* 46.

66. *Territorial Enterprise,* 16 March 1876.

67. Annals of St. Mary in the Mountains, Annals, MS, copy, SMMA. The sister is anonymous in the account, indicating only that she left the Virginia City mission in 1874, thus eliminating Sister Frederica from the list of people possibly charged with the late night duty.

68. Annals, "Unidentified Sister's Reminiscence of Sister Frederica and the Journey from Maryland to Nevada," MS, copy, SMMA; "Remarks on Sister Alice McGrath, Who Died April 18th, 1913, at Mount St. Joseph's Infant Asylum, San Francisco, California, 88 Years of Age, 60 of Vocation," Annals, copy, MS, SMMA; *Daily Territorial Enterprise,* 17 August 1897.

69. McCarthy letters, 24 August 1897, Mamie McCarthy to brother Joe McCarthy (on file, Nevada State Historic Preservation Office).

8. Divination on Mount Davidson

1. *Territorial Enterprise,* 23 August 1887.

2. Daniel Cohen, *The Magic Art of Foreseeing the Future* (New York: Dodd and Mead and Company, 1973), 15, 26, 30, 32.

3. Whitney R. Cross, *The Burned-Over District: The Social and Intellectual History of Enthusiastic Religion in Western New York, 1800–1850* (New York: Harper and Row, 1965).

4. Slater Brown, *The Heyday of Spiritualism* (New York: Hawthorn Books, 1970), 98.

5. Russel M. R. Goldfarb and Clare R. C. Goldfarb, *Spiritualism and Nineteenth-Century Letters* (Cranbury, N.J.: Associated University Presses, Inc., 1978), 131.

6. See, for example, Barbara Welter, "The Cult of True Womanhood: 1820–1860," *American Quarterly* 18 (summer 1966): 151–74.

7. Alex Owen, *The Darkened Room* (Philadelphia: University of Pennsylvania Press, 1990), 9; Ann Braude, *Radical Spirits* (Boston: Beacon Press, 1989), 201.

8. Walter Van Tilburg Clark, ed., *The Journals of Alfred Doten,* 3 vols. (Reno: University of Nevada Press, 1973), 1:784.

9. Owen, *Darkened Room,* 291.

10. Braude, *Radical Spirits,* 121.

11. Gold Hill *Evening News,* 5 February 1866.

12. Owen, *Darkened Room,* 95.

13. Clark, *Journals of Alfred Doten,* 2:953.

14. Ibid., 965.

15. R. Goldfarb and C. Goldfarb, *Nineteenth-Century Letters,* 86.

16. Clark, *Journals of Alfred Doten,* 2:967.

17. Ibid., 2:969.

18. Ibid., 2:967.

19. Braude, *Radical Spirits,* 195.

20. *Daily Safeguard,* 6 October 1868.

21. *Virginia and Truckee Railroad Directory,* 1873–74, 88.

22. *Virginia Evening Chronicle,* 15 November 1872.

23. Allen Kardec, *The Book on Mediums: A Guide for Mediums and Invocators* (1874; reprint, York Beach, Me.: Samuel Weiser, 1970), 107.

24. *Territorial Enterprise,* 13 November 1872.

25. As quoted in the *Virginia Evening Chronicle*, 19 November 1872.

26. Ibid., 16 November 1872.

27. Ibid., 15 November 1872.

28. Ibid., 16 January 1875.

29. Storey County Court Records on file at the Storey County Courthouse, Virginia City, Nevada.

30. *Virginia Evening Chronicle*, 16 January 1875.

31. Storey County Court Records on file at the Storey County Courthouse, Virginia City, Nevada.

32. Gold Hill *Evening News*, 1 July 1867, 10 July 1867.

33. Clark, *Journals of Alfred Doten*, 2:882.

34. Dan De Quille, *The Washoe Giant* (Reno: University of Nevada Press, 1941), 94.

35. Ibid., 90.

36. Ibid., 90.

37. Ibid., 92.

38. *Territorial Enterprise*, 16 March 1878.

39. Ibid., 3 January 1875.

40. *Virginia Evening Chronicle*, 1 November 1876.

41. Ibid., 6 July 1877.

42. Ibid., 12 February 1876, 28 August 1876, 1 November 1876, 2 June 1877, 7 January 1880.

43. Mary McNair Mathews, *Ten Years in Nevada: Or Life on the Pacific Coast* (1880; reprint, Lincoln: University of Nebraska Press, 1985) 230–31.

44. *Territorial Enterprise*, 23 August 1887.

9. *"The Advantages of Ladies' Society"*

1. Gold Hill *News*, 13 October 1863. For a discussion of women and newspapers on the Nevada mining frontier, see Barbara Cloud, "Images of Women in the Mining-Camp Press," *Nevada Historical Society Quarterly* 36, no.3 (fall 1993): 194–207.

2. In 1860 the ratio of women to men in Storey County, the district in which the Comstock was located, was lopsided, with females comprising 5 percent of the population. As the population increased, the percentage of women increased, but females were still underrepresented, at 16 percent in 1861, 15 percent in 1862, and 31 percent in 1870. It was, however, typical for mining communities in the West to have drastically uneven male-female ratios. A territorial census and a special census provide figures for 1861 and 1862, but numbers are not available again until a U.S. census was conducted in 1870. See the chapter by James and Fleiss in this volume for further detail. Percentages compiled by Jean Ford from Waller H. Reed, *Population of Nevada: Counties and Communities, 1860–1980* (Reno: Nevada Historical Society, 1983–1984).

3. The term "sphere," although it carries its own rhetorical baggage, is used throughout this chapter. It was employed in the nineteenth century to describe physical and metaphorical spaces in the lives of both men and women. As an example, Frances Willard used it to describe her efforts to achieve temperance, and Anna Fitch referred to a "true woman's sphere" in her criticism of women in the public arena.

4. The belief in their moral superiority went beyond local civic activity for a number of middle-class Protestant women. There were institutionalized efforts in the West to reform women who had lapsed and to rescue those who were victimized. For a discus-

sion of this activity in the late nineteenth and early twentieth centuries, see Peggy Pascoe, *Relations of Rescue: The Search for Female Moral Authority in the American West, 1874–1939* (New York: Oxford University Press, 1990).

5. Barbara Welter, "The Cult of True Womanhood: 1820–1860," *American Quarterly* 18, no. 2 (summer 1966): 151–74; Lori D. Ginzberg, *Women and the Work of Benevolence: Morality, Politics, and Class in the Nineteenth-Century United States* (New Haven: Yale University Press, 1990), 16. Regarding women moving from private to public activity, see Mary P. Ryan, *Women in Public: Between Banners and Ballots, 1825–1880* (Baltimore: Johns Hopkins University Press, 1990). The concept of a domestic sphere, particularly regarding western settlement, has been a subject of debate among historians. See Robert L. Griswold, "Anglo Woman and Domestic Ideology in the American West in the Nineteenth and Early Twentieth Centuries," in *Western Women: Their Land, Their Lives,* ed. Lillian Schlissel, Vicki Ruiz, and Janice J. Monk (Albuquerque: University of New Mexico Press, 1988).

6. For a discussion of the flexibility of gender roles see Karen Lystra, *Searching the Heart: Women, Men, and Romantic Love in Nineteenth-Century America* (New York: Oxford University Press, 1989).

7. For an analysis of the rhetoric and imagery of the process of civilizing the wilderness and the western frontier, see Carolyn Merchant, "Reinventing Eden: Western Culture as a Recovery Narrative," in *Uncommon Ground: Toward Reinventing Nature,* ed. William Cronon (New York: W. W. Norton and Company, 1995).

8. Ginzberg, *Women and the Work of Benevolence,* 12.

9. See Carl N. Degler, *At Odds: Women and the Family in America from the Revolution to the Present* (New York: Oxford University Press, 1980); and Steven Mintz and Susan Kellogg, *Domestic Revolutions: A Social History of American Family Life* (New York: The Free Press, 1988) for an overview of changes in American family relations and expectations.

10. Eleanor Flexner, *A Century of Struggle* (New York: Atheneum, 1972), 107; Sam P. Davis, *History of Nevada* (Reno: Elms Publishing, 1913), 268–70; and Sandra L. Myres, *Westering Women and the Frontier Experience, 1800–1915* (Albuquerque: University of New Mexico Press, 1982), 218. The Sanitary Commission in Nevada is best known for a fifty-pound sack of flour that Reuel Gridley carried around Nevada and California in 1864. The flour was first auctioned in Austin, then donated for resale, and eventually raised $175,000. The cities of the Comstock raised $23,000 for the Sanitary Commission with the flour sack auctions.

11. *Territorial Enterprise,* 10 January 1863.

12. J. Ross Browne, *A Peep at Washoe and Washoe Revisited* (1863, 1864, 1869; reprinted, Balboa Island, Calif.: First Paisano Press, 1959), 64.

13. *Territorial Enterprise,* 3 March 1867.

14. Ibid., 6 March 1867.

15. Ibid., 3, 6, March 1867.

16. Mathews came to Virginia City in 1869 to settle the estate of her younger brother. A widow, she brought her young son with her and stayed in Nevada for almost a decade. Mathews supported herself in a variety of ways and had contact with a broad range of the individuals who peopled Virginia City. Her book about her life on the Comstock is a valuable source of detail about daily life in the bustling mining towns. Mary McNair Mathews, *Ten Years in Nevada: Or Life on the Pacific Coast* (1880; reprint, Lincoln: University of Nebraska Press, 1985), 268–81. The Bonanza Ring to which Math-

ews referred were four miners and stockbrokers—Mackay, Fair, Flood, and O'Brien—who controlled the wealth produced by a rich ore deposit, the "Big Bonanza"; after 1875 the Bonanza firm established economic dominance on the Comstock. See Russell Elliott, *History of Nevada* (Lincoln: University of Nebraska Press, 1973), 132–37.

17. *Territorial Enterprise,* 21 July 1870.

18. Ibid., 26 June 1869.

19. Elliott, *History of Nevada,* 148.

20. Davis, *History of Nevada,* 690–93; Nevada *State Journal,* February 11, 1894. See Anne Butler's chapter in this volume for activity of Daughters of Charity in Virginia City, and Ronald M. James's chapter regarding the Irish on the Comstock.

21. Marion S. Goldman, *Gold Diggers and Silver Miners: Prostitution and Social Life on the Comstock* (Ann Arbor: University of Michigan Press, 1981), 84; Katherine Hillyer and Katherine Best, *The Amazing Story of Piper's Opera House in Virginia City, Nevada* (Virginia City, Nev.: Enterprise Press, 1953), 3–11.

22. *Virginia Evening Chronicle,* 9 January 1875.

23. Louise Palmer, "How We Live in Nevada," *Overland Monthly* (May 1869): 457–62.

24. Ibid, 462.

25. *History of the First Presbyterian Church, Virginia City, Nevada* (Philadelphia: Presbyterian Historical Society, 1876), quoted in Charles Jeffrey Garrison, "'How the Devil Tempts Us to Go Aside from Christ,'" *Nevada Historical Society Quarterly* 36, no. 1 (spring 1993): 27.

26. The religious activity of western women is discussed in Myres, *Westering Women.*

27. Dr. Charles Meigs quoted in Welter, "The Cult of True Womanhood," 152.

28. Myron Angel, ed., *History of Nevada* (Oakland: Thompson and West, 1881), 206.

29. Information on Episcopal church from Verna Paterson Papers, 1860–1962, Special Collections, Getchell Library, University of Nevada, Reno.

30. John P. Marshall, "Jews in Nevada: 1850–1900," *Journal of the West* 23, no. 4 (January 1984): 62–72.

31. An early church history notes that the Baptist church building was "filled with rooms to lodge strangers." First Baptist Church manuscript collection, Special Collections, Getchell Library, University of Nevada, Reno.

32. *Territorial Enterprise,* 8 October 1868; Virginia City *Daily Safeguard,* 8 October 1868; Gold Hill *Daily News,* 5 March 1867.

33. *Webster's New World Dictionary* defines a "mite" as an English coin worth half a farthing, a small amount of money. In Luke 21:2–4 of the King James version of the Bible, there is reference to the "widow's mites." A widow gave little compared to the rich men, but the amount she contributed was significant, because it was all she had. It is likely that the term for the church organizations is based on these references.

34. *Territorial Enterprise,* 5, 16, February 1871.

35. Ibid., 8 October 1868; Virginia City *Daily Safeguard,* 8 October 1868. See also the chapter by Anne Butler in this volume.

36. Mathews, *Ten Years,* 172.

37. Nevada *State Journal,* 24 December 1870.

38. *Territorial Enterprise,* 3 March 1867.

39. "St. Mary's Asylum and School, Virginia City, Nevada," manuscript in files of Northern Nevada Children's Home, Nevada State Archives, Carson City, Nevada.

40. John Taylor Waldorf, *A Kid on the Comstock: Reminiscences of a Virginia City Childhood* (Palo Alto: American West Publishing, 1970; reprint, Reno: University of Nevada Press, 1991), 110; William Breault, *The Miner Was a Bishop* (Rancho Cordova, Calif.: Landmark Enterprises, 1970), 65.

41. Phoebe Hansford, *Women of the Century* (Augusta, Me.: True and Co., 1882), 483–87; Rev. John O. Foster, *Life and Labors of Mrs. Maggie Newton Van Cott* (Cincinnati: Hitchcock and Walden, 1872).

42. *Territorial Enterprise,* 2 December 1873.

43. Ibid., 11–12 February 1875; Leonard J. Arrington and Davis Bitton, *The Mormon Experience: A History of the Latter-day Saints* (New York: Alfred A. Knopf, 1979), 230. The issue of polygamy was widely discussed and debated in the West, particularly among women. See Pascoe, *Relations of Rescue,* chapter 2, for an examination of this conflict. "Bluestocking" is a derogatory term referring to a female intellectual who has lost her femininity in her quest for education.

44. Gold Hill *Daily News,* 5 March 1867.

45. *Territorial Enterprise,* 8 October 1868.

46. Paul Fatout, *Mark Twain in Virginia City* (Bloomington: Indiana University Press, 1964), 196–213; emphasis and spelling in original, Gold Hill *Daily News,* 30 May 1864.

47. Ginzberg, *Women and the Work of Benevolence,* thoroughly and effectively explores this topic.

48. Julie Roy Jeffries, *Frontier Women: The Trans-Mississippi West, 1840–1880* (New York: Hill and Wang, 1979), 184–85.

49. Elko *Independent,* 23 June 1869.

50. Nevada *State Journal,* 11 February 1871.

51. An example of their weekly notice is found in the *Territorial Enterprise,* 31 March 1863.

52. Wells Drury, *An Editor on the Comstock Lode* (Reno: University of Nevada Press, 1985), 122.

53. Mathews, *Ten Years,* 116.

54. Ibid.

55. *Evening Chronicle,* 26 January 1875.

56. Carolyn De Swarte Gifford, "Protection: The WCTU's Conversion to Woman Suffrage," in *Gender, Ideology, and Action,* ed. Janet Sharistanian (New York: Greenwood Press, 1986), 96.

57. *Territorial Enterprise,* 29 October 1874.

58. Jeffries, *Frontier Women,* 192.

59. See Jean Ford and James W. Hulse, "The First Battle for Women's Suffrage in Nevada: 1869–1871—Correcting and Expanding the Record," *Nevada Historical Society Quarterly* 38, no. 3 (fall 1995): 174–88.

60. "Woman Suffrage. Speech of Hon C. J. Hillyer, Delivered in the Assembly of the State of Nevada, Tuesday, February 16, 1869," Appendix to the *Journal of the Assembly,* Fourth Session, Legislature of the State of Nevada, 1869 (Carson City: State Printer, 1869).

61. *Territorial Enterprise,* 25 April 1869.

62. Ibid.

63. Carroll Smith-Rosenberg, *Disorderly Conduct: Visions of Gender in Victorian America* (New York: Oxford University Press, 1985), 206, 258–60.

64. *Territorial Enterprise,* 25 April 1869.

65. Reese River *Reveille,* 25 November 1867.

66. Ibid., 20 June 1868.

67. Mathews, *Ten Years,* 78.

68. Elko *Independent* as quoted by *Carson Appeal,* 8 July 1870.

69. Walter Van Tilberg Clark, ed., *The Journals of Alfred Doten,* 3 vols. (Reno: University of Nevada Press, 1973), 2:1100.

70. *Territorial Enterprise,* 23 July 1870.

71. Reno *Crescent,* 23 July 1870.

72. Clark, *Journals of Alfred Doten,* 2:1101.

73. Nevada *State Journal,* 4 February 1871.

74. *Journal of the Assembly,* 1871, 150, 154.

75. Clark, *Journals of Alfred Doten,* 2:1145.

76. Jeffries, *Frontier Women,* 190–91.

10. Their Changing World

1. I am indebted to the Nevada Humanities Committee, which, through an exemplary grant from the National Endowment for the Humanities, made possible much of this research.

2. For more details on this topic, see Ronald Takaki, *A Different Mirror: A History of Multicultural America* (Boston: Little, Brown and Company, 1993), Chapter 8.

3. This has been detailed in an excellent study by Judy Yung, *Unbound Feet: A Social History of Chinese Women in San Francisco* (Berkeley: University of California Press, 1995), chapter 1. See also June Mei, "Socioeconomic Origins of Emigration: Guangdong to California, 1850–1882," *Modern China* 5 (October 1979): 463–501; and David Chuenyan-Lai, *Chinatowns: Towns Within Cities in Canada* (Vancouver, B.C.: University of British Columbia Press, 1988).

4. In the 1870s White Pine County, for example, was apparently regarded as a safer community, and the census manuscript for 1870 shows a high ratio (three to one) of Chinese men to Chinese women. Many of these women are listed as married and "keeping house." Undoubtedly some of the husbands of these women were miners who were working nearby.

5. Census manuscript (1860), Nevada portion of Utah Territory, searched by "China" as place of birth.

6. Nevada State Legislature, "Census of the Inhabitants of the State of Nevada, 1875," *Appendix to the Journals of the Senate and Assembly in the Eighth Session of the Legislature of the State of Nevada* (Carson City: John K. Hill State Printer, 1877), vol. 3. The figures are not based upon the summary but on a recalculation of the manuscripts performed by Sue Edwards.

7. Census manuscript (1870), Nevada, Storey County.

8. Alvin Yiu-Cheong So, "Ethnic Doctors in Los Angeles's Chinatown," *Journal of Ethnic Studies* 11, no. 4 (winter 1984): 75–82.

9. For information about Dr. Lock and King Lee, see the *Territorial Enterprise* and the Gold Hill *News,* 18–23 February 1871. On the subject of Chinese organizations, see Him Mark Lai, "Historical Development of the Chinese Consolidated Benevolent Association/*Huiguan* System," in *Chinese America: History and Perspectives 1987* (San Francisco: Chinese Historical Society of America, 1987), 13–51.

10. The photograph, "Mary, wife of a physician, her son, and her friends Nellie and Susie," can be found at the Nevada Historical Society and the Special Collections Library of the University of Nevada, Reno. If Dr. Lock had a son, Mary might be his wife, but the census manuscript shows no children, and the boy shows up as an older child in photographs with Euro-Americans.

11. Patricia Buckley Ebrey, *The Inner Quarters: Marriage and the Lives of Chinese Women in the Sung Period* (Berkeley: University of California Press, 1993), 218, and Yung, *Unbound Feet,* 37–41. See also, Maria Jaschok, *Concubines and Bondservants: The Social History of a Chinese Custom* (London: Zed, 1988): and Margery Wolf, *Women and the Family in Rural Taiwan* (Stanford: Stanford University Press, 1972).

12. *Directory of Chinese Business Houses* (San Francisco: Wells Fargo and Company, 1878), 84–86.

13. *Silver State* (Winnemucca), 25 February 1875; Wilbur Shepperson, "Immigrant Themes in Nevada Newspapers," *Nevada Historical Society Quarterly,* 12, no. 2 (summer, 1969): 16; and Beth Amity Au, "Home Means Nevada: The Chinese in Winnemucca, Nevada, 1870–1950, A Narrative History" (M.A. thesis, University of California, Los Angeles, 1993).

14. This figure is based on the census manuscript and on soundex cards and therefore includes, for example, Wha Lee, female, age eighteen, who is not in the census manuscript, and Sam Lee, male, age thirty—both laundry workers in Gold Hill.

15. For more information, see Rubie S. Watson, "Wives, Concubines, and Maids: Servitude and Kinship in the Hong Kong Region, 1900–1940," in *Marriage and Inequality in Chinese Society,* ed. Rubie S. Watson and Patricia Buckley Ebrey (Berkeley: University of California Press, 1993), 231–55. See also, Jaschok, *Concubines;* Benson Tong, *Unsubmissive Women: Chinese Prostitutes in Nineteenth-Century San Francisco* (Norman: University of Oklahoma Press, 1994), 71; and Huping Ling, "Surviving on the Gold Mountain: Chinese-American Women and Their Lives" (Ph.D. diss., Miami University, 1992).

16. See, for example, Anthony B. Chan, *Gold Mountain: The Chinese in the New World* (Vancouver, B.C.: New Star Books, 1983); and Shih-shan Henry Tsai, *The Chinese Experience in America* (Bloomington: Indiana University Press, 1986). See also Yung, *Unbound Feet,* 26–37, 73–77.

17. See Kang Chao, *The Development of Cotton Textile Production in China* (Cambridge: Harvard University Press, 1977); Yan Zhongping, *A Draft on the History of the Chinese Cotton Textile, 1289–1937* (Beijing: Kexue Chubanshe, 1963), 20–17; Bobby Siu, *Women of China: Imperialism and Women's Resistance, 1900–1949* (London: Zed, 1982), 38–43; Ronald M. James, introduction to this volume; Paul P. C. Siu, *The Chinese Laundryman: A Study in Social Isolation* (New York: New York University Press, 1987); and Paul M. Ong, "An Ethnic Trade: The Chinese Laundries in Early California," *Journal of Ethnic History* 11, no. 3 (spring 1992): 41–67.

18. Lucie Cheng Hirata, "Free, Indentured, Enslaved: Chinese Prostitutes in Nineteenth-Century America," *Signs* 5, no. 1 (autumn 1979), 3–29. See also Marion S. Goldman, *Gold Diggers and Silver Miners: Prostitution and Social Life on the Comstock Lode* (Ann Arbor: University of Michigan Press, 1981), whose statistics are inaccurate, thus resulting in misleading conclusions; see also her two articles, "Prostitution and Virtue in Nevada," *Nevada Historical Society Quarterly* 10, no. 4 (November/December 1978), and "Sexual Commerce on the Comstock Lode," *Nevada Historical Society Quarterly* 21,

no. 2 (summer 1978), 72–89; and Anne M. Butler, *Daughters of Joy, Sisters of Misery: Prostitutes in the American West, 1865–90* (Urbana: University of Illinois Press, 1985).

19. The most comprehensive study of Chinese American prostitutes is Tong, *Unsubmissive Women*. See also James Francis Warren, *Ah Ku and Karayuki-san: Prostitution in Singapore, 1870–1940* (New York: Oxford University Press, 1993); and Sue Gronewold, *Beautiful Merchandise: Prostitution in China, 1860–1936* (New York: Haworth Press, Inc., 1982). See also George M. Blackburn and Sherman I. Richards, "The Prostitutes and Gamblers of Virginia City, Nevada: 1870," *Pacific Historical Review* 48, no. 2 (May 1979): 239–58, and "The Chinese of Virginia City, Nevada: 1870," *Amerasia* 7, no. 1 (1980): 51–71; James and Fliess in this volume; and Loren B. Chan, "The Chinese in Nevada: An Historical Survey, 1856–1970," *Nevada Historical Society Quarterly* 25, no. 4 (winter 1982): 266–314.

20. For an excellent background on Chinese prostitution, see Gail Hershatter, "Prostitution and the Market in Women in Early Twentieth-Century Shanghai," in *Marriage and Inequality,* ed. Watson and Ebrey, 256–85. According to Herschatter, in a survey of 500 prostitutes conducted in 1948, two-thirds were unmarried, a fifth were widows, and more than 9 percent had living spouses. The latter category could not divorce their husbands or were so poor that engaging in prostitution was their only means of survival.

21. Herschatter, "Prostitution and the Market in Women" in *Marriage and Inequality,* ed. Watson and Ebrey, 260–63.

22. Walter Van Tilburg Clark, ed., *The Journals of Alfred Doten,* 3 vols. (Reno: University of Nevada Press, 1973), 2:867.

23. Opium was legal until 1909 in the United States; see Tong, *Unsubmissive Women,* 65, and Warren, *Prostitution in Singapore,* 42, 293, 306, 310, and 365 for more information on the connection between prostitution and opium.

24. Clark, *Journals of Alfred Doten,* 2:817, 839, 866–67.

25. Tong, *Unsubmissive Women,* 6; and Yung, *Unbound Feet,* 31, 33–34.

26. Hirata, "Free, Indentured, Enslaved," 17.

27. In 1869 in Silver Peak, Esmeralda County, Chinese employed in the mine and mill earned one dollar per day with board. See Hugh A. Shamberger, *The Story of Silver Peak, Esmeralda County, Nevada* (Carson City: Department of Conservation and Natural Resources, 1976), 11. Central Pacific Railroad workers between 1866 and 1867 averaged thirty-five dollars for a twenty-six-day month, but after they had deducted board and broker fee netted only twenty to twenty-five dollars per month. See *Alta California* (San Francisco) 16 November 1867, 9 November 1868.

28. *Territorial Enterprise,* 22 June 1867; Clark, *Journals of Alfred Doten,* 2:877.

29. Tong, *Unsubmissive Women,* 114.

30. Goldman, *Gold Diggers,* 134–35.

31. *Territorial Enterprise,* 18 February 1871.

32. Ibid., 2 October 1874.

33. Hill Gates, "The Commoditization of Chinese Women," *Signs: Journal of Women in Culture and Society* 14, no. 4 (1989): 799.

34. Richard Kock Dare, "The Economic and Social Adjustment of the San Francisco Chinese for the Past Fifty Years" (M.A. thesis, University of California, Berkeley, 1959), 23.

35. Mary McNair Mathews, *Ten Years in Nevada: Or Life on the Pacific Coast* (1880; reprint, Lincoln: University of Nebraska Press), 257.

36. The Chinese Exclusion Act, 22 Stat. 58 (May 6, 1882) and its supplements, the Scott Act, 25 Stat. 504 (October 1, 1888), Geary Act, 17 Stat. 15 (May 5, 1892), Chinese Exclusion Laws, 33 Stat. 428 (1904), Immigration Act, 39 Stat. 874 (February 5, 1917), and Immigration Act, 43 Stat. 153 (May 26, 1924) are reproduced in Bill Ong Hing, *Making and Remaking Asian America Through Immigration Policy, 1850–1990* (Stanford: Stanford University Press, 1993), 203–14. See also, Sharon M. Lee, "Asian Immigration and American Race-relations: From Exclusion to Acceptance?" *Ethnic and Racial Studies* (Great Britain) 12, no. 3 (1989): 368–90; and Sucheng Chan, "Exclusion of Chinese Women, 1870–1943," in *Entry Denied: Exclusion and the Chinese Community in America, 1882–1943,* ed. Sucheng Chan (Philadelphia: Temple University Press, 1991), 94–146; and George Anthony Pfeffer, "Forbidden Families: Emigration Experiences of Chinese Women under the Page Law, 1875–1882," *Journal of American Ethnic History* 6 (1986): 28–46.

37. *Territorial Enterprise,* 9 October 1866.

38. Prison Records, Nevada State Archives, Carson City.

39. *The Eureka Daily Sentinel,* 7 January 1882, reported the kidnapping of the ten- or eleven-year-old daughter of wealthy Winnemucca Chinese merchant, Tong Ting.

40. See, for example, Virginia *Evening Bulletin,* 31 July 1863.

41. Hsiung, "Constructed Emotions," 87–117.

42. See, for example, taped interview with Frank Chang, 1992, Special Collections, James R. Dickinson Library, University of Nevada, Las Vegas.

43. *Nevada Statutes,* 1865, Chapter CXLV Section 50 and Nevada Statutes, 1867, Chapter LII Section 21.

44. This was the result of the Stoutmeyer case of 1872, which involved African American parents who wanted their children to attend public school. See Elmer Rusco's forthcoming work on the Chinese and Nevada law.

45. Daniels, *Asian America,* 82–83.

46. June 28, 1882.

47. Storey County, *Marriage License Stubs,* ST-013 for 1874–1879, ST-014 for 1879–1902. Nevada Museum and Historical Society, Las Vegas. All licenses mentioned were found in the above. I am indebted to David Millman of the Nevada Museum and Historical Society for his assistance.

48. *Territorial Enterprise,* 28 May 1875.

49. Ibid., 19 September 1877.

50. For more information about Chinese community organizations, see Him Mark Lai, *Cong Huaren dao Huaqiao [From Overseas Chinese to Chinese American]* (Hong Kong: Joint Publishing, Inc., 1992), chapter 2.

51. For more information, see Tong, *Unsubmissive Women,* 134.

52. The *Territorial Enterprise,* 12 January 1878, actually referred to it as Mongolian rather than Chinese.

53. On this process, see Elliott R. Barkan, "Race, Religion, and Nationality in American Society: A Model of Ethnicity—From Contact to Assimilation," *Journal of American Ethnic History* 14, no. 2 (winter, 1995): 38–75.

54. Yung, *Unbound Feet,* chapter 2; Tong, *Unsubmissive Women,* 30.

55. Marlon K. Hom, "Some Cantonese Folksongs on the American Experience," *Western Folklore* 42, no. 2 (1983): 130.

56. Folk song translated in Marlon K. Hom, ed. and trans., *Songs of Gold Mountain: Cantonese Rhymes from San Francisco Chinatown* (Berkeley: University of California

Press, 1987), 146, and Hakka folk song in Tin-Yuke Char, *The Sandalwood Mountains: Readings and Stories of the Early Chinese in Hawaii* (Honolulu: University of Hawaii Press, 1975), 67.

57. Both men probably held one of the three traditional Chinese degrees; the average age of men holding the top level degree was in the mid-thirties, and from the age of these two men, it seems that they probably held at least the lowest (i.e., bachelor's) degree.

58. Gung and Mung were probably the same surname in Chinese.

59. This statistical analysis was performed by Sue Edwards and is based upon the 1870 and 1880 census manuscript and the 1875 Nevada census.

60. S. Chan, "Exclusion of Chinese Women," and Pfeffer, "Forbidden Families."

61. *Nevada Statutes,* 1861, Chapter 32; *Nevada Statutes,* 191, Chapter 72; see also a forthcoming work by Elmer Rusco on aliens and the law in Nevada, and Philip I. Earl, "Blood Will Tell: A Short History of Nevada's Miscegenation Laws," *Nevada Public Affairs Review* 2 (1987): 82–86. The law was not repealed until 1959. See also Megumi Dick Osumi, "Asians and California's Anti-Miscegenation Laws," in *Asian and Pacific American Experiences: Women's Perspectives,* ed. Nobuya Tsuchida (Minneapolis: Asian/Pacific American Learning Resources Center, University of Minnesota, 1982), 1–37.

62. Carson City *Morning Appeal,* 21 June 1880.

63. Eureka *Weekly Sentinel,* 16 February 1889.

64. Bureau of the Census, 1870, 1880.

65. Oral interview with Juanita Pontoon, granddaughter, in Reno, July 1994. Documents about the marriage and subsequent children are in Special Collections, #93-21, James R. Dickinson Library, University of Nevada, Las Vegas.

66. For more information on the repeal of Chinese exclusion, see S. Chan, "Exclusion of Chinese Women," and Hing, *Making and Remaking.*

67. Her name can be mistakenly rendered as "Gung Gow." Based on her *huiguan* association, her surname was probably Yee, in this case Romanized as You.

68. This is a secret society organization. For information on these organizations, see Him Mark Lai, Cong Huaren, and Douglas W. Lee, "Sacred Cows and Paper Tigers: Politics in Chinese America, 1880–1900," *Annals of the Chinese Historical Society of the Pacific Northwest* 3 (1985–1986): 86–103.

69. See H. Lai, *Cong Huaren,* 29, for more information on these *huiguan.* Yung Wa is more commonly Romanized as Young Wo.

70. This was not the case for Euro-American prostitutes. See Tong, *Unsubmissive Women,* 175.

71. The 1900 and 1910 census information was provided by Mary K. Rusco. The women are discussed earlier in this study.

72. This was not unique to the Chinese. See Arodays Robles and Susan Cotts Watkins, "Immigration and Family Separation in the U.S. at the Turn of the Twentieth Century," *Journal of Family History* 18, no. 3 (1993): 191–211.

73. Marlon K. Hom, "Some Cantonese Folksongs," 136.

74. Census manuscript (1900) Nevada, Storey County, Virginia City, and U.S. Immigration and Naturalization File No. 13561/141 in the National Archives, San Bruno, California. I am indebted to Mary and Elmer Rusco for the latter document.

75. Census manuscript (1900) Nevada, Storey County, Virginia City.

76. There is a tremendous amount of literature on the subject of acculturation and assimilation. See, for example, Raymond H. C. Teske, Jr. and Bardin H. Nelson,

"Acculturation and Assimilation: A Clarification," *American Ethnologist* 1, no. 2 (May 1974): 351–68.

11. *"And Some of Them Swear Like Pirates"*

1. Julian H. Steward, *Basin-Plateau Aboriginal Sociopolitical Groups* (Smithsonian Institution Bureau of American Ethnology Bulletin, No. 120, 1938; reprinted Salt Lake City: University of Utah Press, 1970), 44.

2. Herbert Barry III and Alice Schlegel, "Cross-Cultural Codes on Contributions by Women to Subsistence," *Ethnology* 21, no. 2 (1982): 165–88, Table 1, No. 137. Omer C. Stewart, "Culture Element Distributions: xiv Northern Paiute," *University of California Anthropological Records* 4, no. 3 (1941): 361–446).

3. Margaret M. Wheat, *Survival Arts of the Primitive Paiutes* (Reno: University of Nevada Press, 1967), 29–39.

4. Catherine S. Fowler, *In the Shadow of Fox Peak: An Ethnography of the Cattail-Eater Northern Paiute People of Stillwater Marsh* (U.S. Fish and Wildlife Service, Region 1, Cultural Resource Series No. 5. Washington D.C.: Government Printing Office, 1992), 58–59.

5. Wheat, *Survival Arts*, 12–15.

6. Albert D. Richardson, *Beyond the Mississippi* (Hartford: American Publishing Co., 1869), 512.

7. F. Dodge, "Report of the Carson Valley Agency," *Report of the Commissioner of Indian Affairs for the Year 1859* (Washington, D.C.: Government Printing Office, 1860), 373–77. Dan De Quille, *The Big Bonanza* (Hartford: American Publishing Co., 1876; reprint, Alfred A. Knopf, 1947), 20–21.

8. De Quille, *Big Bonanza*, 155.

9. Myron Angel, ed., *History of Nevada* (Oakland: Thompson and West, 1881; reprint, Berkeley: Howell-North Books, 1958), 169.

10. Gold Hill *Evening News*, 2 June 1864; *Territorial Enterprise*, 6 July 1879.

11. Gold Hill *Evening News*, 2 June 1864.

12. Ibid., 13, 18 May 1875.

13. Eugene M. Hattori, *Northern Paiutes on the Comstock: Archaeology and Ethnohistory of an American Indian Population in Virginia City, Nevada* (Carson City: Nevada State Museum, Occasional Paper, No. 3, 1975), 20–22; *Territorial Enterprise*, 6 October 6, 1877.

14. *Territorial Enterprise*, 20 January 1872, 18 January 1879, 30 August 1881.

15. De Quille, *The Big Bonanza*, 214–15; *Territorial Enterprise*, 20 July 1881.

16. *Territorial Enterprise*, 9 August 1868.

17. John T. Waldorf, *A Kid on the Comstock: Reminiscences of a Virginia City Childhood* (1970; reprint, Reno: University of Nevada Press, 1991), 34.

18. Mary McNair Mathews, *Ten Years in Nevada: Or Life on the Pacific Coast* (1880; reprint, Lincoln: University of Nebraska Press, 1985), 289; *Territorial Enterprise*, 30 August 1881.

19. J. Ross Browne, *A Peep at Washoe and Washoe Revisited* (1863; 1864; 1869; reprint, Balboa Island, Calif.: Paisano Press), 189; Mathews, *Ten Years*, 287; Waldorf, *A Kid on the Comstock*, 33; *Territorial Enterprise*, 15 August 1872.

20. *Territorial Enterprise*, 14 June 1872.

21. Ibid., 11 October 1872.

22. De Quille, *Big Bonanza*, 215; *Territorial Enterprise*, 6 January 1872.

23. Stewart, *Culture Element Distributions*, 440.

24. Alice Schlegel, "Comments to Charles Callender and Lee M. Kochems, The North American Berdache," *Current Anthropology* 24, no. 4 (1983): 462–63.

25. *Territorial Enterprise*, 11 October 1872.

26. U.S. Bureau of the Census, tenth census (1880), manuscript census for Gold Hill, Storey County, Nevada.

27. *Territorial Enterprise*, 18 March 1871, 21 May 1872.

28. Mathews, *Ten Years*, 288; Hattori, *Northern Paiutes on the Comstock*, 58, 65.

29. *Territorial Enterprise*, 26 August 1866, 9 August 1868, 6 October 1877; Mathews, *Ten Years*, 286.

30. *Territorial Enterprise*, 25 October 1876; De Quille, *Big Bonanza*, 217.

31. Mathews, *Ten Years*, 287.

32. *Territorial Enterprise*, 14 July 1871.

33. U.S. Bureau of the Census Office, Report on Indians Taxed and Indians Not Taxed in the United States at the Eleventh Census: 1890, 1894; (Washington, D.C.: Government Printing Office, 1894).

34. Joy Leland, "Population," in *Handbook of North American Indians*, vol. 11 [Great Basin], ed. Warren d'Azevedo, William C. Sturtevant, Catherine S. Fowler, Jesse D. Jennings, Don D. Fowler, and William H. Jacobsen Jr. (Washington, D.C.: Smithsonian Press, 1986), 608.

35. *Territorial Enterprise*, 7 August 1870, 15 August 1872, 20 September 1877; see also Gold Hill *Evening News*, 2 June 1864.

36. *Territorial Enterprise*, 6 October 1877.

37. U.S. Bureau of the Census Office, tenth census (1880).

38. Fowler, *In the Shadow of Fox Peak*, 152.

39. *Territorial Enterprise*, August 7, 1870; De Quille, *Big Bonanza*, 216.

40. *Territorial Enterprise*, 8 June 1872.

41. *Territorial Enterprise*, 10 January 1871, 30 January 1872, 21 May 1872; Catherine S. Fowler and Sven Liljeblad, "Northern Paiute," in *Handbook of North American Indians*, vol. 11, 460.

42. *Territorial Enterprise*, 15 April 1871.

43. Ibid., 13 August 1881.

44. Ibid., 9 August 1868; De Quille, *Big Bonanza*, 209–10.

45. Sarah Winnemucca Hopkins, *Life Among the Piutes: Their Wrongs and Claims* (Boston: Cupples, Upham & Co., 1883), 64, 79; Gold Hill *Evening News*, 22 September 1864.

46. *Territorial Enterprise*, 18 February 1879.

47. De Quille, *Big Bonanza*, 216–217.

12. Erin's Daughters on the Comstock

1. John Taylor Waldorf, *A Kid on the Comstock: Reminiscences of a Virginia City Childhood* (1970; reprint, Reno: University of Nevada Press, 1991), 115.

2. Information on the Cronins is based on the tenth U.S. manuscript census (1880) and on the Murphy family Bible, courtesy of Susan Murphy Kastens and Eldeane

Murphy of Carson City, Nevada. See also Walter van Tilburg Clark, ed., *The Journals of Alfred Doten: 1849–1903*, 3 vols. (Reno: University of Nevada Press, 1973), 2:1497; and Gold Hill *Evening News*, 28 July 1875.

3. This study classifies anyone with at least one parent born in Ireland as Irish American. Such determinations are not possible in the eighth and ninth U.S. censuses of 1860 and 1870, because parents' nativity was not recorded during those years. Asserting that the child of an Irish parent would choose an Irish American identity is problematic. Nevertheless, studies in ethnicity have demonstrated that with Irish heritage, this is typically the case. See Mary C. Waters, *Ethnic Options: Choosing Identities in America* (Berkeley: University of California Press, 1990), 34–37.

4. Two of the most important sources are Kerby A. Miller, *Emigrants and Exiles: Ireland and the Irish Exodus to North America* (New York: Oxford University Press, 1985): and Hasia R. Diner, *Erin's Daughters in America: Irish Immigrant Women in the Nineteenth Century* (Baltimore: Johns Hopkins University Press, 1983). Miller argues that the immigrant perceived himself as an involuntary traveler, exiled from his beloved homeland. His analysis includes women, but questions remain as to how applicable his approach is for women. Diner sees Irish immigrant women as ambitious, hard workers intent on exploiting opportunities in the New World. Her sources are, however, largely eastern. See also Maria Luddy and Cliona Murphy, *Women Surviving: Studies in Irish Women's History in the 19th and 20th Centuries* (Dublin: Poolbeg Press, 1989); Robert Arthur Burchell, *San Francisco Irish, 1848–1880* (Berkeley: University of California Press, 1980); David M. Emmons, *The Butte Irish: Class and Ethnicity in an American Mining Town, 1875–1925* (Urbana: University of Illinois Press, 1989); James P. Walsh, *The San Francisco Irish, 1850–1976* (San Francisco: The Irish Literary and Historical Society, 1978); and Laurie K. Mercier, "'We are Women Irish': Gender, Class, Religious, and Ethnic Identity in Anaconda, Montana," *Montana: The Magazine of Western History* 44, no. 1 (winter 1994). See the special issue of the *Journal of the West* 31, no. 2 (April 1992) for some excellent essays on Irish in the West, including in particular Mary Murphy, "A Place of Greater Opportunity: Irish Women's Search for Home, Family, and Leisure in Butte, Montana," and Nancy J. Emmick, "Biographical Essay on Irish-Americans in the West," 87–94.

5. James P. Walsh, "The Irish in the New America: 'Way Out West,'" in *America and Ireland, 1776–1976: The American Identity and the Irish Connection*, ed. David Noel Doyle and Owen Dudley Edwards (Westport, Conn.: Greenwood Press, 1980).

6. Actually, two of the Comstock Irish women in 1860 were living in Silver City, which is now in Lyon County. This county is not part of statistical overviews used here based on subsequent census manuscripts, because only Storey County has been computerized to date.

7. The number 111 refers exclusively to that part of the Comstock later represented by Storey County. Only twelve Irish women appear in that part of the census, making them roughly 10 percent of the whole.

8. Bridget Deobin, the servant discussed in chapter 2 of this volume, was apparently an Irish American. The eighth U.S. census does not record parents' nativity, but ethnicity can be surmised by the subject's name.

9. It is not possible to assess the numbers of Irish Americans from the ninth U.S. manuscript census of 1870, because parents' nativity is not included.

10. Tenth U.S. manuscript census (1880); Sanborn Perris Fire Insurance Map for Storey County (New York, 1890).

11. The method used is to locate names in the 1870 document that are present in the 1880 document. Pages that exhibit higher degrees of repetition of the same names can be taken to indicate that the enumerator was recording the same street as is indicated in 1880. The problem with this methodology is that without recorded streets and addresses, there is no way to be certain that the enumerator did not jump back and forth from street to street. Source criticism for a conclusion reached about geography in 1870 is consequently necessary.

12. Storey County Assessor Book of Records, Storey County Courthouse, Virginia City, Nevada.

13. Ninth and tenth U.S. manuscript censuses, Storey County, Nevada, 1870 and 1880, respectively.

14. Diner, *Erin's Daughters,* 79–94, provides a comparative analysis from the East.

15. One house in the midst of the four did not have a servant. The circumstance of Weltmore and her clerk employer may be questionable, since they were the only two occupants of the household and both were in their early twenties; it is possible that they had a relationship that went beyond the economic bond of employer-servant. See the tenth U.S. manuscript census (1880).

16. Mary McNair Mathews, *Ten Years in Nevada: Or Life on the Pacific Coast* (1880; reprint, Lincoln: University of Nebraska Press, 1985), 252. Mathews was actually quoting another woman with this opinion, but clearly Mathews agreed with the sentiment.

17. Diner, *Erin's Daughters,* 80–82.

18. Ronald M. James, Richard D. Adkins, and Rachel J. Hartigan, "Competition and Coexistence in the Laundry: A View of the Comstock," *Western History Quarterly* 25, no. 2 (summer 1994): 165–84.

19. Tenth U.S. manuscript census (1880).

20. Tenth U.S. manuscript census (1880). The reason for the higher percentage of widows among the Irish could relate to a reluctance to remarry or to the exposure to danger of their spouses. Mining employed fewer Irish men per capita than it did Cornish men, for example, but the former may have performed more dangerous tasks associated with the occupation. The ninth U.S. manuscript census (1870) fails to list many widows. See Ronald M. James, "Defining the Group: Nineteenth-Century Cornish on the North American Mining Frontier," in *Cornish Studies: Second* Series [No. 2], ed. Philip Payton (Exeter: University of Exeter, 1994).

21. Anne M. Butler, *Daughters of Joy, Sisters of Misery: Prostitution in the American West* (Urbana: University of Illinois Press, 1985), 14.

22. It is difficult to discuss ethnic exogamy/endogamy rates before 1880, because earlier enumerators did not record parental nativity.

23. Diner, *Erin's Daughters,* 50–51, observes similar exogamy rates for eastern communities. For stereotypes about Irish men, see John J. Appel, "From Shanties to Lace Curtains: The Irish Image in Puck, 1876–1910," *Comparative Studies in Society and History* 13, no. 4 (October 1971); and for a Comstock example of ethnic humor pointed at Irish men see De Quille, *Big Bonanza,* 69–70.

24. This conclusion is based on a comparison of the location of residence based on the tenth U.S. manuscript census with the Irish neighborhoods identified in Map 1.

25. For comparative purposes, men must be used here, because it is not possible to identify Cornish women reliably. See James, "Defining the Group," 38–40.

26. Gold Hill *Evening News,* 28 January 1865, 3:1.

27. Vincent A. Lapomarda, S.J., "Saint Mary's in the Mountains: The Cradle of

Catholicism in Western Nevada," 58-62. *Nevada Historical Society Quarterly* 35, no. 1 (spring 1992): 58–62; Rachel J. Hartigan, "Looking for a Friend among Strangers: Virginia's Religious Institutions as Purveyors of Community" (undergraduate thesis, Yale University, 1993). The Great Fire of October 26, 1875, left the church in ruins. Workers were apparently able to reuse the brick walls in its rebuilding. The church is also called St. Mary's in the Mountains, but St. Mary in the Mountains is generally preferred.

28. John B. McGloin, S.J., "Patrick Manogue, Gold Miner and Bishop," *Nevada Historical Society Quarterly* 14, no. 2 (summer 1971): 25–31; William Breault, S.J., *The Miner was a Bishop: The Pioneer Years of Patrick Manogue* (Rancho Cordova, Calif.: Landmark Enterprises, 1988). Gold Hill's Catholic church dedicated to St. Patrick also had a large Irish congregation, but that parish has long since disappeared. Records from that facility are fewer than are those from St. Mary's in Virginia City, and so conclusions about the Gold Hill Catholics are more difficult to reach.

29. Tenth U.S. manuscript census (1880). See also the chapter on the Daughters of Charity by Anne Butler in this volume.

30. Gold Hill *Evening News*, 10 August 1864, 28 January 1865. *Territorial Enterprise*, 18 December 1868; 20, 22, 23 February, 1870; 9, 13 March 1870; 27, 29, 30 October 1870; 26 November 1870. See also the ninth and tenth U.S. census manuscripts from 1870 and 1880, respectively.

31. Clark, *Journals of Alfred Doten*, vol. 3: 1801–1802; Henry Bergstein, M.D., "Medical History," in Sam P. Davis, ed., *The History of Nevada* (Reno: The Elms Publishing Company, 1913), 622.

32. See the Virginia *Daily Union*, 9, 28 February 1864, and 17 March 1864. And see ibid., 19 March 1866, for a mention of women participating in Saint Patrick's Day celebrations. Unfortunately, there are few earlier accounts of St. Patrick's Day celebrations, owing in part to the fact that newspapers from the month of March of previous years are largely nonextant.

33. *Territorial Enterprise*, 2 June 1869, 2:6.

34. R. V. Comerford, "Patriotism as Pastime: The Appeal of Fenianism in the mid-1860s," in *Reactions to Irish Nationalism* (London: The Hambledon Press, 1987), 30.

35. Miller, *Emigrants and Exiles*.

36. See the O'Connell tombstone in the Virginia City Catholic cemetery.

37. Miller, *Emigrants and Exiles*.

13. Girls of the Golden West

1. Bernard DeVoto, *Mark Twain's America & Mark Twain at Work* (Cambridge, Mass.: The Fellows of Harvard College, 1932), 11.

2. Loosely quoted from a dinner conversation between Edna O'Brien and the author in Dublin, Ireland, ca. 1980.

3. Bernard DeVoto, "Brave Days in Washoe," *American Mercury* 17 (June 1929) and *Mark Twain's America*, 11. See the chapter "Washoe," 162–63, and 310. DeVoto was among the first people Beebe and Clegg chose for the editorial staff of their *Territorial Enterprise*. Born in Ogden, Utah, in 1897, he earned his degree from Harvard, joining the faculty in 1929. In 1935 he became an editor and columnist for *Harpers* and an editor of the *New England Quarterly*.

4. DeVoto obit, Boston Daily *Globe*, 14 November 1995, Harvard Alumni File at the Pusey Library. "Seething resolution to write" is quoted in an article about Charles

Townsend Copeland, the Boylston Professor of Rhetoric and Oratory, who taught, among others, Lucius Beebe, Robert Hillyer, Walter Lippman, Max Perkins, and De-Voto. This article is also included in Copeland's Harvard file.

5. Lucius Beebe, *Boston and the Boston Legend* (New York: Appleton-Century, 1935).

6. Lucius Beebe, "Stars Speeding to Virginia City for Film Debut," and "Premiere of 'Virginia City' Held in Rollicking Frontier Fashion," from the New York *Herald Tribune,* 16, 17 March 1940.

7. See Lucius Beebe, "High Jinks at Film Premiere—Beebe Recalls Virginia City Wingding," San Francisco *Chronicle,* September 3, 1960.

8. Ibid.

9. The only nineteenth-century writer who wrote of the Comstock with a sense of place was Clarence King; see his *Report of the Geological Exploration of the Fortieth Parallel,* edited by U.S. Exploration of the Fortieth Parallel, 1867–1881 (Washington, D.C.: Government Printing Office, 1900); the *Clarence King Memoirs* (New York: G. P. Putnam's Sons, 1904); and *Mountaineering in the Sierra Nevada* (1872; reprint, Lincoln: University of Nebraska Press, 1997). Other period writers, including Eliot Lord in his *Comstock Mining and Miners* (Washington, D.C.: Government Printing Office, 1883), focus less on the setting of the Comstock.

10. Walter Van Tilburg Clark, *The Ox-Bow Incident* (New York: Random House, 1940) and *The Track of the Cat* (New York: New American Library, 1949; reprint, Reno: University of Nevada Press, 1993).

11. *Nevada Highway & Parks,* special edition (fall 1971): 8–9.

12. DeVoto's review of Walter Van Tilburg Clark's *The Watchful Gods and Other Stories* (New York: Random House, 1950) appeared in the *New York Times Book Review,* 24 September 1950).

13. Butterfield—author, journalist, and historian—wrote about the West for national magazines and served as "staffer" on Beebe and Clegg's *Territorial Enterprise,* as well as *Life,* the *Saturday Evening Post,* and other Curtis group magazines, such as *Gentlemen's Quarterly.* Beebe's letters to Butterfield, 1950–1954, are at the Nevada Historical Society in Reno.

14. Mike Thomas, ed., *Literary Las Vegas* (New York: Henry Holt and Company, 1995).

15. See the letter from Alfred Lunt to Lucius Beebe, Graham Hardy Collection, California State Railroad Museum, Sacramento, California.

16. Jan Morris, "Virginia City Spirit is Bloodied but Alive," *Comstock Chronicle,* 15 January 1993.

17. During the 1920s and 1930s, because of the notoriety associated with Reno divorces, many romance novelists based their plots in northern Nevada. Clare Booth Luce's 1936 play, *The Women,* was probably the best of this genre. It was subsequently filmed in Reno in 1939.

18. Abbott Joseph Liebling, "Our Far-flung Correspondents: Out Among the Lamisters," *New Yorker,* 30 (March 27, 1954): 71–86; and "A Reporter at Large: The Mustanging Buzzers," *New Yorker,* 30 (April 3, 1954): 35ff; 30 (April 10, 1954), 66ff.

19. Vardis A. Fisher, *City of Illusion* (New York: Harper & Bros., 1941).

20. J. H. Jackson, "Book," San Francisco *Chronicle,* 30 March 1941, 5.

21. This is a favorite story of Kyle K. Wyatt, assistant curator of the Nevada State Railroad Museum, Carson City.

22. Margaret Mitchell, *Gone with the Wind* (New York: The Macmillan Company, 1936).

23. Duncan Emrich, *It's an Old West Custom* (New York: Vanguard, 1949), 117.

24. Duncan Emrich, "In the Delta Saloon: Conversations with Residents of Virginia City, Nevada, Reno Oral History Program, University of Nevada, Reno. Transcribed in 1990 and 1991 Conversations with Residents of Virginia City, Nevada" (Reno: Oral History Program, University of Nevada, Reno, transcribed in 1990 and 1991).

25. Beebe's letter to Roger Butterfield, dated March 20, 1952. Letters on file at the Nevada Historical Society, Reno, Nevada.

26. Katherine Hillyer and Katherine Best wrote *Julia Bulette and Other Red Light Ladies* (Sparks, Nev.: Western Printing, 1959), not to be confused with Douglas McDonald's *The Legend of Julia Bulette and the Red Light Ladies of Nevada* (Reno: Nevada Publications, 1980). Zeke Daniels, *The Life and Death of Julia C. Bulette* (Virginia City, Nev.: Lamppost, 1958).

27. Lucius Beebe and Charles Clegg, "Legends of the Fair but Frail," in *Legends of the Comstock Lode* (Stanford: Stanford University Press, 1954), 15–22. "Julia Bulette: The Comstock's First Cyprian," was posthumously published in *The Lucius Beebe Reader* (Garden City, N.Y.: Doubleday, 1967), 320–26.

28. Mahoney letter to H. M. Gorham, 10 November 1937, in Gorham file, MSS 837, California Historical Society.

29. "Nevada Historians Amazed by NBC's Version of Virginia City's History," a Beebe editorial, *Territorial Enterprise*, 12 December 1952, 4.

30. For books about Bowers, see Gloria Millicent Mapes, et al., *Bowers Mansion* (Reno: Bowers Mansion Restoration Group, 1952); and Barbara J. Jeffers, *Sandy Bowers's Widow: The Biography of Allison Eilley Bowers* (Reno: Barringer Historical Books, 1993).

31. Beebe and Clegg, *Legends of the Comstock Lode*, 23–28.

32. Ross's Nevada novels are *Bonanza Queen* (Indianapolis: Bobbs-Merrill, 1949), *Reno Crescent* (Indianapolis: Bobbs-Merrill, 1951), and *Tonopah Lady* (Indianapolis: Bobbs-Merrill, 1950), 28.

33. DeVoto, *Mark Twain's America*, 162–63.

34. Louis L'Amour, *Comstock Lode* (New York: Bantam Books, 1981).

35. Beebe and Clegg, "Nabobs in Broadcloth," in *Legends of the Comstock Lode*, 67. It is surprising that Beebe, who appreciated excess and would have felt comfortable in the court of Louis XVI, became such an ardent critic of Louise Mackay. It was, he pronounced, her "naive determination" to display her wealth that irked him so. Such ostentation Beebe somehow found unconscionable. Of course, Beebe's considered opinion was in direct opposition to the Comstock newspapers of Mackay's day, which never failed to carry news about her. Mackay herself kept all that was written about her, good or bad; Berlin's research notes were kindly given to Special Collections, Getchell Library, University of Nevada, Reno.

36. Lucius Beebe's review of *Silver Platter* appeared in the *Territorial Enterprise and Virginia City News*, 20 December 1957, 7. Ellin Berlin, *Silver Platter* (Garden City, N.Y.: Doubleday & Co., 1957).

37. Henry M. Gorham, *My Memories of the Comstock* (Los Angeles: Suttonhouse, 1939); John Taylor Waldorf, *A Kid on the Comstock: Reminiscences of a Virginia City Childhood* (Palo Alto: American West, 1970); Miriam Michelson, *The Wonderlode of Sil-*

ver and Gold (Boston: Stratford, 1934); David Belasco, *Gala Days of Piper's Opera House and the California Theater* (Sparks, Nev.: Falcon Hill Press, 1991).

38. "Nevada Historians Amazed by NBC's Version of Virginia City's History," *Territorial Enterprise*, 12 December 1952, 4.

14. Gender and Archaeology on the Comstock

1. David Lowenthal, *The Past is a Foreign Country* (Cambridge: Cambridge University Press, 1985), 249.

2. See, for example, Joan Gero and Margaret Conkey, eds., *Engendering Archaeology: Women and Prehistory* (Oxford: Basil Blackwell, 1991); Elizabeth Scott, ed., *Those of Little Note: Gender, Race, and Class in Historical Archaeology* (Tuscon: University of Arizona Press, 1994); Donna J. Seifert, ed., *Gender in Historical Archaeology*, edited issue of *Historical Archaeology* 25, no. 4 (1991); Dale Walde and Noreen D. Willows, eds., *The Archaeology of Gender: Proceedings of the 22nd (1989) Annual Chacmool Conference* (Calgary: The Archaeological Association of the University of Calgary, 1991); and Diana diZerega Wall, *The Archaeology of Gender: Separating the Spheres in Urban America* (New York: Plenum Press, 1994).

3. Gero and Conkey, *Engendering Archaeology*, 23.

4. Catherine H. Blee, *Sorting Functionally-Mixed Artifact Assemblages with Multiple Regression: A Comparative Study in Historical Archaeology* (Ph.D. diss., University of Colorado, Boulder, 1991); Donald L. Hardesty, "Gender Roles on the American Mining Frontier: Documentary Models and Archaeological Strategies," Paper presented at the 22nd Annual Chacmool Conference, the Archaeological Association of the University of Calgary, 1989) and "Class, Gender Strategies, and Material Culture in the Mining West," in *Those of Little Note: Gender, Race, and Class in Historical Archaeology*, ed. Elizabeth M. Scott (Tuscon: University of Arizona Press, 1994), 129–45; Margaret Kennedy, "Houses with Red Lights: The Nature of Female Households in the Sporting Subculture Community," in *Households and Communities Proceedings of the 21st (1988) Annual Conference of the Archaeological Association of the University of Calgary*, ed. David J. W. Archer and Richard D. Gavin (Calgary: University of Calgary Archaeological Association, 1989), 93–100; and Alexy Simmons, *Red Light Ladies: Settlement Patterns and Material Culture on the Mining Frontier*, Anthropological Northwest No. 4 (Corvallis: Department of Anthropology, Oregon State University, 1989).

5. James Deetz, "Public Aesthetics Versus Personal Experience: Worker Health and Well-Being in 19th-Century Lowell, Massachusetts," *Historical Archaeology* 27, no. 2 (1993): 90–105; Robert Schuyler, "Archaeological Remains, Documents, and Anthropology: A Call for a New Culture History," *Historical Archaeology* 22, no. 1 (1988): 6–42.

6. Pierre Bourdieu, *Distinctions: A Social Critique of the Judgement of Taste*, trans. Richard Nice (Cambridge: Harvard University Press, 1984).

7. James Deetz, "Introduction: Archaeological Evidence of Sixteenth- and Seventeenth-century Encounters," in *Global Perspectives on Historical Archaeology*, ed. Lisa Falk (Washington, D.C.: Smithsonian Institution Press, 1991), 2–3.

8. Arjun Appadurai, ed., *The Social Life of Things: Commodities in Cultural Perspective* (Cambridge: Cambridge University Press, 1986); Bourdieu, *Distinctions;* Mary Douglas and Baron Isherwood, *The World of Goods* (New York: Basic Books, 1979).

9. Jill Derby, "Cattle, Kin and the Patrimonial Imperative: Social Organization on Nevada Family Ranches" (Ph.D. diss., University of California, Davis, 1988).

10. Daniel E. Sutherland, *The Expansion of Everyday Life, 1860–1876* (New York: Harper and Row, 1989), xii.

11. Vincent P. DeSantis, *The Shaping of Modern America: 1877–1920* (Arlington Heights, Ill.: Forum Press, 1989); Thomas Schlereth, *Victorian America, 1876–1915* (New York: Harper Collins, 1991); Alan Trachtenberg, *The Incorporation of America* (New York: Hill and Wang, 1982).

12. Mary P. Ryan, Cradle of the Middle Class: *The Family in Oneida County, New York, 1790–1865* (New York: Cambridge University Press, 1981).

13. See chapter 9 by Watson, Ford, and White for ways that this ideology was manifested in nineteenth-century Comstock communities.

14. Paula Petrik, *No Step Backward, Women and Family on the Rocky Mountain Mining Frontier, Helena, Montana 1865–1900* (Helena: Montana Historical Society Press, 1987).

15. Ibid., 25–58.

16. Wall, *The Archaeology of Gender*, 158.

17. See also Karen Halttunen, *Confidence Men and Painted Women* (New Haven: Yale University Press, 1982).

18. Wall, *The Archaeology of Gender*, see chapters 5 and 6.

19. Andrew Jackson Downing, *The Architecture of Country Houses* (New York: D. Appleton, 1850), 23.

20. Wall, *The Archaeology of Gender*, 160.

21. Downing, *Architecture of Country Houses*, 24.

22. Lee Virginia Chambers-Schiller, *Liberty a Better Husband: Single Women in America: The Generations of 1780–1840* (New Haven: Yale University Press, 1984).

23. Frances B. Cogan, *All American Girl: The Ideal of Real Womanhood in Mid-Nineteenth Century America* (Athens: University of Georgia Press, 1989).

24. Ibid., 4.

25. Suzanne Spencer-Wood, "Towards an Historical Archaeology of Materialistic Domestic Reform," in *The Archaeology of Inequity*, ed. Robert Paynter and Randall McGuire (Oxford: Basil Blackwell, 1991), 231–86.

26. Ibid., 250.

27. Ibid., 252.

28. Ibid., 255.

29. Suzanne Spencer-Wood, "Diversity and Nineteenth-Century Domestic Reform: Relationships among Classes and Ethnic Groups," in *Those of Little Note*, ed. Elizabeth Scott, 175–208.

30. Mary McNair Mathews, *Ten Years on the Comstock: Or Life on the Pacific Coast* (Lincoln: University of Nebraska Press, 1985).

31. See, for example, Russell M. Magnaghi, "Virginia City's Chinese Community, 1860–1880," *Nevada Historical Society Quarterly* 24 (1981): 130–57. And see the chapter by Sue Fawn Chung in this volume.

32. Blee, *Sorting Functionally-Mixed Artifact Assemblages*.

33. Priscilla Wegars, "Besides Polly Bemis: Historical and Artifactual Evidence for Chinese Women in the West, 1848–1930," in *Hidden Heritage*, ed. Priscilla Wegars (Amityville, N.Y.: Baywood Publishing Company, 1993) 229–54.

34. Simmons, *Red Light Ladies*, 62–67.

35. Marion S. Goldman, *Gold Diggers and Silver Miners: Prostitution and Social Life on the Comstock* (Ann Arbor: University of Michigan Press, 1981); Simmons, *Red Light Ladies*, 64.

36. Goldman, *Gold Diggers;* Maribeth Hamby, "Women of the Comstock and Material Culture" (professional paper, University of Nevada, Reno, 1980); and Simmons, *Red Light Ladies.*

37. Elliott West, *The Saloon on the Rocky Mountain Mining Frontier* (Lincoln: University of Nebraska Press, 1981), 26.

38. Blee, "Sorting Functionally-Mixed Artifact Assemblages."

39. Donna J. Seifert, "Within Sight of the White House: The Archaeology of Working Women," *Historical Archaeology* 25, no. 4 (1991): 104.

40. William Rathje and Cullen Murphy, *Rubbish! The Archaeology of Garbage* (New York: Harper Collins, 1992).

41. Mary C. Beaudry, "Public Aesthetics Versus Personal Experience: Worker Health and Well-Being in 19th-Century Lowell, Massachusetts," *Historical Archaeology* 27, no. 2 (1993): 90.

42. Stephen A. Mrozowski, "Landscapes of Inequality," in *The Archaeology of Inequality,* ed. Randall H. McGuire and Robert Paynter (Oxford: Basil Blackwell, 1991), 79–101.

43. Robert Z. Melnick, *Cultural Landscapes: Rural Historic Districts in the National Park System* (Washington, D.C.: U.S. Department of the Interior, National Park Service, 1984).

44. Donald L. Hardesty, Steven Mehls, and Priscilla Mecham, "A Class III Cultural Resource Inventory of the 493 Acre Bodie Study Area, Mono County, California: Historic Resources, Volume 1" (Report prepared for Bodie Consolidated Mining Company, Bridgeport, California, 1990).

45. J. Ross Browne, "A Trip to Bodie Bluff and the Dead Sea of the West," *Harper's New Monthly Magazine* (August 1865): 276.

46. Frank S. Wedertz, *Bodie, 1859–1900* (Bishop, Calif.: Chalfant Press, 1969), 5.

47. Ibid., 28.

48. Ibid., 32, 35.

49. Simmons, *Red Light Ladies,* 98–125.

50. Virginia *Evening Bulletin,* 20 August 1863.

51. Margaret Purser, "Several Paradise Ladies are Visiting Town: Gender Strategies in the Early Industrial West," *Historical Archaeology* 25, no. 4 (1991): 9–13.

52. Mathews, *Ten Years.*

53. Cedric E. Gregory, *A Concise History of Mining* (New York: Pergamon Press, 1980), 221–22.

54. William G. White and Ronald M. James, "Little Rathole on the Big Bonanza: Historical and Archaeological Assessment of an Underground Resource" (Carson City: State Historic Preservation Office, 1991).

Angel, Myron, ed. *History of Nevada with Illustrations and Biographical Sketches of its Prominent Men and Pioneers.* Oakland: Thompson and West, 1881. Reprint, Berkeley: Howell-North Books, 1958.

Armitage, Susan, and Elizabeth Jameson. *The Women's West.* Norman: University of Oklahoma Press, 1987.

Arnold, Emmett L. *Gold Camp Drifter: 1906–1910.* Reno: University of Nevada, 1973.

Arrington, Leonard J., and Davis Bitton. *The Mormon Experience: A History of the Latter-day Saints.* New York: Alfred A. Knopf, 1979.

Banner, Lois. *American Beauty.* New York: Alfred A. Knopf, 1983.

Barber, Elizabeth W. *Women's Work: The First 20,000 Years.* New York: W. W. Norton, 1994.

Barnhart, Jacqueline Baker. *The Fair but Frail: Prostitution in San Francisco, 1849–1900.* Reno: University of Nevada Press, 1986.

Barry, Herbert, III, and Alice Schlegel. "Cross-Cultural Codes on Contributions by Women to Subsistence." *Ethnology* 21, no. 2 (1982): 165–88.

Barth, Paul G. *Bitter Strength: A History of the Chinese in the United States, 1850–1870.* Cambridge: Harvard University Press, 1964.

Beebe, Lucius, and Charles Clegg. *Legends of the Comstock Lode.* Stanford: Stanford University Press, 1954.

———. "Julia Bulette: The Comstock's First Cyprian." In *The Lucius Beebe Reader,* edited by Charles D. Emrich. Garden City, N.Y.: Doubleday, 1967.

———. "Legends of the Fair but Frail." In *The Lucius Beebe Reader,* edited by Charles D. Emrich. Garden City, N.Y.: Doubleday, 1967.

Belasco, David. *Gala Days of Piper's Opera House and the California Theater.* Sparks, Nev.: Falcon Hill Press, 1991.

Berlin, Ellin. *Silver Platter.* Garden City, N.Y.: Doubleday & Co., 1957.

Blackburn, George M., and Sherman I. Richards. "The Prostitutes and Gamblers of Virginia City, Nevada: 1870." *Pacific Historical Review* 48, no. 2 (May 1979): 239–59.

Blee, Catherine H. "Sorting Functionally-Mixed Artifact Assemblages with Multiple Regression: A Comparative Study in Historical Archaeology." Ph.D. diss., University of Colorado, Boulder, 1991.

Blum, Stella. *Victorian Fashions and Costumes from Harper's Bazaar: 1867–1898.* New York: Dover Publications, 1974.

Brandt, Brenda. "Arizona Clothing: A Frontier Perspective." *Dress* 15 (1989): 69–76.

Braude, Ann. *Radical Spirits.* Boston: Beacon Press, 1989.

Breault, William. *The Miner Was a Bishop: The Pioneer Years of Patrick Manogue, California-Nevada 1854–1895.* Rancho Cordova, Calif.: Landmark Enterprises, 1988.

Brewer, Eileen Mary. *Nuns and the Education of American Catholic Women, 1860–1920.* Chicago: Loyola University Press, 1987.

Brown, Slater. *The Heyday of Spiritualism.* New York: Hawthorn Books, 1970.

Browne, J. Ross. *A Peep at Washoe and Washoe Revisited.* 1863, 1864, 1869. Reprint, Balboa Island, Calif.: First Paisano Press, 1959.

Burchell, Robert Arthur. *San Francisco Irish, 1848–1880.* Berkeley: University of California Press, 1980.

Butler, Anne M. *Daughters of Joy, Sisters of Misery: Prostitutes in the American West, 1865–90.* Urbana: University of Illinois Press, 1985.

Castañeda, Antonia I. "Women of Color and the Rewriting of Western History: The Discourse, Politics, and Decolonization of History." *Pacific Historical Review* 61, no. 4 (November 1992): 501–33.

Chambers-Schiller, Lee Virginia. *Liberty a Better Husband: Single Women in America: The Generations of 1780–1840.* New Haven: Yale University Press, 1984.

Chan, Anthony B. *Gold Mountain: The Chinese in the New World.* Vancouver, B.C.: New Star Books, 1983.

Chan, Loren B. "The Chinese in Nevada: An Historical Survey, 1856–1970." *Nevada Historical Society Quarterly* 25, no. 4 (winter 1982): 266–314.

Chan, Sucheng. *Asian Americans: An Interpretive History.* Boston: Twayne, 1991.

———, ed. *Entry Denied: Exclusion and the Chinese Community in America, 1882–1943.* Philadelphia: Temple University Press, 1991.

Clark, Walter Van Tilburg, ed. *The Journals of Alfred Doten, 1849–1903.* 3 vols. Reno: University of Nevada Press, 1973.

Cloud, Barbara. "Images of Women in the Mining Camp Press." *Nevada Historical Society Quarterly* 36, no. 3 (fall 1993): 194–207.

Cogan, Frances B. *All American Girl: The Ideal of Real Womanhood in Mid-Nineteenth Century America.* Athens: University of Georgia Press, 1989.

Cohen, Lizbeth A. "Embellishing a Life of Labor: An Interpretation of the Material Culture of American Working-Class Homes, 1885–1915." In *Material Culture Studies in America,* edited by Thomas J. Schlereth. Nashville: American Association of State and Local History, 1982.

Conlin, Joseph. *Bacon, Beans and Galantines: Food and Foodways on the Western Mining Frontier.* Reno: University of Nevada Press, 1987.

Coray, Michael S. "Influences on Black Family Household Organization in the West, 1850–1860." *Nevada Historical Society Quarterly* 31, no. 1 (spring 1933): 1–31.

———. "African-Americans in Nevada." *Nevada Historical Society Quarterly* 35, no. 4 (winter 1992): 239–57.

Dailey, Christie. "A Woman's Concern: Millinery in Center Iowa, 1870–1880." *Journal of the West* 21 2 (1982): 20–32.

Daniels, Roger. *Asian America: Chinese and Japanese in the United States Since 1850.* Seattle: University of Washington Press, 1992.

Daniels, Zeke. *The Life and Death of Julia C. Bulette.* Virginia City, Nev.: Lamppost, 1958.

Davis, Sam P. *History of Nevada.* Reno: Elms Publishing, 1913.

d'Azevedo, Warren D., William C. Sturtevant, Catherine S. Fowler, Jesse D. Jennings, Don D. Fowler, and William H. Jacobsen Jr. *Handbook of North American Indians.* Vol. II (Great Basin). Washington, D.C.: Smithsonian Institution, 1986.

Degler, Carl N. *At Odds: Women and the Family in America from the Revolution to the Present.* New York: Oxford University Press, 1980.

De Graaf, Lawrence B. "Race, Sex, and Region: Black Women in the American West, 1850–1920." *Pacific Historical Review* 49, no. 2 (May 1980): 285–313.

De Quille, Dan. *The Big Bonanza.* 1876. Reprint, New York: Alfred A. Knopf, 1947.

Derby, Jill. "Cattle, Kin and the Patrimonial Imperative: Social Organization on Nevada Family Ranches." Ph.D. diss., University of California, Davis, 1988.

Diner, Hasia R. *Erin's Daughters in America: Irish Immigrant Women in the Nineteenth Century.* Baltimore: John Hopkins University Press, 1982.

Dosch, Henry Ernst. *Vigilante Days at Virginia City: A Personal Narrative.* Facsimile Reproduction. Seattle: Shorey Book Store, 1967.

Drury, Wells. *An Editor on the Comstock Lode.* New York, Toronto: Farrar & Rinehart, 1936. Reprint, Reno: University of Nevada Press, 1984.

Dwyer, John T. *Condemned to the Mines: The Life of Eugene O'Connell, 1815–1891, Pioneer Bishop of Northern California and Nevada.* New York: Vantage Press, 1976.

Dwyer, Richard A., and Richard E. Lingenfelter. *Dan De Quille, The Washoe Giant: A Biography and Anthology.* Reno: University of Nevada Press, 1990.

Earl, Phillip I. "Blood Will Tell: A Short History of Nevada's Miscegenation Laws." *Nevada Public Affairs Review* 2 (1987): 82–86.

Elliott, Russell. *History of Nevada.* Lincoln: University of Nebraska Press, 1973.

Ewens, Mary. *The Role of the Nun in Nineteenth-Century America.* New York: Arno Press, 1978.

Faragher, John Mack. *Women and Men on the Overland Trail.* New Haven: Yale University Press, 1979.

Fatout, Paul. *Mark Twain in Virginia City.* Bloomington: Indiana University Press, 1964.

Fernandez, Nancy Page. "'If a Woman Had Taste . . .': Home Sewing and the Making of Fashion 1850–1910." Ph.D. diss., University of California, Irvine, 1987.

Fisher, Vardis. *City of Illusion.* Caldwell, Ida.: The Caxton Printers, Ltd., 1941.

Flexner, Eleanor. *A Century of Struggle.* New York: Atheneum, 1972.

Ford, Bonnie L. "Women, Marriage, and Divorce in California, 1849–1872." Ph.D. diss., University of California, Davis, 1985.

Ford, Jean, and James W. Hulse. "The First Battle for Women's Suffrage in Nevada: 1869–1871—Correcting and Expanding the Record." *Nevada Historical Society Quarterly* 38, no. 3 (fall 1995): 174–88.

Fowler, Catherine S. *In the Shadow of Fox Peak.* U.S. Fish and Wildlife Service, Region 1, Cultural Resource Series, No. 5, 1992.

Francke, Bernadette. "The Neighborhood and Nineteenth-Century Photographs: A Call to Locate Undocumented Historic Photographs of the Comstock Region." *Nevada Historical Society Quarterly* 35, no. 4 (winter 1992): 258–69.

Garrison, Charles Jeffrey. "'How the Devil Tempts Us to Go Aside from Christ.'" *Nevada Historical Society Quarterly* 36, no. 1 (spring 1993): 13–34.

Ginzberg, Lori D. *Women and the Work of Benevolence: Morality, Politics, and Class in the Nineteenth-Century United States.* New Haven: Yale University Press, 1990.

Glass, Mary Ellen. "Nevada's Census Taker: A Vignette." *Nevada Historical Society Quarterly* 19, no. 4 (winter 1966): 1–12.

Goldfarb, Russel M., and Clare R. Golfarb. *Spiritualism and Nineteenth-Century Letters.* Cranbury, N.J.: Associated University Presses, Inc., 1978.

Goldman, Marion S. *Gold Diggers and Silver Miners: Prostitution and Social Life on the Comstock Lode.* Ann Arbor: University of Michigan Press, 1981.

———. "Sexual Commerce on the Comstock Lode." *Nevada Historical Society Quarterly* 21, no. 2 (summer 1978): 99–139.

Gorham, Henry M. *My Memories of the Comstock.* Los Angeles: Suttonhouse, 1939.

Gorman, Thomas. *Seventy-Five Years of Catholic Life in Nevada.* Reno: Journal Publishing, 1935.

Griswold, Robert L. "Apart but Not Adrift: Wives, Divorce, and Independence in California, 1850–1890." *Pacific Historical Review* 49, no. 2 (1980): 265–83.

———. *Family and Divorce in California, 1850–1890: Victorian Illusions and Everyday Realities.* Albany: State University of New York Press, 1982.

Halem, Lynne, and Nelson Blake. *The Road to Reno: A History of Divorce in the United States.* New York: McMillan, 1962.

Hannefin, Daniel. *Daughters of the Church: A Popular History of the Daughters of Charity in the United States, 1809–1987.* Brooklyn: New City Press, 1989.

Hansford, Phoebe. *Women of the Century.* Augusta, Me.: True and Co., 1882.

Hardesty, Donald L. *The Archaeology of Mining and Miners: A View from the Silver State.* The Society for Historical Archaeology: Special Publication Series, No. 6, 1988.

———. "Gender Roles on the American Mining Frontier: Documentary Models and Archaeological Strategies." Paper presented at the 22nd Annual Chacmool Conference, the Archaeological Association of the University of Calgary, 1989.

———. "Class, Gender Strategies, and Material Culture in the Mining West." In *Those of Little Note: Gender, Race, and Class in Historical Archaeology,* edited by Elizabeth M. Scott. Tucson: University of Arizona Press, 1994.

Hareven, Tamara K., ed. *Family and Kin in Urban Communities, 1700–1930.* New York: New Viewpoints, 1977.

Hartigan, Rachel J. "Looking for a Friend among Strangers: Virginia City's Religious Institutions as Purveyors of Community." A.B. thesis, Yale University, 1993.

Hattori, Eugene M. *Northern Paiutes on the Comstock: Archaeology and Ethnohistory of an American Indian Population in Virginia City, Nevada.* Nevada State Museum Occasional Paper, No. 3, 1975.

Hillyer, Katharine, and Katherine Best. *The Amazing Story of Piper's Opera House in Virginia City, Nevada.* Virginia City, Nev.: Enterprise Press, 1953.

———. *Julia Bulette and Other Red Light Ladies.* Sparks, Nev.: Western Printing, 1959.

Hirata, Lucie Cheng. "Free, Indentured, Enslaved: Chinese Prostitutes in Nineteenth-Century America." *Signs* 5, no. 1 (autumn 1979): 3–29.

Holliday, J. S. *The World Rushed In: The California Gold Rush Experience.* New York: Simon and Schuster, 1981.

Hulse, James W. *The Nevada Adventure.* 6th ed. Reno: University of Nevada Press, 1990.

James, Ronald M. "A Plan for the Archeological Investigation of the Virginia City Landmark District." Nevada Comprehensive Preservation Plan: Addendum. Carson City, Nev.: Nevada State Historic Preservation Office, 1992.

———. "Women of the Mining West: Virginia City Revisited." *Nevada Historical Society Quarterly* 36, no. 3 (fall 1993): 153–77.

———. "Defining the Group: Nineteenth-Century Cornish on the North American Mining Frontier." In *Cornish Studies: Second Series* [No. 2], edited by Philip Payton. Exeter: University of Exeter, 1994.

James, Ronald M., Richard D. Adkins, and Rachel J. Hartigan. "Competition and Coexistence in the Laundry: A View of the Comstock." *Western Historical Quarterly* 25, no. 2 (summer 1994): 165–84.

Jeffers, Barbara J. *Sandy Bowers's Widow: The Biography of Allison Eilley Bowers.* Reno: Barringer Historical Books, 1993.

Jensen, Joan M., and Darlis A. Miller. "The Gentle Tamers Revisited: New Approaches to the History of Women in the American West." *Pacific Historical Review* 49 (1980): 173–213.

Johnson, David Alan. *Founding the Far West: California, Oregon, and Nevada, 1840–1890.* Berkeley: University of California Press, 1992.

Johnson, Susan Lee. "'A Memory Sweet to Soldiers': The Significance of Gender in the History of the 'American West.'" *Western Historical Quarterly* 24 (1993): 495–517.

Lapomarda, Vincent A. "Saint Mary's in the Mountains: The Cradle of Catholicism in Western Nevada." *Nevada Historical Society Quarterly* 35, no. 1 (spring 1992): 58–62.

Levy, Jo Ann. *They Saw the Elephant: Women in the California Gold Rush.* Hamden, Conn.: Archon Books, 1990.

Lewis, Marvin. *Martha and the Doctor: A Frontier Family in Central Nevada.* Reno: University of Nevada Press, 1977.

Lewis, Oscar. *Silver Kings: The Lives and Times of Mackay, Fair, Flood, and O'Brien, Lords of the Nevada Comstock Lode.* New York: Alfred A. Knopf, 1947. Reprint, Reno: University of Nevada Press, 1986.

Limerick, Patricia Nelson, Clyde A. Milner II, and Charles E. Rankin, eds. *Trails Towards a New Western History.* Lawrence: University Press of Kansas, 1991.

Ling, Huping. "Surviving on the Gold Mountain: Chinese-American Women and Their Lives." Ph.D. diss., Miami University, 1992.

Locke, Mary Lou. "Out of the Shadows and into the Western Sun: Working Women of the Late Nineteenth-Century Urban Far West." *Journal of Urban History* 16 (1988): 175–204.

Lord, Eliot. *Comstock Mining and Miners.* 1883. Reprint, Berkeley: Howell-North, 1959.

Lystra, Karen. *Searching the Heart: Women, Men, and Romantic Love in Nineteenth-Century America.* New York: Oxford University Press, 1989.

Madsen, Carol Cornwall. "'At Their Peril': Utah Law and the Case of Plural Wives, 1850–1900." *The Western Historical Quarterly* 21, no. 4 (1990): 425–43.

Magnaghi, Russell M. "Virginia City's Chinese Community, 1860–1880." *Nevada Historical Society Quarterly* 24 (1981): 130–57.

Mann, Ralph. *After the Gold Rush: Society in Grass Valley and Nevada City, California, 1849–1870.* Palo Alto: Stanford University Press, 1982.

Marshall, John P. "Jews in Nevada: 1850–1900." *Journal of the West* 23, no. 4 (January 1984): 62–72.

Mathews, M. M. (Mary McNair). *Ten Years in Nevada: Or Life on the Pacific Coast.* Buffalo: Baker, Jones, 1880. Reprint, Lincoln: University of Nebraska Press, 1985.

McDonald, Douglas. *The Legend of Julia Bulette and the Red Light Ladies of Nevada.* Las Vegas: Nevada Publications, 1980.

————. *Virginia City: The Silver Region of the Comstock Lode.* Las Vegas: Nevada Publications, 1982.

McGloin, John Bernard. "Patrick Manogue, Gold Miner and Bishop." *Nevada Historical Society Quarterly* 14, no. 2 (summer 1971): 25–31.

Mercier, Laurie K. "'We are Women Irish': Gender, Class, Religious, and Ethnic Identity in Anaconda, Montana." *Montana: The Magazine of Western History* 44, no. 1 (winter 1994): 28–41

Michelson, Miriam. *The Wonderlode of Silver and Gold.* Boston: Stratford, 1934.

Mintz, Steven, and Susan Kellogg. *Domestic Revolutions: A Social History of American Family Life.* New York: The Free Press, 1988.

Morrisey, Katherine C. "Engendering the West." In *Under an Open Sky: Rethinking America's Western Past,* edited by George Miles and Jay Citlin. New York: W. W. Norton, 1992.

Moynihan, Ruth B., Susan Armitage, and Christiane Fischer Dichamp, eds. *So Much to Be Done: Women Settlers on the Mining and Ranching Frontier.* Lincoln: University of Nebraska Press, 1990.

Murphy, Mary. "Bootlegging Mothers and Drinking Daughters: Gender and Prohibition in Butte, Montana," *American Quarterly* 46, no. 2 (June 1994): 174–94.

————. *Recreating Butte: Gender, Work, and Leisure in a Western Mining City, 1914–41.* Champaign: University of Illinois Press, 1996.

Myres, S. L. *Westering Women and the Frontier Experience, 1800–1915.* Albuquerque: University of New Mexico Press, 1982.

Nylander, Jane C. *Our Own Snug Fireside.* New Haven: Yale University Press, 1992.

Osumi, Megumi Dick. "Asians and California's Anti-Miscegenation Laws." In *Asians and Pacific American Experiences: Women's Perspectives,* edited by Nobuya Tsuchida. Minneapolis: Asian/Pacific American Learning Resources Center, University of Minnesota, 1982.

Paine, Swift. *Eilley Orrum: Queen of the Comstock.* Palo Alto: Pacific Books, 1949.

Palmer, Cynthia. *Shaman Woman, Mainline Lady.* New York: Quill, 1982.

Pascoe, Peggy. *Relations of Rescue: The Search for Female Moral Authority in the American West, 1874–1939.* New York: Oxford University Press, 1990.

Paul, Rodman W. *Mining Frontiers of the Far West, 1848–1880.* New York: Holt, Rinehart and Winston, 1963. Reprint, Albuquerque: University of New Mexico Press, 1974.

Paynter, Robert, and Randall McGuire, eds. *The Archaeology of Inequity.* Oxford: Basil Blackwell, 1991.

Peck, Gunther. "Manly Gambles: The Politics of Risk on the Comstock Lode, 1860–1880." *Journal of Social History* 26 (summer 1993): 701–23.

Petrik, Paula. *No Step Backward: Women and Family on the Rocky Mountain Mining Frontier, Helena, Montana 1865–1900.* Helena: Montana Historical Society Press, 1987.

Pfeffer, George Anthony. "Forbidden Families: Emigration Experiences of Chinese Women under the Page Law, 1875–1882." *Journal of American Ethnic History* 6 (1986): 28–46.

Purser, Margaret. "Several Paradise Ladies Are Visiting Town: Gender Strategies in the Early Industrial West." *Historical Archaeology* 25, no. 4 (1991): 9–13.

Reed, Waller H. *Population of Nevada: Counties and Communities, 1860–1980.* Reno: Nevada Historical Society, 1983–1984.

Riley, Glenda. *The Female Frontier: A Comparative View of Women on the Prairies and the Plains.* Lawrence: University Press of Kansas, 1988.

———. *A Place to Grow: Women in the American West.* Arlington, Ill.: Harlan Davidson, 1992.

Rusco, Elmer R. *"Good Times Coming?": Black Nevadans in the Nineteenth Century.* Westport, Conn.: Greenwood Press, 1975.

Schlereth, Thomas J., ed. *Victorian America, 1876–1915.* New York: HarperCollins, 1991.

Schlissel, Lillian. *Women's Diaries of the Westward Journey.* New York: Schocken Books, 1982.

Schlissel, Lillian., Vicki L. Ruiz, and Janice Monk, eds. *Western Women: Their Land, Their Lives.* Albuquerque: University of New Mexico Press, 1988.

Scrugham, James G., ed. *Nevada: A Narrative of the Conquest of a Frontier Land.* 3 vols. Chicago: American Historical Society, 1935.

Seifert, Donna J., ed. *Gender in Historical Archaeology.* Edited issue of *Historical Archaeology* 25, no. 4 (1991).

Shamberger, Hugh A. *The Story of Silver Peak, Esmeralda County, Nevada.* Carson City, Nev.: Department of Conservation and Natural Resources, 1976.

Shepperson, Wilbur S. "Immigrant Themes in Nevada Newspapers." *Nevada Historical Society Quarterly* 12, no. 2 (summer 1969): 4–46.

———. *Restless Strangers: Nevada's Immigrants and Their Interpreters.* Reno: University of Nevada Press, 1970.

Shepperson, Wilbur S., with Ann Harvey. *Mirage-Land: Images of Nevada.* Reno: University of Nevada Press, 1992.

Shinn, Charles H. *The Story of the Mine: As Illustrated by the Great Comstock Lode of Nevada.* New York: D. Appleton and Co., 1896. Reprint, Reno: University of Nevada Press, 1992.

Simmons, Alexy. *Red Light Ladies: Settlement Patterns and Material Culture on the Mining Frontier.* Anthropological Northwest No. 4. Corvallis: Department of Anthropology, Oregon State University, 1989.

Smith, Grant. *The History of the Comstock Lode: 1850–1920.* Reno: Nevada Bureau of Mines, 1943.

Strasser, Susan. *Never Done: A History of American Housework.* New York: Pantheon, 1982.

Tong, Benson. *Unsubmissive Women: Chinese Prostitutes in Nineteenth-Century San Francisco.* Norman: University of Oklahoma Press, 1994.

Tsuchida, Nobuya, ed. *Asian and Pacific American Experiences: Women's Perspectives.* Minneapolis: Asian/Pacific American Learning Resources Center, University of Minnesota, 1982.

Twain, Mark. *Roughing It.* 1871. Reprint, New York: Harper and Brothers, Publishers, 1913.

Walde, Dale, and Noreen D. Willows, eds. *The Archaeology of Gender: Proceedings of the 22nd (1989) Annual Chacmool Conference.* Calgary: The Archaeological Association of the University of Calgary, 1981.

Waldorf, John Taylor. *A Kid on the Comstock: Reminiscences of a Virginia City Childhood.* Palo Alto: American West Publishing, 1970. Reprint, Reno: University of Nevada Press, 1991.

Walsh, James P. *The San Francisco Irish, 1850–1976.* San Francisco: The Irish Literary and Historical Society, 1978.

————. "The Irish in the New America: 'Way Out West.'" In *America and Ireland, 1776–1976: The American Identity and the Irish Connection,* edited by David Noel Doyle and Owen Dudley Edwards. Westport, Conn.: Greenwood Press, 1980.

Wegars, Priscilla, ed. *Hidden Heritage.* Amityville, N.Y.: Baywood Publishing Company, 1993.

Weisenburger, Francis P. "God and Man in a Secular City: The Church in Virginia City, Nevada." *Nevada Historical Society Quarterly* 14 (summer 1971): 3–23.

Welter, Barbara. "The Cult of True Womanhood: 1820–1860." *American Quarterly* 18, no. 2 (summer 1966): 151–74.

West, Elliot. "Beyond Baby Doe: Child Rearing on the Mining Frontier." In *The Women's West,* edited by Susan Armitage and Elizabeth Jameson. Norman: University of Oklahoma Press, 1987.

————. *Growing up with the Country: Childhood on the Far-Western Frontier.* Albuquerque: University of New Mexico Press, 1989.

White, William G., and Ronald M. James. "Little Rathole on the Big Bonanza: Historical and Archeological Assessment of an Underground Resource." Unpublished report, Carson City, Nev.: State Historical Preservation Office, 1991.

Winnemucca Hopkins, Sarah. *Life Among the Piutes: Their Wrongs and Claims.* Boston: Cupples, Upham & Co., 1883. Reprint, Reno: University of Nevada Press, 1994.

Yung, Judy. *Chinese Women of America: A Political History.* San Francisco: Chinese Cultural Foundation of San Francisco, 1986.

————. *Unbound Feet: A Social History of Chinese Women in San Francisco.* Berkeley: University of California Press, 1995.

Zanjani, Sally. *Goldfield: The Last Gold Rush on the Western Frontier.* Athens: Ohio University Press, Swallow Press, 1992.

CONTRIBUTORS

ANNE M. BUTLER is editor of the *Western Historical Quarterly* and professor of history at Utah State University. The author of *Gendered Justice in the American West: Women Prisoners in Men's Penitentiaries* and *Daughters of Joy, Sisters of Misery: Prostitution in the American West,* she has published extensively on the subject of women in western history. Currently, she is working on a monograph about Roman Catholic Sisters in the American West.

SUE FAWN CHUNG is associate professor and former chair of the history department at the University of Nevada, Las Vegas, and is a specialist in Asian and Asian-American history. She has prepared a book manuscript, "Beyond Gum San: A History of the Chinese in Nevada," based on her research for an exhibit that has been on display at the state museums in Carson City and Las Vegas. In 1996 she was given the "Outstanding Nevadan Award" by the Nevada Humanities Committee.

ANDRIA DALEY TAYLOR is the chair of the Comstock Historic District Commission and northern Nevada representative to the National Trust for Historic Preservation. A long-time Comstock resident living in the Piper-Beebe House, Daley Taylor is currently writing a biography of Lucius Beebe and Charles Clegg. Her numerous articles have appeared in regional publications.

KENNETH H. FLIESS is associate professor of anthropology at the University of Nevada, Reno. He has published widely on the historical demography of Texas and the southwestern United States. His current projects include computerizing the Federal Manuscript Census for Nevada from 1860 through 1920 and examining civil death records for Storey County (the Comstock) from 1865 through 1920.

JEAN E. FORD is a former member of the Nevada State Assembly and Senate. She is currently teaching Nevada's women's history at the University of Nevada, Reno. Ford is co-founder and chair of the Nevada Women's History Project. She has played an instrumental role in elevating the study of women's issues in Nevada and developing an archive to document the role women have played in shaping the destiny of the state.

Bernadette Smith Francke served as the Inspector-Clerk for the Comstock Historic District Commission for six years until 1996. Her articles on Comstock history have appeared in the *Nevada Historical Society Quarterly*. Francke is also the author of a booklet on Virginia City cemeteries. Her current research interests include the history of nineteenth-century photography in the West.

Donald L. Hardesty is professor of anthropology at the University of Nevada, Reno. He has been president of the Society for Historical Archaeology and a member of the executive council of the Mining History Association. He has published widely in the field of archaeology.

Eugene M. Hattori is an archaeologist with the Nevada State Historic Preservation Office. He also is a research associate for the California Academy of Sciences and an adjunct associate professor of Anthropology for the University of Nevada, Reno. He specializes in Great Basin and California prehistory and historic archaeology.

Ronald M. James is the Nevada State Historic Preservation Officer and serves on the Comstock Historic District Commission. His book manuscript on the Bonanza of the mining West is in press with the University of Nevada Press. James has also published on the subjects of European and western history and folklore.

Janet I. Loverin is curator of clothing and textiles at the Nevada State Museum, Marjorie Russell Clothing and Textiles Research Center in Carson City. Her extensive research on Comstock needle workers spans several years. Previous publications have dealt with the subject of clothing textiles, and she is the co-author of *To Clothe Nevada Women*.

Sharon Lowe is a member of the Arts and Science Department at Truckee Meadows Community College where she teaches American history and western traditions. Her interest in social history of the Comstock began as a graduate student at the University of Nevada, Reno, which led to research on drug abuse in western mining frontiers. Most recently, she has been researching the role of women and their connection to drug usage in these communities.

Julie Nicoletta is assistant professor of public history at the University of Washington Tacoma and holds a Ph.D. in the history of art from Yale University. She has published *The Architecture of the Shakers* and is currently co-authoring a book with Ronald M. James on the buildings of Nevada.

Robert A. Nylen is curator of history at the Nevada State Museum in Carson City. He has published articles on the state's brewing industry, early frontier baseball, and cowboy chuck wagon cooking. He is currently writing a history of the Nevada mining town of Belmont.

C. Elizabeth Raymond is associate professor and chair of the history department at the University of Nevada, Reno. The author of books and essays on several aspects of Nevada history, she specializes in the study of regional landscape perception and the sense of place.

Contributors

KATHRYN DUNN TOTTON has published widely in Nevada history on topics ranging from the military of Fort Churchill to Hannah Clapp, one of the state's earliest educators. She is photograph curator in the Special Collections Department, University of Nevada Libraries, and teaches Nevada history for the University's Continuing Education Program and Truckee Meadows Community College. Divorce on the Comstock is a long-standing research interest.

ANITA ERNST WATSON earned a Ph.D. from the University of Nevada, Reno, focusing on the history of divorce, and of the family. In 1996 she completed *Reflection, Recollection and Change,* a history of the Nevada State Board of Medical Examiners. The three-year project included eighteen oral histories, primarily of Nevada physicians, and reflects her interest in oral history.

LINDA WHITE has a Ph.D. in Basque studies and is currently the Assistant Coordinator of the Basque Studies Program at the University of Nevada, Reno. She wrote her dissertation on Basque women writers of the twentieth century. She also teaches graduate classes in Basque literature and undergraduate classes in Basque language.